Inhalation Studies

Inhalation Studies

Foundations and Techniques

Second Edition

by

Robert F. Phalen

School of Medicine
University of California, Irvine
Irvine, California, USA

informa
healthcare

New York London

First published in 2009 by Informa Healthcare, Telephone House, 69-77 Paul Street, London EC2A 4LQ, UK.

Simultaneously published in the USA by Informa Healthcare, 52 Vanderbilt Avenue, 7th Floor, New York, NY 10017, USA.

Informa Healthcare is a trading division of Informa UK Ltd. Registered Office: 37–41 Mortimer Street, London W1T 3JH, UK. Registered in England and Wales number 1072954.

A CIP record for this book is available from the British Library.

Library of Congress Cataloging-in-Publication Data available on application

ISBN-13: 9780849314001

Orders may be sent to: Informa Healthcare, Sheepen Place, Colchester, Essex CO3 3LP, UK
Telephone: +44 (0)20 7017 5540
Email: CSDhealthcarebooks@informa.com
Website: http://informahealthcarebooks.com/

For corporate sales please contact: CorporateBooksIHC@informa.com
For foreign rights please contact: RightsIHC@informa.com
For reprint permissions please contact: PermissionsIHC@informa.com
Printed in the UK

Preface

The 1984 edition of *Inhalation Studies: Foundations and Techniques* was a success because it presented essential information for inhalation toxicologists and other health professionals. Prior to preparing the second edition, suggestions were received from a group of anonymous reviewers. They were unified in recommending that the second edition should both focus on the essentials, and update the key developments. The author has learned to take the advice of reviewers.

In the years since the first edition was published, numerous scientific developments have occurred. Some of the more important to those who perform inhalation studies are:

- the explosion of knowledge in the genetic and molecular realms,
- the introduction of a large number of transgenic animal models,
- the emergence of aerosol medicine,
- the increased threat of aerosol-related terrorism,
- the realization that low levels of ultrafine particles may have health effects,
- the increased appreciation of nonpulmonary effects of inhaled substances,
- the use of medical scanning techniques to study respiratory tract structure,
- the introduction of new inhalation exposure systems,
- the emergence of aerosol concentrators for use in air pollution studies,
- the application of computational fluid dynamics for modeling inhaled aerosols,
- the introduction of new devices for production and characterization of aerosols, and
- the need for greater security for biomedical laboratories.

As this list is only a sampling of the relevant developments, this edition represents a significant, as opposed to a minor, update. Several new sections, and hundreds of new references have been added. Key older references and descriptions of early studies have been preserved when they still have useful information or show the evolution of modern concepts.

The topics covered are broad, representing dozens of specialties. The terminology, symbols, and units of these specialties have been used without any attempt to harmonize across all chapters.

Countless suggestions were offered by expert reviewers, including William Hinds, Richard Mannix, Michael Kleinman, Kathryn Osann, and Melanie Fabian. Ms. Leslie Owens expertly word-processed and edited the book, and performed many administrative functions; the author is eternally grateful for her dedication and expertise. Artists Tuan Nguyen, Robert Olide, and Joshua Bracks contributed to the illustrations. Katherine Phalen checked the references. Still, the author is solely responsible for inaccuracies and omissions.

This book was possible because of the efforts of scientists who published their work. They are acknowledged with the deepest gratitude.

Finally, this book is dedicated to Kayla, Joseph, and Samuel, young children who did not receive all of the attention they deserved from parents and grandparents who worked on this book: May they have happy and prosperous futures.

Robert F. Phalen, Ph.D
March, 2008

Contents

Introduction

Humans breathe several thousand times more volumes of air each day than the volumes of food and water consumed. Thus the potential for injury from inhaled particles and gases is ever-present. Thousands of substances in the home, workplace, and outdoor air, along with aerosol medicines and intentionally toxic aerosols, must be extensively studied in toxicology laboratories. *Inhalation Studies: Foundations and Techniques, Second Edition* describes why and how such studies are performed. This thorough and richly illustrated treatment represents a significant update of the widely used, original 1985 edition: Each chapter has been revised and key references updated. Some older material has been retained where it demonstrates basic principles or essential techniques. The book is organized into 10 chapters that cover: (1) aerosols and gases; (2) respiratory tract anatomy and physiology; (3) generation of experimental atmospheres; (4) characterization of exposures; (5) inhalation exposure systems; (6) testing for toxicity; (7) experimental designs; (8) facilities and supporting functions; (9) animal models; and (10) regulations and guidelines. New material, supported by over 300 new references, covers recent developments including: new animal models; nonpulmonary effects of inhaled materials; ultrafine and nanotechnology-related aerosols; aerosol concentrators and other new exposure systems; dosimetry developments including computational fluid dynamics deposition models; and new requirements for facilities. Although intended primarily for active researchers and graduate students, the material is presented in a manner that is understandable by other professionals in the medical, engineering, regulatory, and environmental communities. Those who conduct, support, or use inhalation research, as well as those who are interested in aerosol medicine, air pollutants, or aerosol bioterror/emerging airborne infections, will find this book to be an important reference.

1
Aerosols and Gases

INTRODUCTION

An *aerosol* is a relatively time-stable, two-phase system consisting of finely divided and condensed particulate matter in a suspending gaseous medium. The particulate phase may consist of liquids, solids, or both. The condensed phase particles are small, having dimensions in the 0.001–100 μm range. The behavior of the aerosol is influenced by factors associated with the particles, the surrounding gas, the containment, and external forces (e.g., gravity, electrical fields, and radiation).

Whether by intent or otherwise, aerosol particles are always present in atmospheres studied in inhalation experiments. Air can be considered relatively particle-free when the mass of suspended particles exists in trace amounts of about 1 ng [10^{-9} g] /m^3 of air. A nearly particle-free condition is obtained by the use of a clean, relatively inert containment system and by filtration of the air.

A standard for particle-free air, the *class 100 clean room*, must contain no more than 100 particles of 0.5 μm diameter or larger per cubic foot of air. Assuming an average particle specific gravity of 2, this implies an airborne particle mass concentration of about 0.5 ng/m^3 of air. Since the density of dry air at standard room conditions is 1.2 kg/m^3, the particles constitute less than $0.5 \times 10^{-10}\%$ of the mass of the aerosol–gas system. In contrast, ordinary room air may contain 10,000 to 50,000 particles per cubic centimeter of air, even when no unusual sources, such as a burning cigarette, are present. People are usually totally unaware of such levels of airborne particles.

As the intent of this chapter is to provide a foundation for dealing with aerosols in inhalation experiments, only a few selected properties of aerosols are presented. Such properties that relate to generating, controlling, and understanding the response of exposed subjects include particle size, shape, density, electrical charge, hygroscopicity, surface area, settling behavior, diffusion, inertial properties, coagulation, and rate of dissolution in fluids such as are found in the lung.

The Impact of Aerosols

The impact of aerosols on our daily lives is large, as our activities are performed in an atmospheric sea containing gases and particles (Table 1.1). The particles, liquid and solid, organic and inorganic, viable and nonviable, influence the environment. Natural particle phenomena include cloud formation, the role of particles in the water cycle, the shaping of

Table 1.1 Some Particles Commonly Found in Air, Their Sizes and Impacts on Natural Phenomena and Human Health

Particle	Typical diameter range (µm)	Impact
Viruses	0.01–0.45	Some produce infection
Bacteria	0.2–30	Some produce infection
Fungal spores	2–100	Some are allergens
Moss spores	6–30	Propagation of plants
Fern spores	20–60	Propagation of plants
Pollen	10– >200	Some are allergens
Coal dust	3–30	Can produce lung diseases
Natural fog	2–80	Contributes to smog
Tobacco smoke	0.05–5	Can produce lung diseases
Metal fumes	0.01–100	Can produce lung diseases
Fly ash	0.5 and up	Unknown
Plant and insect bits	5–100 and up	Some are allergens
Molecular clusters (gaseous ions)[a]	0.001–0.005	Centers of droplet condensation

[a]Not true particles; do not persist if uncharged.

land by wind, pollination of plants, and the distribution of seeds and spores. Human uses of aerosols include the atomization of fuels prior to combustion, the application of paints, cosmetics, medicines, insecticides, and lubricants; and scientific uses.

Unfortunately, aerosols often cause problems which resist eradication. Among these are infectious diseases including the common cold, influenza, viral pneumonia, measles, mumps, and tuberculosis. Other diseases in which inhaled particles often play a central role are bronchitis, pulmonary emphysema, asthma, diffuse interstitial fibrosis, alveolitis, silicosis, anthracosilicosis, berylliosis, farmers lung, byssinossis, lung cancer, and nasal cancer.

Inhaled particles can induce disease states in many tissues or organ systems when they, or their metabolic products, are systemically distributed via blood or lymph. Examples include liver necrosis, aplastic anemia (bone marrow failure), hemolytic anemia, leukopenia, fluorosis, bone cancer, headache, dizziness, insomnia, irritability, and muscle weakness. This list of aerosol-related diseases is by no means complete.

Aside from these adverse effects on health, aerosols are implicated in the following: damage to crops and other plants; deterioration of works of art and structural materials; dust explosions; reduction of visibility; soiling of mirrors, lenses, windows, painted surfaces, clothing, skin, hair, food, and water; damage to air pumps, motors, and electronics; reduction of the solar constant at the earth's surface; production of air inversions; and the formation of smog.

Size Regimes

The great diversity in particle size, shape, and composition makes it impossible to describe aerosol behavior simply. As a starting point, one can divide aerosols into *regimes* (Table 1.2). These regimes, which encompass given size ranges, are each associated with sets of equations that describe the physical behavior of aerosols. An important dimensionless parameter, the *Knudsen number*, Kn, which relates the particle radius, r_p, to the molecular mean free-path of the suspending gas, λ_g, is given by:

$$\mathrm{Kn} = \frac{\lambda_g}{r_p} \qquad \text{(eq. 1.1)}$$

Table 1.2 The Major Particle Regimes and the Dependence of Various Properties on Particle Radius

	Regime			
	Free molecule	Transition	Slip flow	Continuum
Knudsen number	> 10	10 to 0.3	0.3 to 0.1	< 0.1
Particle radius	< 0.005	0.005 to 0.2	0.2 to 0.65	> 0.65
Resistance to motion	Proportional to r^2	Transitional		Proportional to r
Evaporation rate	Proportional to r^2	Transitional		Proportional to r
Light scattering	Proportional to r^6	Transitional		Proportional to r^2
Coagulation rate	Function of r	Transitional		Independent of r

Source: Adapted from Hesketh (1977), Chapter 1.

The molecular mean free-path represents the average distance traveled by a molecule of gas between successive collisions with other gas molecules. For air at standard laboratory conditions, the molecular mean free-path is about 0.065 μm, which is about 20 times the average distance between gas molecules and about 200 times the diameter of an average air molecule. Particles with Knudsen numbers greater than 10 are small with respect to the spaces between gas molecules and therefore "experience" the surrounding gas molecules as individual, rapidly moving, bombarding entities (in the *Free Molecule Regime* particle motion is dominated by diffusion). At small Knudsen numbers, the particles are large enough so that the surrounding gas acts as a continuous medium (in the *Continuum Regime* particle motion is dominated by inertial forces). Between these two extremes, *Transition* and *Slip Flow Regimes* can be described in which particle behavior must be treated by using corrections to the equations of the two other regimes. Figure 1.1 provides a scale drawing of a 0.01-μm diameter particle in air.

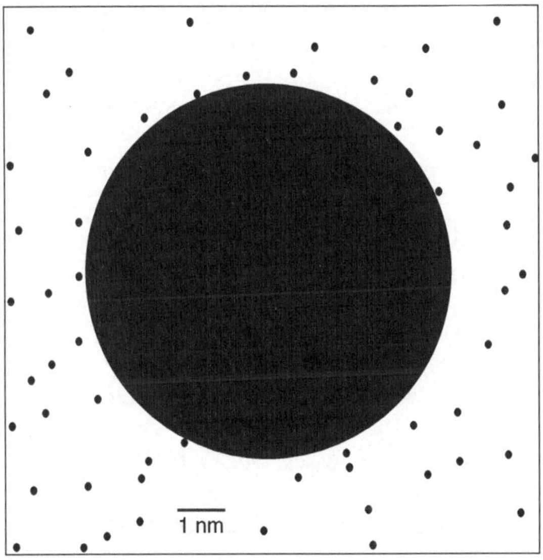

Figure 1.1 Scale depiction of a 0.01-μm diameter particle surrounded by air molecules. The Knudsen number is 13, so the particle is in the *free molecule regime*.

Aerosol Terminology

That aerosols of various types affect our lives in many ways is evident by the large number of terms used to refer to various aerodisperse systems. Examples of terms commonly used to describe aerosols include: air contaminants, air pollutants, Aitken nuclei, aerocolloids, aerosols, ash, clouds, colloids, condensation nuclei, dispersoids, droplets, dusts, emissions, exhausts, fallout, fine particles, floculates, fogs, fumes, hazes, lapilli, mists, motes, nanoparticles, nuclei, particles, plumes, powders, smogs, smokes, soots, sprays, and ultrafines. Definitions of selected terms are given below.

Aitken nuclei, Condensation nuclei—Particles that are detected by their tendency to serve as centers for condensation of water vapor under supersaturated conditions in the approximate relative humidity range of 200–300%. Such particles are usually in the diameter range of about 0.01–0.2 μm.

Aerosol, Aerocolloid—(1) A disperse system in air. According to Drinker and Hatch (1936), the term *aerosol* was first introduced by Gibbs in 1924; (2) a relatively time-stable suspension of small liquid and/or solid particles in a gas. The diameter size range of aerosol particles is about 0.001–100 μm.

Cloud—Any free (not spatially confined) aerosol system with a definite overall shape and size. Rain clouds and smoke rings are examples.

Colloid—A dispersion of liquid or solid particles in a gas, liquid, or solid medium that has all of the following properties: slow settling, large surface to volume ratio, invisibility to the unaided eye, and producing scattering of a light beam. Examples include smoke, milk, and gelatin.

Dust—Dry particles dispersed in a gas as a by-mechanical disruption of a solid or powder.

Fine particles—Particles having aerodynamic equivalent diameters from 2.5 μm to 0.1 μm.

Fume—An agglomerated aerosol consisting of clusters of smaller primary particles. Fumes form by condensation and usually resist disruption into free, individual, primary particles.

Mist—Traditionally a liquid droplet aerosol of particles having diameters greater than about 20 μm, but the term has been used to describe all liquid aerosols even in the submicrometer diameter range.

Nanoparticles—Particles smaller than ultrafine particles. Dimensions are usually 1–50 nm, but sometimes larger.

Particle—A small piece of matter which may or may not be suspended in a liquid or a gas.

Particulate—This term is usually an adjective, meaning "in the form of separate particles," but it can be used as a noun meaning "particles" (both solid and liquid).

Smog—A highly variable mixture of aerosol particles and gases found in the air in or downwind from urban centers. The term smog, originally meaning smoke and fog, is now associated with air pollution in general.

Smoke—Any of a variety of concentrated, visible aerosols formed in large part by condensation of supersaturated vapors. Smokes usually result from combustion of organic materials and may contain a variety of solids, liquids, and gases. Due to their high gas and particle concentrations, smokes often exhibit cloud behavior.

Ultrafine particles—Particles having geometric diameters less than 0.1 μm.

The general lack of agreement on the precise particle size ranges that typify the above aerosols arises from specialization of contributors to the scientific literature; such specialization includes atmospheric chemistry, industrial hygiene, engineering, inhalation toxicology, combustion technology, and medicinal therapy. Each specialty has its own terminology.

Several reference books on aerosols have been published. The basic theoretical reference is a work by Nicholai A. Fuchs (1964) entitled *The Mechanics of Aerosols*, which was translated from Russian into English by R.E. Daisley and Marina Fuchs and edited by C.N. Davies. A variety of additional books, some general and some specialized, are presented in Table 1.3. Although not exhaustive, the listed references cover most problems that arise in studies with aerosols.

DIRECT OBSERVATION OF AEROSOL PARTICLES

The commonly accepted upper limit of diameter of an aerosol particle, about 100 μm, is near the lower limit of resolution of the human eye. The Rayleigh condition for resolving two points of equal brightness is that the centers of the points are separated by a distance at least as great as the radius of the central disk of the diffraction pattern. Using the Rayleigh criterion, the normal eye at close range should just resolve two objects whose separation is about 70 μm. This separation subtends about 1′ of arc, that is, 3 cm separation at 100 m. The unaided eye is inadequate for resolving most individual aerosol particles.

The resolution of a high-power optical microscope is about one-half the wavelength of the light used for viewing, or about 0.2 μm, providing a resolution 350 times smaller than the unaided eye. This resolution is achieved by filling the space between the specimen and the objective lens with an oil that has a refractive index greater than that of air (n_{air} = 1.00). The index of refraction of typical microscope oil is about 1.5. In this medium, the wavelength of light is less than that in the air, resulting in improved resolving power. Therefore, examination by optical microscope is appropriate for particles with diameters down to about 0.3 μm. When particles smaller than this are viewed using the optical microscope, they are likely to be missed. Although this situation is improved by use of short wavelength or dark field illumination, one must be cautious when sizing particle samples with a light microscope. Errors due to inadequate resolution must always be expected when the particle size distribution has a falloff value in the diameter range near or just above the limit of resolution. This situation is illustrated in Figure 1.2, which shows a hypothetical particle distribution and the renormalized distribution obtained using a light microscope. This principle applies to any sizing device or method with a finite limit of resolution that is greater than the smallest particle in the sample.

An improved limit of resolution is available in the electron microscope. By using electrons generated from a hot filament, the practical limit of resolution is near 0.001 μm. The wavelength of the electron, a function of its velocity, v, is given by the de Broglie equation:

$$\lambda = h/mv \qquad \text{(eq. 1.2)}$$

where h is Planck's constant and m the mass of the electron. With an accelerating potential difference of 50,000 volts, the electron wavelength is 0.25 Å units (0.25×10^{-10} m). In reality, the resolution of the electron microscope is limited by factors other than the wavelength of the electrons. This limitation is of little consequence in aerosol technology, since the practical limit of resolution is smaller than the smallest aerosol particle. However, several artifacts occur when sizing particles using the electron microscope. Assume that a representative sample suitable for viewing has been obtained—no simple

Table 1.3 Selected References on Aerosols

Author/title/publisher/date	Comments
Brown, L.M., Collings, N., Harrison, R.M., Maynard, A.D., and Maynard, R.L., Eds., *Ultrafine Particles in the Atmosphere*, Imperial College Press, London, 2000	Has 16 chapters on ultrafine aerosol physics, sources, analysis, and health effects
Cohen, B.S. and McCammon, C.S. Jr., Eds. *Air Sampling Instruments, 9th Ed.*, ACGIH ® (American Conference of Governmental Industrial Hygienists), Cincinnati, OH, 2001	Twenty-three chapters prepared by various experts covering the rationale and methods for sampling aerosols, and instrumentation; geared toward industrial hygiene applications
Cox, C.S. and Wathes, C.M., Eds., *Bioaerosols Handbook*, Lewis Publishers, Boca Raton, FL, 1995	Contains 21 chapters by experts on bioaerosol physics, sampling, size distributions, generation, analysis, environmental problems, and laboratory safety and containments
Davies, C.N., *Recent Advances in Aerosol Research*, Macmillan, New York, 1964	A bibliographical review of publications on aerosol acoustics, adhesion, reactions, coagulation, diffusion, combustion, size and shape, evaporation and condensation, filtration, generation, nucleation and growth, electrical properties, sampling, phoresis, sedimentation, radioactivity, and deposition
Davies, C.N., Ed., *Aerosol Science*, Academic Press, London and New York, 1966	Contains 12 chapters on aerosol generation, filtration, charge, measurement, adhesion, and deposition
Dennis, R., *Handbook on Aerosols*, U.S. Energy Research and Development Administration, Oak Ridge, Tenn., 1976	A practical guide to aerosol generation, sampling, sizing, optical properties, and dynamic behavior in air
Drinker, P. and Hatch, T., *Industrial Dust*, McGraw-Hill, New York, 1936	Slanted toward dust hazards, the book covers basic aerosol properties, effects on humans, practical measurement of size, concentration, and composition as well as dust control methods
Einstein, A., *Investigations on the Theory of Brownian Movement*, Dover, New York, 1956	A translation of five papers written between 1905 and 1908, covering thermally induced particle motion and its contribution to various physical phenomena
Finlay, W.H., *The Mechanics of Inhaled Pharmaceutical Aerosols: An Introduction*, Academic Press, New York, 2001	Covers particle-size distributions and particle physics, plus information on respiratory tract deposition and medical aerosol generators and medical aerosols
Friedlander, S.K., *Smoke, Dust and Haze, 2nd Ed.*, Oxford University Press, New York, 2000	Textbook covering the atmosphere, aerosols, air pollution, and transport models
Fuchs, N.A., *The Mechanics of Aerosols*, Dover Publications Inc., New York, 1964	A basic reference on particle physics covering size, steady and nonuniform motion, Brownian motion, diffusion, coagulation, and dispersal
Fuchs, N.A. and Sutugin, A.G., *Highly Dispersed Aerosols*, Ann Arbor Science, Ann Arbor, 1970	Covers characterization, generation, and properties of particles with diameters below 1 μm

Gehr, P. and Heyder, J., Eds., *Particle–Lung Interactions*, Marcel Dekker, Inc., New York, 2000	A comprehensive treatment in 19 invited chapters covering environmental, industrial, and medical aerosols: inhalation, clearance, biological research, and health consequences
Green, H.L. and Lane, W.R., *Particulate Clouds: Dust, Smokes and Mists, 2nd Ed.*, Van Nostrand, New York, 1964	Thorough treatment of aerosol physics, generation, sampling, collection, health hazards, and industrial applications, with some spectacular aerosol photography
Hesketh, H.E., *Fine Particles in Gaseous Media*, Ann Arbor Science, Ann Arbor, MI., 1977	Theoretical treatment of size; size measurement; motion; effects of forces such as electrostatic, magnetic; and acoustic; and particle collection
Hickey, A.J., Ed., *Inhalation Aerosols: Physical and Biological Basis for Therapy, 2nd Ed.*, Informa Healthcare U.S.A., New York, 2007	Nearly 40 expert contributors provide chapters covering aerodynamic behavior, biological considerations, and pharmaceutics
Hidy, G.M. and Brock, J.R., *The Dynamics of Aerocolloidal Systems*, Pergamon Press, Elmsford, NY, 1970	An engineering and physical chemistry approach containing sections on aerosol dynamics, heat and mass transfer, diffusion, generation, nucleation, and coagulation
Hidy, G.M., *Aerosols: An Industrial and Environmental Science*, Academic Press, Orlando, FL, 1984	Covers aerosol dynamics, generation, measurements, applications, environmental and health effects, and regulation
Hinds, W.C., *Aerosol Technology: Properties, Behavior and Measurement of Airborne Particles, 2nd Ed.*, John Wiley & Sons, New York, 1999	A college-level textbook for persons with a background in chemistry, physics, and mathematics. Covers basic properties, respiratory tract deposition, dust explosions, size measurement, and generation techniques. Has problems and answers
Irani, R.R. and Callis, C.F., *Particle Size: Measurement Interpretation, and Application*, John Wiley & Sons, New York, 1963	Collection and sizing techniques including sedimentation, microscopy, sieving, and several other methods
Liu, B.Y.H., Ed., *Fine Particles*, Academic Press, New York, 1976	A symposium proceedings with 34 papers on aerosol generation, sampling, measurement, and analysis
Lundgren, D.A., Harris, F.S. Jr., Marlow, W.H., Lippmann, M., Clark, W.E., and Durham, M.D., Eds., *Aerosol Measurement*, University Press of Florida, Gainesville, FL, 1979	Fifty-seven papers covering centrifuges, cyclones, impactors, optical counters, electrical analyzers, condensation nuclei counters, and diffusion batteries
Marple, V.A. and Lui, B.Y.H., Eds., *Aerosols in the Mining and Industrial Work Environments*, 3 Vols, Ann Arbor Science, Ann Arbor, MI, 1983	Over 1200 pages covering 81 papers from a comprehensive international symposium on workplace aerosols, their properties, sampling, analysis, and inhalation
Mercer, T.T., *Aerosol Technology in Hazard Evaluation*, Academic Press, New York, 1973	Geared toward instrumentation and hazard analysis, covers size distributions, basic aerosol properties, production of test aerosols, and measurement of concentration, size, and respirable fraction
Mercer, T.T., Morrow, P.E., and Stober, W., Eds., *Assessment of Airborne Particles*, Charles C. Thomas, Springfield, IL, 1972	Proceedings of a symposium on aerosol fundamentals, generation and measurement, analysis, deposition, and hazard assessment. Has 28 separate papers
Murphy, C.H., *Handbook of Particle Sampling and Analysis Methods*, Verlag Chemie International, Deerfield Beach, NJ, 1984	Textbook format covering particle characteristics, sampling, and several analytical techniques

Continued

Table 1.3 Selected References on Aerosols—cont'd

Author/title/publisher/date	Comments
Ruzer, L.S. and Harley, N.H., Eds., *Aerosols Handbook: Measurement, Dosimetry and Health Effects*, CRC Press, Boca Raton, FL, 2005	Has 24 chapters by leading researchers on a broad range of topics from basic aerosol science and medical aerosols to health effects of environmental and radioactive aerosols
Salem, H., and Katz, S.A., Eds., *Inhalation Toxicology 2nd Ed.*, Taylor & Francis, Boca Raton, FL, 2006	Contains 40 chapters (1034 pages) by experts on inhalation toxicology methods, measurements, and in-depth material on asbestos, toxic gases, cigarette smoke, and bioaerosols
Seinfeld, J.H., and Pandis, S.N., *Atmospheric Chemistry and Physics, 2nd Ed.*, Wiley, New York, 2006	Textbook covering the atmosphere, aerosols, air pollution, and transport models
Silverman, L., Billigs, C.E., and First, M.W., *Particle Size Analysis in Industrial Hygiene*, Academic Press, New York, 1971	Covers particle behavior, sampling and sizing methods, size distribution analysis, and field applications of particle sizing
Vincent, J.H., *Aerosol Science for Industrial Hygienists*, Elsevier Science, Tarrytown, NY, 1995	Covers aerosols and gases including physical behavior, sampling, inhalation, and control in the workplace
Wen, C.S., *The Fundamentals of Aerosol Dynamics*, World Scientific, Singapore, 1996	Theoretical treatments of aerosol motion, sedimentation, coagulation heat transfer, and interaction in concentrated systems
Willeke, K., Ed., *Generation of Aerosols and Facilities for Exposure Experiments,* Ann Arbor Science, Ann Arbor, MI, 1980	A symposium proceedings with 28 papers on aerosol generation, characterization, deposition, dissolution, health effects, charge effects, deliquescence, exposure techniques, and other topics
Willeke, K. and Baron, P.A., Eds., *Aerosol Measurement: Principles Techniques and Applications,* Van Nostrand Reinhold, New York, 1993	Contains 38 chapters by numerous experts covering aerosol behavior, sampling, measurement, instrumentation, and applications.

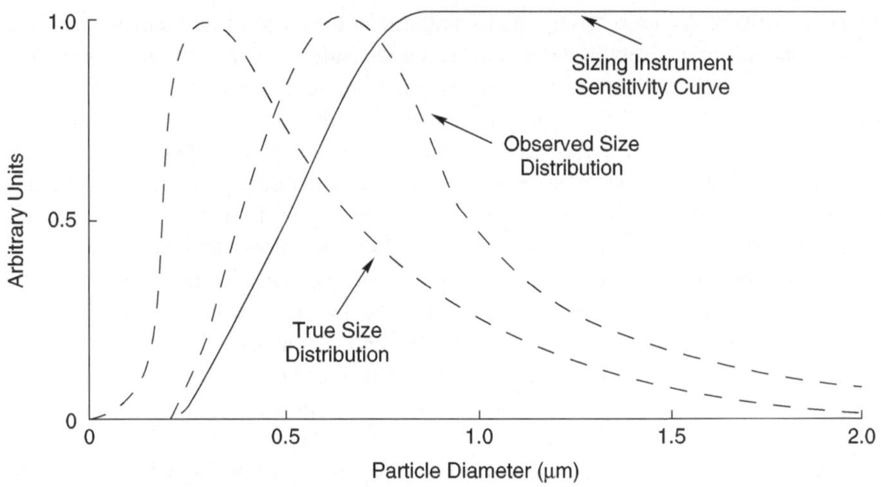

Figure 1.2 Hypothetical size distribution and artifactual measured distribution due to inadequate instrumental size resolution. The distributions have been normalized.

feat itself. Inside the microscope, the specimen is subjected to a high vacuum, 10^{-4} atm or thereabouts. Many materials evaporate rapidly in this condition. In the focused electron beam the temperature (600°C or more) increase may evaporate particles that are normally stable under vacuum. It is not uncommon to see the sample disappear within seconds. In addition, particles can become charged by the beam and fly off of the collection substrate if the sample and surface are not electrically conductive. At times, particles grow due to condensation of vapors on the sample. Oils, greases, and other organic materials can carbonize on contact with the hot particles in the viewing beam and form a coating which may rapidly reach a thickness of 1–2 nm. The coating may introduce appreciable error when sizing tiny particles. Control of these artifacts is not trivial, and special techniques (such as sample cooling) must be considered.

CIGARETTE SMOKE: A FAMILIAR AEROSOL SYSTEM

A familiar aerosol system, cigarette smoke, can be used to illustrate some important properties of aerosols. The combustion of tobacco at about 1000°C leads to the formation of a large variety of inorganic and organic gases, liquids, and solids (Baker, 1974). Ignoring side-stream smoke, which is that produced between puffs, the hot mixture flows through the unburned tobacco undergoing filtration, dilution with residual gases and fresh air, and enrichment with additional vaporized materials. Both the particle size and number concentration of the exiting smoke depend on the unburned butt length—the longer the butt, the fewer and larger the particles (Keith and Derrick, 1960; Ishizu et al., 1978). This effect is apparently primarily due to the action of filtration and preferential removal of smaller particles. Fresh, undiluted smoke may contain several billion particles per cubic centimeter of air, with droplets predominant in the 0.1–1.0 μm diameter range and solids predominant above and below that size. The gaseous components in fresh smoke, too numerous to list completely, include water vapor, carbon monoxide, carbon dioxide, nitric oxide, hydrogen sulfide, isoprene, acetone, toluene, acetaldehyde, and hydrogen cyanide

(Jenkins et al., 2000). As these components undergo dilution with ambient air, several things occur including a drop in temperature, condensation of vapors onto particles, chemical reactions, evaporation of volatiles, coagulation, sedimentation, and diffusional transport of particles. Depending on the dilution ratio with fresh air, the smoke may be dense enough to exhibit *cloud aerodynamic behavior* (Phalen et al., 1994a; Hinds et al., 2002) or dilute enough so that each particle moves independently without significant influence from surrounding particles. Coagulation of fresh cigarette smoke particles can be rapid, and in less than 1 s the particle number per cubic centimeter can fall to one-half of the original number. This coagulation tends to increase the particle size, but when one actually measures the particle diameter as a function of time in free air, it is often seen to decrease. This implies that evaporation of particles and formation of new particles can reduce the average size more than it is increased by coagulation.

If the smoke is inhaled, deposition will occur in the respiratory system. Although the breathing pattern, including whether or not breath-holding occurs, will modify the deposition efficiency and pattern, typically one measures deposition rates of about 50–90% of the inhaled mass (Landahl and Tracewell, 1957; Hinds et al., 1983; Martonen, 1992; Phalen et al., 1994a; Hofmann et al., 2001). This value is greater than one would expect for inert particles of the same median particle diameter (between 0.1 and 1.0 μm), and is evidence that in addition to the usual particle deposition mechanisms other phenomena are occurring. Several mechanisms are at play including distillation of volatiles to the respiratory tract walls, cloud behavior in which the particle–particle interactions keep the cloud relatively intact so that it deposits more or less as a very large low-density object, Raleigh-Taylor instability caused by settling of the suspending gas (Hinds et al., 2002), and enhanced deposition due to electrical charges on the smoke.

Once deposited, the persistence times of various smoke components in the respiratory tract will be variable. Some rapidly dissolving components will enter the body fluids and be removed from lung tissue. Other components may resist dissolution or other clearance mechanisms and persist in the respiratory tract for years. Most components will clear with intermediate rates.

In the foregoing example, one sees how a multitude of physical and chemical properties of an aerosol are relevant to the inhalation toxicologist. Aerosol technology is one of the inseparable foundations of inhalation toxicology. The material that follows in this and other chapters covers in more detail the concepts that were introduced in this example with cigarette smoke.

PARTICLE SIZE

The issue of particle size arises when one considers aerosol particles. Perhaps the most misunderstood property used to describe aerosols is the *diameter*. The only physical object with a unique geometrical diameter is a smooth sphere. Several factors complicate the determination of the geometrical diameter. Aerosol systems usually consist of a great number of particles of differing size and shape, which necessitates use of statistical concepts of geometrical size. For spherical particles, measurements of the diameters of a representative number of particles can be combined and used to estimate a mean or median diameter and an associated estimate of the range of sizes, such as a standard deviation. When individual particles are not spherical, several measurements of diameter can be made on each particle or on a large number of randomly oriented particles and the data summarized statistically.

A useful statistical diameter that is applicable to globular-shaped particles is the *projected area diameter*. This is the diameter of a circle that has the same cross-sectional area as the projected two-dimensional image of the particle. In practice, the projected area diameter is obtained by fitting the particle with a circular overlay such that the particle area excluded equals the excluded area of the circle. Instruments exist for conveniently and rapidly measuring the projected area diameters of particles from photographs. The Endter Gebauer Analyzer (Zeiss® TGZ3, Zeiss, Germany) is perhaps the best known of such instruments. But this technology has largely been replaced by image analysis software.

Diameter can also be defined with respect to any of several measurable properties. Instruments for such measurements are usually calibrated with spherical aerosol particles, and calibration curves are obtained relating instrument response to geometrical diameter of the calibration aerosols. In this case, the measure is termed an *equivalent diameter*. Common equivalent diameters are based on measurement of light scatter, aerodynamic behavior, surface area, diffusional excursions, and mobility in an electrical field. Since these characteristics often depend on particle properties such as index of refraction, shape, density, and surface roughness, the geometrical sizes of two *equivalent* particles may differ considerably. Such size conventions based upon physical properties are useful because they relate directly to the ways in which particles interact with the environment. When visibility of distant objects is of interest, diameter conventions based on light scatter are appropriate. In inhalation toxicology, the *equivalent aerodynamic diameter* has proven to be very useful. This is usually defined as the diameter of a spherical particle of unit standard density (1 g/cm^3) that has the same terminal settling velocity, with respect to still air, as the particle in question. The aerodynamic diameter, which is generally applicable to particles whose diameters are greater than 0.5 μm and thus not strongly influenced by Brownian bombardment, determines important inertial properties such as inability to follow air streams and rate of settling. Such properties are major determinants of deposition of inhaled particles in the lung. Less than about 0.5 μm in geometrical diameter, approaching the mean free-path between collisions of air molecules, particle motion is strongly influenced by diffusional forces. In this size range, particles no longer experience the air as a continuous fluid, and randomly uneven molecular bombardment causes the particle to wander in random directions, making invalid the concept of a steady terminal settling velocity.

SIZE DISTRIBUTIONS

The individual particles in an aerosol are not identical to one another; variation being an inherent characteristic. Thus, when a physical property such as diameter that applies to an individual particle is described, a distribution of values exists for the particle population. It is useful to deal with mathematical representations of size distributions rather than the distribution data set itself for a variety of reasons including compactness, lack of ambiguity, and the ease with which new parameters can be derived. For example, if a lognormal distribution function provides a good fit to sizing data of an aerosol sample, then only two numbers, a median and a geometric standard deviation, define the size distribution. From these two values one can reconstruct a facsimile of the original count distribution. If desired, the distributions of volume and surface can also be derived with the aid of suitable equations, provided that the particles' geometrical shapes are simple, for example, spherical or cubical.

A variety of mathematical size distributions have been successfully applied to particle size data, and relatively complete descriptive summaries of these distributions can be found in several of the references listed in Table 1.3. For most purposes, lognormal distributions provide reasonably good fits to commonly encountered particle size data.

The lognormal distribution function is similar in form to the normal distribution function in that the log of a particle property, say diameter, is normally distributed. Normalized equations giving the fraction of the sample present as a function of diameter are shown below for the normal and the lognormal distributions.

$$\text{Normal Distribution } f(D) = \frac{1}{\sigma(2\pi)^{1/2}} \exp\left(-[D-\bar{D}]^2/2\sigma^2\right) \qquad \text{(eq. 1.3)}$$

$$\text{Lognormal Distribution } F(D) = \frac{1}{D \ln \sigma_g (2\pi)^{1/2}} \exp\left(-[\ln D - \ln D_g]^2/2 \ln^2\sigma_g\right) \qquad \text{(eq. 1.4)}$$

where $\bar{D}$ is the mean value of the diameter, σ the standard deviation, D_g the geometric mean (or count median diameter), and σ_g the geometric standard deviation. Two parameters, one measuring the central tendency and the other the spread of the distribution, uniquely describe each curve.

Considering the lognormal distribution function further, the cumulative count distribution function, giving the fraction of particles below a given diameter D, is

$$C(D) = \int_o^D \frac{1}{D \ln \sigma_g (2\pi)^{1/2}} \exp\left(-[\ln D - \ln D_g]^2/2\ln^2\sigma_g\right) d\ln D \qquad \text{(eq. 1.5)}$$

When C(D) is plotted versus D on log-probability graph paper, a straight line results. The point at which C(D) equals 0.5 determines the count median diameter, D_g. The geometric standard deviation σ_g is found from the values of D for C(D) = 0.5, 0.16, and 0.84 by the relationships:

$$\sigma_g = \frac{D_{50}}{D_{16}} = \frac{D_{84}}{D_{50}} \qquad \text{(eq. 1.6)}$$

For convenience, aerosol size distribution parameters are often determined from cumulative plots of sizing data on log-probability graph paper.

Numerical computational methods are also useful for estimating D_g and σ_g. If the particle sizing data are organized into pairs of numbers, one being the midpoint of a diameter interval D_i and the other being the number of particles in that diameter interval N_i, then estimators for D_g and σ_g are:

$$\ln CMD = \sum_i \frac{N_i \ln D_i}{N} \qquad \text{(eq. 1.7)}$$

$$(\ln \sigma_g)^2 = \sum_i \frac{N_i(\ln D_i - \ln CMD)^2}{N-1} \qquad \text{(eq. 1.8)}$$

where k is the number of intervals, i = 1→k, and N is the total number of particles such that:

$$N = \sum N_i \qquad \text{(eq. 1.9)}$$

The equations of Hatch and Choate (1929) allow one to estimate the volume median diameter (VMD) and the surface median diameter (SMD) for a particle population that is log-normally distributed with a known D_g and σ_g:

$$\ln VMD = \ln CMD + 3 \ln^2 \sigma_g \quad \text{(eq. 1.10)}$$

$$\ln SMD = \ln CMD + 2 \ln^2 \sigma_g \quad \text{(eq. 1.11)}$$

Both the volume and surface distributions have a σ_g theoretically identical to that of the count distribution.

In order to aid in the understanding of geometric standard deviation, it is useful to consider some collections of hypothetical particles, each collection with the same count median diameter, but having different geometric standard deviations. The collections of dark circles depicted in Figure 1.3 represent geometric standard deviations of 1.1 (essentially monodisperse), 2.0, and 3.0. These values span the typically encountered geometric standard deviations for aerosols generated by a single source.

AEROSOL PROPERTIES

Shape

As with macroscopic objects, aerosol particles exist in a large variety of shapes (Fig. 1.4). For practical purposes, four categories of shape appear to be adequate for describing particles of interest in inhalation studies. The first three categories are defined in terms of three mutually perpendicular axes convergent at the center of the particle. If the particle boundary extends to about the same distance along each axis, the particle can be classified as *globular* or roughly spherical in appearance. Liquid particles are typically spherical as are many viable particles and particles formed by condensation of supersaturated vapors or evaporation of droplets.

If one axis of the particle is much shorter than the other two, a plate-like, flat shape is obtained. Examples of this shape include particles of graphite, talc, mica, and

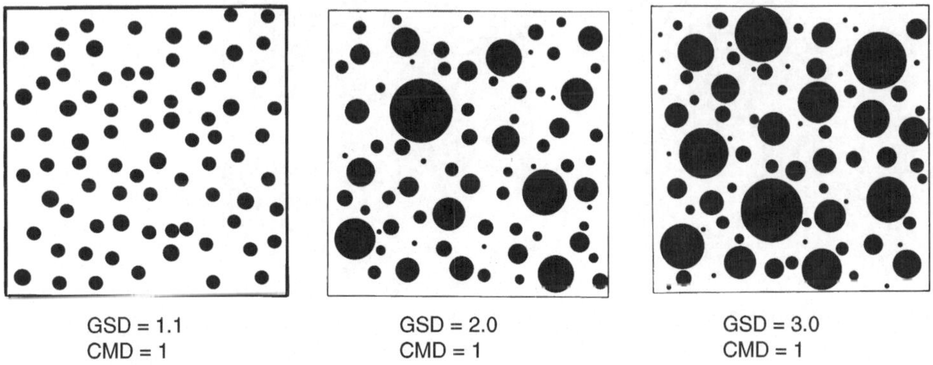

Figure 1.3 Collections of circles each having a count median diameter (CMD) of 1 and various geometric standard deviations (GSDs). *Left*: GSD = 1.1; *Center*: GSD = 2.0; *Right*: GSD = 3.0.

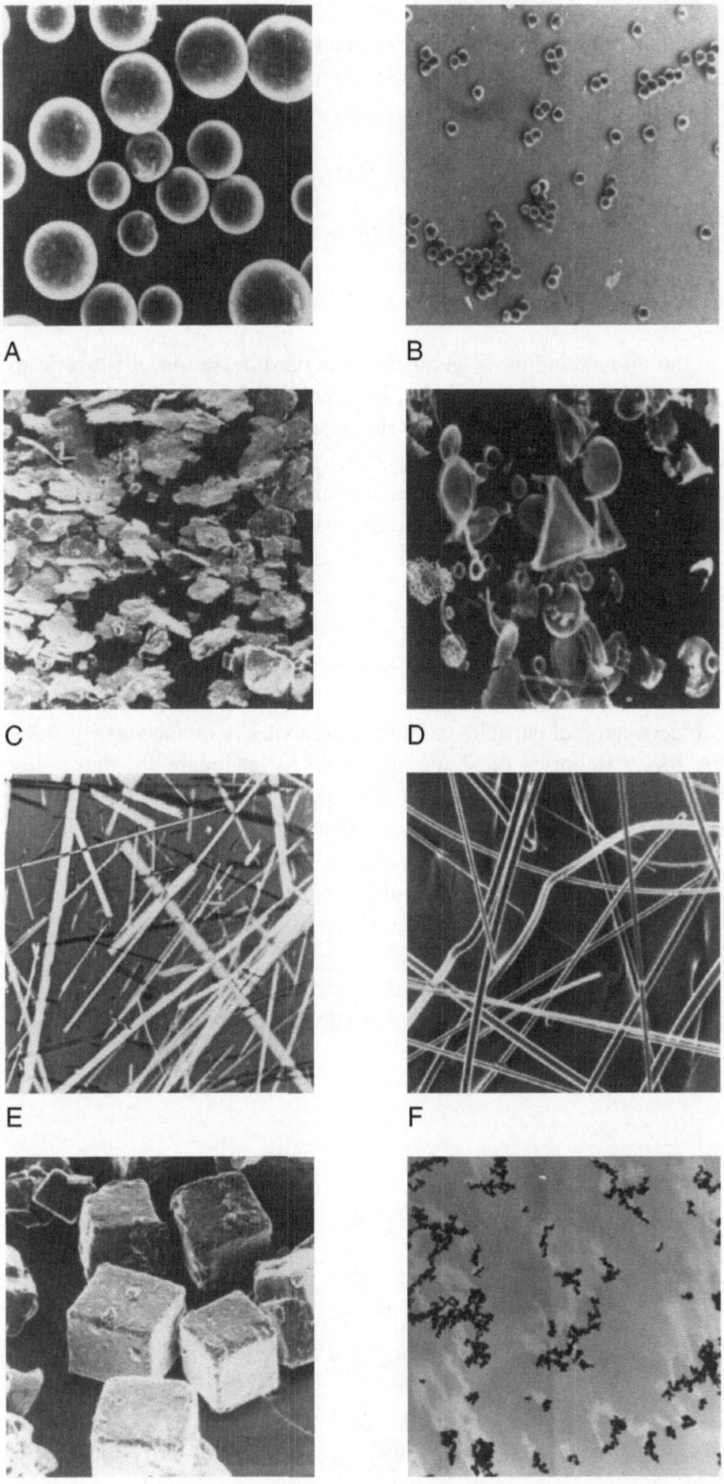

Figure 1.4 Examples of common particle shapes: (**A**) polystyrene-latex spheres; (**B**) ragweed pollen; (**C**) bronze powder; (**D**) diatoms; (**E**) asbestos fibers; (**F**) mineral wool fibers; (**G**) sodium chloride crystals; and (**H**) metal (silver) fume agglomerates. *Source*: (**A–G**) from *The Particle Atlas*, Vols. 2 and 3, 1973, with kind permission of the McCrone Institute.

insect scales. Plate-like particles are usually formed by mechanical means, for example, abrasion.

Fibers represent an important shape of particle for which length along one axis is three or more times greater than along the other two. Asbestos, glass fibers, mineral wool, hair fragments, and various plastic fibers are examples. These particles tend to line up in a stream of moving air with their long axis parallel to their direction of motion, and therefore can have aerodynamic diameters that are strongly influenced by their smallest dimensions rather than their lengths. For this reason, fibers can deposit more deeply in the respiratory tract than globular particles having the same volume. Also, particles of this shape may resist engulfment by mobile cells in the lung and remain in the body for long periods.

A fourth class of particles is those whose shapes are so irregular as to exclude them from the other categories. Examples are the star-like particles of zinc oxide smoke and the clusters of small primary particles that comprise soots, fly-ash, and metal fumes. Chain agglomerate particles which may consist of hundreds of small primary units are commonly encountered. Their shapes are fractal (Moskal et al., 2006), often approximating those of distorted spiders, having long, often branched, chain-like limbs. Such particles may have very large, specific surface areas (surface area per unit mass) and correspondingly low values (as small as 10% of the parent material) of effective particle density. The large surface and low densities of branched-chain agglomerates permit them to follow airstreams and remain suspended in air for long periods. They also can penetrate deeply into the lung and can deliver relatively large amounts of materials to alveolar surfaces. The persistence of such particles in the lung is often brief, since a large surface is available for dissolution. Metal fumes of lead that are inhaled may have half-times of a few hours in the lung, with solubilized lead appearing rapidly in the bloodstream. The distinction between primary particles and agglomerates, that is, clusters, of such particles should be emphasized. Typically, agglomerates are quite stable, being held together by electrical and molecular forces. The aerodynamic behavior of agglomerates may differ considerably from the properties of their individual primary particles.

Density

Particle density, or the amount of mass per unit volume of a particle, influences behavior in many important ways. For example, two spherical particles of the same diameter that differ with respect to their densities will have different masses and different terminal settling velocities in still air. If their geometrical diameters are above about 1 μm, the settling velocity of each will be directly proportional to density over a wide range of densities. Conversely, if two spherical particles have the same settling velocity, the one with lower density will have greater mass, and hence can potentially deliver a greater amount of toxicant if deposited in the lung. Also, for two spherical particles of equal mass, the one with greater density will be geometrically smaller and its motion in air influenced more by Brownian bombardment.

The values for densities of various materials that are found in handbooks are seldom applicable to aerosol particles. Most aerosol particles have apparent densities that are significantly less than the corresponding bulk material values. This is due to porosity, that is the presence of voids in the particles. Beeckmans (1964) has shown that it is not uncommon for up to 90% of a particle volume to be void. Values in Table 1.4 are typical for aerosol particle densities. Measurement of particle density is not a simple matter; methods used include flotation tests in various fluids, and the simultaneous determination of geometrical diameter and aerodynamic diameter followed by calculation of density.

Table 1.4 Densities of Aerosol Particles in Comparison With Densities of Bulk Material of the Same Composition

Material	Method of formation	Particle density (g/cm^3)	Bulk density (g/cm^3)
Au	Electric arc	0.2–8	19.3
Ag	Electric arc	0.64–4.22	10.5
Hg	Heating	0.07–10.8	13.6
MgO	Burning Mg	0.24–3.48	3.6
$HgCl_2$	Heating	0.62–4.3	5.4
CdO	Electric arc	0.17–2.7	6.5

Source: Adapted from Fuchs (1964).

Electrical Charge

Aerosol particles in nature and in the laboratory often possess net electrical charges. In fact, in any aerosol under ordinary conditions a significant fraction of the particles will be charged, even if an attempt has been made to discharge them. The presence of electrical charges on aerosol particles will influence their behavior in several ways including altering their rates of coagulation, their deposition on surfaces, and their trajectories in the presence of electromagnetic fields. This section covers the mechanisms that produce charges on particles, the distribution of electrical charges on particles, the rate of decay of charged aerosols, and methods for producing equilibration of charges in the laboratory.

Mechanisms of Charging

Aerosols usually acquire electrical charges as they are generated. One mechanism involved is the *triboelectric effect*, which is sometimes known as frictional charging. Triboelectric charging occurs when materials with different dielectric constants (a measure of their insulator characteristics) come into contact and are then separated. Although the act of rubbing two dissimilar materials together serves to make and break contact, friction per se does not appear to be necessary for triboelectric charging. In a familiar experiment where a glass rod is rubbed against a silk cloth, the rod acquires a positive charge and the silk a negative one. In general, when two materials are brought into close contact, the better conductor (lower dielectric constant) will become negatively charged.

In aerosol generators, particles may briefly contact surfaces made of metal, rubber, glass, or plastic. This contact can lead to the acquisition of charge. Similarly, when heterogeneous bulk powders are dispersed, differences in dielectric constants between adjacent particles can lead to charges in the resultant aerosol. Even when a powder consisting of all one material is dispersed, some particles will be positively charged and an equal number negatively charged. In the case of dispersion of such powders, Kunkel [cited by Mercer (1973)] found that the average charge per particle, $\overline{q}$, is approximately directly proportional to the particle diameter, D.

$$\overline{q} = 25\ D \qquad \text{(eq. 1.12)}$$

where $\overline{q}$ is expressed in electron charge units and D in μm.

When aerosols are formed by the disruption of ion-containing liquids, the statistically random motion of free ions leads to unequal numbers of positive and negative ions in individual droplets. As the droplets separate, those carrying a net charge will attract ions of the opposite polarity and some loss of charge will occur. Neglecting this latter effect, Smoluchowski, and later Natanson, have given expressions for the average number of charges as a function of formed droplet diameter, D (μm), and the ion concentration in the liquid, N (ions of one sign per cubic centimeter of liquid) (Mercer, 1973).

$$\bar{q} = 8.2 \times 10^{-7} D^{3/2} N^{1/2}, \text{ for } N < 10^{15} \text{ per cm}^3 \qquad \text{(eq. 1.13)}$$

$$\bar{q} = 5.6\, D^{1/2}, \text{ for N between } 10^{18} \text{ and } 10^{20} \text{ per cm}^3 \qquad \text{(eq. 1.14)}$$

The breakup of liquids can lead to either symmetrical charge distributions, or at very high ionic concentrations, unsymmetrical ones.

Another mechanism that can lead to the production of charges on particles is interaction with air ions. Such ions are always present due to natural radioactivity (cosmic radiation, radon gas, and other radioelements) but can be greatly increased by high voltage discharges or large electrical fields. A dissymmetry in the conductivity (current per unit potential difference) of positive and negative air ions can lead to assymmetric charges on aerosol particles due to unequal ionic attachment or discharging rates.

Radioactive particles that emit charged subatomic particles, such as alpha or beta radiations, will acquire net charges. Although the ionization of air due to the emissions will lead to some self-discharging, such radioactive particles will usually acquire net charges. Significant net charges can be expected for particles with nuclear decay rates above about 1 or 2 disintegrations per minute. Radioactive self-charging has been treated theoretically and experimentally by Yeh (1976) and by Yeh et al. (1976).

Attraction of a Charged Particle to a Nearby Conductor

When a particle with charge, q, in free space is at a distance, d, away from an electrically conductive surface, the physical requirement that the conductor's surface have a uniform potential leads to an interesting phenomenon. Charges within the conductor redistribute to produce a counterfield that, in effect, is equivalent to the electrical field that would be produced by a second oppositely charged particle, −q placed behind the conductive surface at a distance d. This second, apparent charge, similar to a virtual image, is called an *image charge*. The two oppositely charged particles then produce fields that just cancel at the surface of the conductor, producing the equipotential condition. Curiously, the real charged particle experiences a force, as if it were attracted to its image counterpart. The magnitude of this attractive force, which tends to attract the particle to the conductive surface as given by Coulomb's law (in c.g.s. units), is:

$$F = \frac{q^2}{d^2} \qquad \text{(eq. 1.15)}$$

Where F is in dynes, q in statculombs, and d in centimeters (Hinds, 1999, Chapter 15). The attractive force is directed along the line joining the particle and its image. Cohen et al. (1996) have shown that image, and space charge, effects can increase inhaled aerosol deposition.

Charge Distributions

Charged aerosol particles attract oppositely charged gaseous ions, and gain and lose charges. Under usual environmental conditions, an equilibrium state will eventually be achieved. The equilibrium charge distribution on an aerosol in the presence of bipolar ions in air (with equal positive and negative conductivities) has a symmetrical form with the most likely charge per particle being zero. In this case, *Boltzmann equilibrium* is established and the average number of charges, positive or negative, $\bar{q}$ (in election charge units), on individual particles is a function of particle diameter, D (in μm).

$$\bar{q} = 2.37\ D^{1/2} \quad \text{(eq. 1.16)}$$

If the positive and negative conductivities of the air (or other gas) are not equal, the particles will acquire a net (nonzero) average charge.

Decay Rate of Charges on Particles

In the laboratory it is often desirable to hasten the discharging rate of aerosols. One device used for this purpose is the 85-Kr aerosol discharger. These units, commercially available (TSI, Shoreview MN, U.S.A.) consist of metal tubes that contain a sealed source of 85-Kr, a beta-emitter that produces both positive and negative air ions within the tube. Liu and Pui (1974) have extensively studied 85-Kr neutralizers. The dischargers studied had a Kr-containing central steel tube (3 mm diameter, 0.076 mm thickness) inside of a larger metal tube through which an aerosol may be passed. They found that Boltzmann equilibrium was achieved for highly charged aerosols when the product (Nt) of ion concentration, N (ions per cubic centimeter of air), and the residence time, t (seconds), for the flowing aerosol exceeded 6×10^6 ion-s/cm^3.

The use of such dischargers, or another source of bipolar ions, is highly desirable for neutralizing freshly generated aerosols in inhalation studies. Aerosols that are not brought to charge equilibrium experience unwanted deposition either within inhalation delivery systems or on external surfaces of the experimental subjects. Table 1.5 shows examples of maximum flow rates for equilibrium discharge of highly charged aerosols in a laboratory discharger. For some applications, a 210-Po neutralizer or a nonradioactive ion generator may be preferable (Covert et al., 1997; Hinds and Kennedy, 2000; Ji et al., 2004).

The loss or gain of particle charge due to interaction with air ions proceeds in an exponential fashion. The higher the level of charge, the more rapid the rate of attraction

Table 1.5 Maximum Flow Rates for Charge Equilibration in an 85-Kr Discharger[a]

Discharger radioactivity (mCi)	Maximum airflow (L/min)
0.5	14
1	17
2	23
10	107

[a]*Note*: For an 85-Kr neutralizer of 1.4 L volume and with the radioactive source in a thin steel tube 0.076 mm in diameter, the Nt product (10^6 ion pair s/cm^3) for equilibrium neutralization of maximally charged particles is achieved by the airflow conditions given; Liu and Pui (1974).

of oppositely charged ions and thus the more rapid the loss of net charge. The rate of charging of uncharged particles can be conveniently expressed by a half-time for reaching the equilibrium charge distribution. Under ordinary conditions in air, this half-time is approximately 400 s (Gunn, 1955). An estimate for the loss rate of charge of charged particles in the atmosphere is difficult to make, but using Liu and Pui's Nt product, about 4 h may be required for a freshly generated aerosol to reach Boltzman equilibrium in the atmosphere.

Light Scattering

Aerosol particles both absorb and scatter incident light. Extinction, or reduction in intensity in the direction of the incident light, is due to both absorption and scattering. Absorption involves the conversion of radiant energy into heat and other forms of energy, including excitation of chemical bonds. Scattering involves the re-radiation of incident light without a change in wavelength. For a single macroscopic homogeneous body, the intensity of the original light beam, I_o, is related to the intensity, I, behind the body by the relationship:

$$I = I_o\, e^{-kx} \qquad \text{(eq. 1.17)}$$

where k is the extinction coefficient, and x the thickness of the body. The ratio of transmitted light intensity to the original intensity, I/I_o, is called the *transmittance*. The product of the extinction coefficient and the thickness, kx, is called the *turbidity*.

For a single particle, the extinction coefficient, or area efficiency factor E, is (Hodkinson, 1966):

$$E = \frac{\text{total energy scattered and absorbed by the particle}}{\text{energy geometrically incident on the particle}} \qquad \text{(eq. 1.18)}$$

For several particles, n per unit volume of air, each having a projected area a (intercepted by the light beam), k equals naE and

$$I = I_o\, e^{-naEx} \qquad \text{(eq. 1.19)}$$

The single particle extinction coefficient, E, varies with particle shape, size, composition, and the wavelength of light.

Scattered light, or more correctly, re-radiated light of the same wavelength, is not in general scattered equally in all directions. The angular dependence of the re-radiated light is quite complex and again dependent upon particle shape, size, composition, and the light wavelength (Hodkinson, 1966).

Several instruments have been developed to exploit aerosol light scattering (see Chapter 4). Aerosol characterization using optical properties has the advantage of minimizing disturbance of the aerosol. Instruments that detect either extinction or scattered intensity over specific angular regions have been used. However, the limitations of such instruments are often severe. Some of these limitations include:

1. Coincidence errors due to overlap of particles in the light beam.
2. Errors due to calibration with particles of one shape or composition followed by use of the instrument with other types of particles.

3. Insensitivity to particles below a given size.
4. Errors due to nonrandom orientations of nonspherical particles in the light beam.
5. Heating and alteration of particles that are easily evaporated.

Hygroscopicity

Hygroscopic aerosol particles, that is those that are soluble in water and grow upon entering a higher humidity environment, comprise a significant fraction of those inhaled. Examples include sea salt, sulfuric acid droplets, tobacco smokes, various medicinals, and a variety of spray-generated aerosols. Since under ordinary conditions the relative humidity in the respiratory tract will exceed that of the gas phase of an inhaled aerosol, water-soluble aerosol particles will grow upon inhalation. Whether or not hygroscopic growth will alter the deposition of these particles is an important question. The Task Group on Lung Dynamics of the International Commission on Radiological Protection (TGLD, 1966) recommended that the equilibrium, or final, size of hygroscopic particles be used for calculating inhalation doses. Computational schemes that take growth rate into account have also been proposed (Landahl, 1972; Ferron, 1977; Martonen, 1982; Ferron and Busch, 1996; Schum and Phalen, 1997; Finlay, 2001). These more sophisticated approaches appear to be warranted, since many hygroscopic particles grow significantly during passage through the airways. Larger particles continue to grow for a longer period of time and smaller particles complete their growth more rapidly. As an example, the calculated growth-time curve for a dry sodium chloride particle of 1 μm initial diameter inhaled into the tracheobronchial tree indicates that the diameter increases at a significant rate during the first 1–5 s, and reaches a final diameter of 4 μm in approximately 10 s. Therefore, more proximal deposition is expected (Ferron and Busch, 1996). Many materials of interest to inhalation toxicologists are hygroscopic. However, even trace quantities of nonhygroscopic substances in particles can greatly alter growth rates (Bell and Ho, 1981), so caution must be used when applying theoretical growth models to real-world aerosols.

Surface Area

The specific surface, or the surface area per unit of mass of small particles, can be astonishingly large. For perfectly smooth spheres, the specific surface area is calculated using the relationship:

$$S_S = \frac{\text{surface}}{\text{mass}} = \frac{6\,\pi\,D^2}{\pi\,D^3\rho} = \frac{6}{D\rho} \qquad \text{(eq. 1.20)}$$

where D is the particle diameter and ρ its density. This equation gives the minimum possible value for a specific surface, and particles that are not spherical with smooth unbroken surfaces will have larger values. Environmental particles can have rough surfaces, pores, cracks, or internal voids that greatly increase their specific surfaces. For such particles, the surface area available for physical or chemical interactions can be obtained by experimental measurement. The fundamental principles that underlie such measurement were described by Brunauer et al. (1938) (the *BET method*). In such measurements, the surface area is derived from knowledge of the amount of a gas that condenses on the particles at various gas partial pressures. The amount of gas adsorbed on the particles can be determined experimentally using a sensitive microbalance or other measurement techniques.

Values of surface areas obtained using these methods are dependent upon the gas used, as some gases penetrate into small crevices better than others. Nitrogen adsorption isotherms are commonly used to determine particle surface areas. Methods and experimental apparatuses appropriate to the measurement of particle-specific surface areas are described by Mercer (1973), and a variety of surface area–measuring instruments are available from Micrometrics Instrument Corporation (Norcross, GA, U.S.A.). Table 1.6 shows the specific surface areas of various particles.

The importance of particle surface areas in inhalation toxicology derives from three phenomena: the ability of aerosols to adsorb gases and thus carry them into the deep lung; the tendency of the particle surface to promote or catalyze chemical reactions, both while airborne and after deposition in the respiratory tract; and the effect of surface area on the rate of dissolution of particles deposited in the lung (Mercer, 1967). The specific surfaces of particles used in inhalation experiments are seldom measured in practice. This often-neglected factor will require more attention in the future.

AEROSOL DYNAMICS

Particle Motion

One of the important properties of aerosol particles is their long suspension times in the air. In order to understand this stability and to make quantitative estimates of aerosol deposition in exposure systems and respiratory systems, it is necessary to examine particle motion, that is *aerosol dynamics*. Three types of forces must be considered to understand the motion of aerosol particles: (1) external, including gravitational and those produced by electrical fields; (2) resistance, which arises from the surrounding gaseous medium; and (3) interaction, among particles (electrical attraction or repulsion for example). Aerosol mechanics usually proceeds from a consideration of the first two types of forces, the third usually being ignored, or included as a correction. For a more thorough presentation of particle motion, see Chapters 3 through 7 in Hinds (1999).

Table 1.6 Specific Surface Areas of Particles of Various Types and Sizes

Type	Diameter (μm)	Specific surface (m^2/g)	Notes
Liquid water	100	0.06	Calculated
	10	0.6	Calculated
	1	6	Calculated
	0.1	60	Calculated
	0.01	600	Calculated
Quartz	100	0.05	Measured[a]
	10	0.4	Measured[a]
	1	3.5	Measured[a]
Silver fume	0.03	16	Measured[b]
Charcoal	—	800	Measured[c]

Sources: [a]Data of Foust et al. in Chapter 1 of Hesketh (1977). [b]Phalen; (1972); [c]Brunauer et al. (1938).

Gravitational and Buoyant Forces

Gravity pulls airborne particles toward the earth with a downward force F_G, known as the particle weight, that is equal to the product of the particle mass M_p, and the gravitational acceleration constant g (980 cm/s^2).

$$F_G = M_p g \quad \text{(eq. 1.21)}$$

Since the particle mass is its volume times its density, ρ_p, one can write:

$$F_G = g\pi D_p^3 \rho_p /6 \quad \text{(eq. 1.22)}$$

This gravitational force is offset by an oppositely directed (given a negative sign) buoyant force, F_B. This force, given by Archimedes' principle is equal to the weight of the fluid displaced by the particle, that is, the particle volume times the fluid density ρ_f.

$$F_G = -g\pi D_p^3 \rho_f /6 \quad \text{(eq. 1.23)}$$

The net force F_{GB} is then the sum of gravitational and buoyant forces.

$$F_{GB} = g\pi D_p^3 (\rho_p - \rho_f)/6 \quad \text{(eq. 1.24)}$$

The magnitude of F_{GB} is quite small compared with F_G, as is seen by substitution of numerical values for a unit density, 1-µm diameter particle.

$$F_{GB} = (980 \text{ cm/s}^2)(3.14)(1 \times 10^{-4} \text{ cm})^3 (1 - 10^{-3} \text{ g/cm}^3)/6 \quad \text{(eq. 1.25)}$$

$$= 5.13 \times 10^{-10} \text{ g cm/s}^2$$

$$= 5.13 \times 10^{-10} \text{ dynes}$$

The Resistance or Drag Force

The magnitude of the drag or resistance force due to the interaction of air with a moving particle will depend upon the size of the particle D_P, its velocity relative to the air U, and the viscosity of the air η. Formulation of this force is straightforward in the size regime where the particle is large with respect to the mean free-path length between collisions among air molecules. In this circumstance, the particle experiences the surrounding medium as a continuous fluid rather than a collection of rapidly moving, individual molecules of gas. Also, if the particle is solid and small enough that its movement does not greatly disturb and accelerate the air fluid, only viscous forces need be considered. When these conditions apply, one is dealing with Stokes' or continuum mechanics and the drag force, F_D, is opposite to the direction of motion and is given by:

$$F_D = 3\pi \eta D_p U \quad \text{(eq. 1.26)}$$

Stokes' law applies to particles in air having diameters between about 1 and 100 µm. As will be seen below, Stokes' law can be extended downward to particles with diameters as small as about 0.05 µm by application of a correction factor (the Cunningham slip correction).

Terminal Settling Velocities

A particle falling under the influence of gravity will steadily accelerate until the air drag, F_D, increases and becomes equal to the net accelerating force, F_{GB}. At this point, the net force on the particle is zero, and it will continue moving at a constant velocity. This constant velocity, called the terminal settling velocity U_t, is found by setting the gravity-buoyant force equal to the drag force at velocity U_t and solving for U_t, as shown below.

$$F_{GB} = F_D \qquad \text{(eq. 1.27)}$$

$$g\pi\, D_p^3(\rho_p - \rho_f)/6 = 3\pi\,\eta\, D_p U_t \qquad \text{(eq. 1.28)}$$

$$U_t = \frac{gD_p^2(\rho_p - \rho_f)}{18\eta} \qquad \text{(eq. 1.29)}$$

In Stokes' mechanics, the terminal settling velocities can be very small, as is seen by substituting numerical values for a unit density, 1-μm diameter sphere.

$$U_t = (980\ \text{cm/s}^2)(1 \times 10^{-4}\ \text{cm})^2(1 - 10^{-3}\ \text{g/cm}^3)/18(184 \times 10^{-6}\ \text{g/s} - \text{cm}) \qquad \text{(eq. 1.30)}$$

$$= 3.0 \times 10^{-3}\ \text{cm/s}$$

$$= 30\ \mu\text{m/s}$$

The strong dependency of settling velocity, U_t on particle diameter can be seen in Table 1.7. Note that the terminal settling velocity is directly proportional to particle density, thus one can multiply the tabulated settling velocity values by the density of nonunit density spheres to obtain an estimate of their settling velocities.

Slip, or Cunningham's Correction

As the particle diameter decreases, the Stokes' assumption of a continuous, isotropic medium becomes less valid. The spaces between molecules of air are no longer negligibly small in relation to the particle. This and other molecular kinetic phenomena lead to a reduction in the drag force, F_D, in a manner that is strongly particle-size dependent. That is:

$$F_D' = F_D/K_s \qquad \text{(eq. 1.31)}$$

$$\text{and}\ \ U_t' = U_t K_s \qquad \text{(eq. 1.32)}$$

where the prime superscript means slip-corrected, and K_s is the unitless Cunningham slip factor, which is always greater than 1. This factor, named after Cunningham for his theoretical treatment, is usually calculated from empirical formulae that are functions of either the molecular mean free-path and particle diameter, or of the atmospheric pressure and particle diameter. One convenient form for calculating K_s was proposed by Davies (1945):

$$K_s = 1 + (2/PD)(6.32 + 2.01\ \exp[-0.1095\ PD]) \qquad \text{(eq. 1.33)}$$

Table 1.7 Terminal Settling Velocity, Uncorrected and Slip Corrected, Cunningham Slip Factors, and Root-Mean-Square Displacement (Brownian) for Various Diameters of Unit Density (1 g/cm^3) Spherical Particles

Particle diameter (μm)	Cunningham slip correction factor	Sedimentation velocity (cm/s)	Corrected sedimentation velocity (cm/s)	r.m.s. Brownian displacement (cm/s)
0.01	22.2	3.01×10^{-7}	6.69×10^{-6}	2.23×10^{-2}
0.05	4.97	7.53×10^{-6}	3.74×10^{-5}	6.84×10^{-3}
0.10	2.87	3.01×10^{-5}	8.63×10^{-5}	3.67×10^{-3}
0.50	1.33	7.53×10^{-4}	1.00×10^{-3}	1.11×10^{-3}
1.00	1.16	3.01×10^{-3}	3.50×10^{-3}	7.40×10^{-4}
5.00	1.03	7.53×10^{-2}	7.77×10^{-2}	3.12×10^{-4}
10.0	1.02	3.01×10^{-1}	3.06×10^{-1}	2.19×10^{-4}
50.0	1.00	7.29	7.31	9.72×10^{-5}
100.0	1.00	2.48×10^{1}	2.48×10^{1}	6.87×10^{-5}

Source: Adapted from Tables for Use in Aerosol Physics, BGI Inc., 1971.

where P is the atmospheric pressure in cm of Hg and D the particle diameter in μm. The slip factors at 1 atm for particle diameters of 1.0 and 0.1 μm are 1.16 and 2.9, respectively. Table 1.7 should be examined in order to appreciate the effect of slip on particle terminal settling velocity.

Brownian Motion

The random bombardment by gas molecules leads to randomly directed changes in the velocities of aerosol particles. This phenomenon leads to a displacement of particles that is superimposed upon their displacements under the action of other forces. Such movement, called *Brownian motion*, is particularly important for small particles less than about 0.5 μm in diameter, and leads to such phenomena as coagulation, spreading of aerosol clouds, and deposition on surfaces of chambers, tubing, and the respiratory tract. Treatment of this phenomenon is found in Fuchs (1964) and Mercer (1967). The average particle velocity due to Brownian bombardment is greater for higher atmospheric temperatures and for smaller particle diameters. Table 1.7 gives the average displacement speeds due to diffusion for a variety of particle sizes at 1 atm air pressure and 20°C temperature. The table also gives slip-corrected displacements. The net effects of diffusion and sedimentation on particle displacements in still air are shown in Figure 1.5. Note the minimum in this curve at a particle diameter of about 0.3 μm. Such particles are minimally influenced by the combined actions of molecular bombardment and gravity, and thus are characterized by relatively long suspension times in air and low deposition efficiencies in the respiratory tract when inhaled.

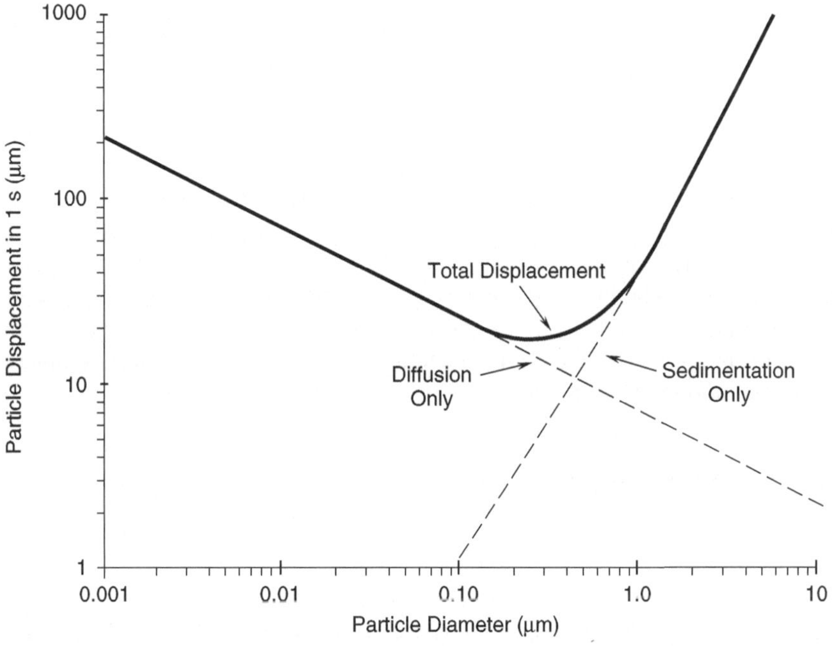

Figure 1.5 Displacement of a particle in 1 s in still air as the sum of diffusional and sedimentational displacements (particle density = 1 g/cm^3).

Coagulation

Airborne particles that come into contact will usually stick together and form a single particle. This mechanism, called *coagulation*, leads to a steady decrease in the number of particles, and an increase in the average particle size.

Coagulation is driven by phenomena that produce net differences in the velocities of individual aerosol particles. Such phenomena include diffusion (due to Brownian motion), aerodynamic polydispersity leading to differences in settling velocities, velocity gradients in air during laminar or turbulent flow, differences in particle electrical mobility, and particle motion produced by external forces such as those produced by light beams. In inhalation studies, aerosol coagulation is usually unwanted, as it can lead to enhanced losses due to sedimentation (but decreased losses due to diffusion), as well as an increase in particle size.

Coagulation involves two particles coming together. In the case where all particles are the same size, it proceeds at a rate proportional to the square of the number of particles, N, per unit volume. In the case of diffusion-produced coagulation of monodisperse particles, the theory of von Smoluchowski (1917) applies. It predicts that the rate of change (negative) in the number of particles per unit volume of air N, is related to N^2 by a constant, K_c, the coagulation coefficient.

$$\frac{dN}{dt} = -K_c N^2 \qquad \text{(eq. 1.34)}$$

K_c, is related to the Cunningham slip factor, K_s, and thus particle size, by:

$$K_c = 3 \times 10^{-10}\ K_s \mathrm{cm^3/s} \qquad \text{(eq. 1.35)}$$

By integration of the differential equation, and imposition of the initial condition, N at time zero equals No, one obtains:

$$N = No/(1 + NoK_c t) \qquad \text{(eq. 1.36)}$$

In this equation one sees that coagulation is a function of the initial number of particles and the coagulation coefficient. Under realistic conditions, the coagulation rate of aerosols will usually differ considerably from that calculated above. This is due to several factors, including any deviation from monodispersity (which will occur as particles coagulate), air currents, electrical charges on particles, and other factors. Thus the formulae given only provide an estimate of the decline in number concentration. For more complete treatments of coagulation see Zebel (1966) or Chapter 12 in Hinds (1999).

PARTICLE SIZE AND TOXICITY

Particle Mass

Many responses to inhaled materials are said to be dose dependent, that is, a greater response is associated with a larger *dose*. Dose is usually expressed in terms of the amount of toxicant present per unit mass of target tissue.

$$\text{Dose} = \frac{\text{Amount of toxicant}}{\text{Mass of target tissue}} \qquad \text{(eq. 1.37)}$$

The "amount" of toxicant is usually the toxicant mass. However, it may also be the particle number, surface area, chemistry or some other property. A list of some potentially toxic characteristics of particulate urban air pollution can be found in Phalen (2002, Table 6.2, p. 72). Recently, evidence has emerged indicating that ultrafine particles may be more toxic per unit mass than larger particles (Oberdorster, 2001).

The deposition of a few particles that have a large mass per particle may have greater impact than the deposition of numerous particles that are each small in mass. This fact is particularly relevant because the mass per particle for aerosols that deposit in the lung can vary over many orders of magnitude. For example, since the mass of a spherical particle is proportional to the cube of the geometric diameter, 1000 particles of 0.1 μm diameter must be deposited in the lung to equal the mass burden from the deposition of a single 1-μm diameter particle.

When particles of different individual masses are deposited in the respiratory tract, the number of cells which each directly affects may vary significantly with respect to mass per particle. A given amount of mass deposited in the respiratory tract may be distributed among numerous small particles or among fewer large particles, and the effect on overall toxicity of these different situations may be striking. Such considerations are probably less important for rapidly dissolved material, which persists as particles only briefly, and most important for material that is resistant to dissolution in the body.

The case of relatively insoluble radioactive particles of alpha-emitting materials deposited in the pulmonary region provides an example of how particle mass might influence toxicity. Since the major direct effect of the particles on the surrounding cells relates to the alpha emissions, each aerosol particle irradiates a small surrounding volume of the tissue. It can be argued that a given mass burden in the lung distributed among a few massive particles is less carcinogenic because the number of cells at risk is limited and those that are irradiated may in fact be over-irradiated and thus be sterilized, preventing development of neoplasia. On the other hand, it can be argued that the distribution of the lung burden in larger particles is more hazardous because of the large local radiation doses received by cells surrounding the particles, and it is less hazardous to have smaller radiation doses which are associated with smaller particles (even though more cells are irradiated). This so-called *hot particle* question bears on the environmental impact of nuclear technology. A similar problem exists with considerations of localized *hot spots* (radioactive or nonradioactive) produced by the nonuniform deposition within the lungs (Phalen et al., 2006).

Aerodynamic Properties

Types of solid particles that can be identified with respect to their shape and concomitant aerodynamic character include: globular particles that approximate spherical shapes; plate-like particles; long, thin particles or fibers; and clusters or agglomerates of particles. For relatively spherical particles of a given aerodynamic diameter, higher density particles have lower total mass. Hollow, or spongy, particles of a homogeneous material will therefore have more mass per particle at a given aerodynamic size. Differences in toxicity with respect to particle density are expected but have not yet been systematically studied.

Long, thin fibers have aerodynamic diameters nearly independent of their length up to a length-to-diameter ratio of about 20 (Timbrell, 1972). For this reason asbestos fibers containing considerable mass can behave like smaller particles aerodynamically and deposit deeply in the lung. Furthermore, such particles are not efficiently removed by phagocytic cells in the lung. These effects are emphasized by the cases of pulmonary disease from the asbestos industry.

An interesting aerosol in inhalation toxicology, the metal fume, consists of chain-like agglomerates of particles smaller than 0.1 μm. Pulmonary injury is known to be associated with the inhalation of metal fumes. This may be due to the large surface area associated with a given mass of fume aerosol. The aerodynamic drag on the large surfaces of fume particles allows them to follow air streams and escape impaction in the upper respiratory tract. As in the case of asbestos fibers, the ability of metal fume particles to penetrate to the deep lung undoubtedly contributes to their hazard.

Surface Area

Consider two categories of toxic particulate materials: those that require dissolution to produce their effects, and those that do not. Materials such as asbestos and quartz which are hazardous as solid particles appear to have toxic shape or surface characteristics. Other materials such as Pb and Mn probably require dissolution in order to be toxic. For both categories of particulate materials, the specific surface, or surface-to-mass ratio, affects their toxicity. The surface-to-mass ratio for smooth spherical particles is equal to $6/\rho D$, where ρ is the physical density and D the geometric diameter. Unit density particles of 1 μm diameter have a specific surface of 6 m^2/g, while 0.01 μm particles have an area of 600 m^2/g. The increased toxicity of finely divided silica, may relate to increased surface area. The mechanism for toxicity appears to involve a tissue reaction to the particle surface (Stober, 1968). Recent research on ultrafine particles indicates that particle surface area, or even particle count, may be an important factor in their toxicity, perhaps more important than particle mass in some cases (Oberdorster, 2001; Kreyling et al., 2006).

A model for dissolution of particles deposited in the lung that correlates well with experimental data on deep lung clearance was proposed by Mercer (1967). The model assumes a rate of dissolution that is proportional to the available surface area of the particles. For materials that are toxic when dissolved, increased surface area tends to enhance toxicity. The dissolution of silver particles (mass median diameter = 0.04 μm) in various aqueous media indicates that even a so-called *insoluble* material can undergo rapid dissolution when in a finely divided state (Fig. 1.6). On the basis of the dissolution rate found in the protein-containing fluid, Mercer's model predicts that these silver particles should essentially completely dissolve in the lung in about 48 h (Phalen, 1972). For a review of methods for measuring particle dissolution rates, see Ansborlo et al. (1999). For information on estimating particle dissolution in cells, see Stefaniak et al. (2005). Systemic toxicants in small particles can be expected to be more rapidly dissolved than in larger particles due to their larger specific surface areas, and hence can be more hazardous when deposited in the lung.

Other Size-Dependent Factors

Aside from the influence of particle size on magnitude and distribution of dose, deposition pattern, and dissolution rate, other size-related factors may bear on toxicity. An optimal particle size of 1.5 μm for efficient uptake of polystyrene spheres by macrophages was suggested by Holma (1967), who gave an upper limit of 8 μm diameter for phagocytic uptake. Kreyling et al. (2006) provide evidence that ultrafine particles are also not efficiently phagocytized, presumably due to their small signal. The question of relative efficiency of uptake by macrophages of the lung for particles in the respirable size range (about 0.01–10 μm) is worthy of further investigation.

The permeability of alveolar membranes to bare particles was reported by Gross and Westrick (1954) and by Tucker et al. (1973). In the earlier study, rats were given small

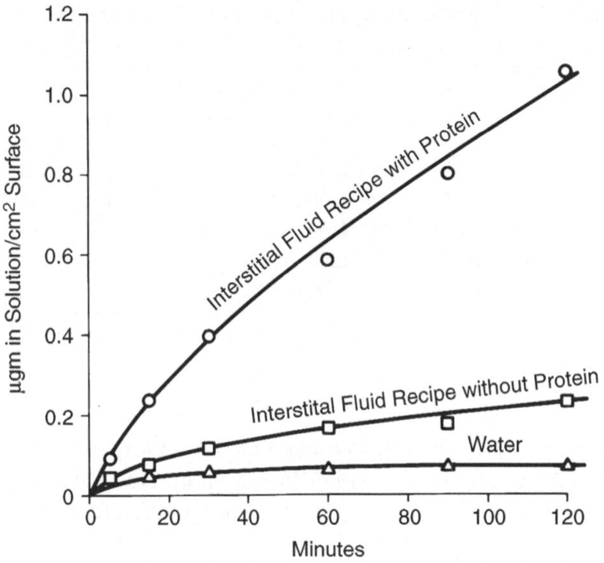

Figure 1.6 Dissolution of metallic silver fume particles in various aqueous media. The increased rate of dissolution with protein present is probably due to the binding of silver ions to protein. The primary particles had a count median diameter of 0.03 μm. *Source*: Redrawn from Phalen (1972).

carbon particles (< 0.2 μm) by intratracheal injection. The particles were found extracellularly in interstitial spaces 19 h later, which was considered by the authors to be proof of membrane penetration by bare particles. In Tucker's experiments, carmine particles ranging from about 5 μm down to below 0.05 μm diameter were inhaled by rats. At 3 h postinhalation, microscopic examination revealed "small aggregates, up to cell size" in the extracellular interstitial spaces (spaces between lung cells). Particulate material in these spaces would presumably either remain, dissolve, undergo transport to lymphatic or blood vessels, or be moved up the respiratory airway and subsequently swallowed. The role of particle size in membrane penetration is a topic of current interest (Kreyling et al., 2006).

Table 1.8 shows values for particle mass, surface area, surface-to-mass ratio, and number of particles in a microgram sample for unit specific gravity spheres. Note that particle mass varies over 15 orders of magnitude!

PROPERTIES OF GASES

Movement from Air into Tissues

What follows will primarily pertain to gases that do not avidly react with tissue. (Additional material on the uptake of inhaled gases may be found in Chapter 2.) The properties of molecular weight and relative equilibrium solubility in air, water, and oil media will largely account for the behavior of these gases upon inhalation.

The movement of a gas from air into tissue involves a series of steps. Inhalation brings the gas into the respiratory tract. Gases that are very soluble in water will be avidly

Table 1.8 Values for Particle Diameter, Mass, Surface Area, and Number of Particles in a Microgram of Material[a]

Diameter (μm)	Mass (g)	Surface (cm^2)	Surface/mass (cm^2/g)	No. /μg
0.001	5.24×10^{-22}	3.14×10^{-14}	6.00×10^{7}	1.91×10^{15}
0.01	5.24×10^{-19}	3.14×10^{-12}	6.00×10^{6}	1.91×10^{12}
0.1	5.24×10^{-16}	3.14×10^{-10}	6.00×10^{5}	1.91×10^{9}
1.0	5.24×10^{-13}	3.14×10^{-8}	6.00×10^{4}	1.91×10^{6}
10.0	5.24×10^{-10}	3.14×10^{-6}	6.00×10^{3}	1.91×10^{3}
100.0	5.24×10^{-7}	3.14×10^{-4}	6.00×10^{2}	1.91

[a]Valid for smooth spheres of unit specific gravity.

absorbed into the wet lining layer of the nose, mouth, and major airways. Gas not trapped in the upper airways will flow into the small airways and diffuse deep into the lung. This diffusion is driven by concentration gradients. If the atmospheric partial pressure of a gas is greater than that in the alveoli, the gas will undergo net movement into the alveolar region. In the alveolus, the gas will move toward equilibration with the lung surface. The initial rate of approach toward equilibrium is dependent upon the partition coefficient, that is, the ratio of relative volume of gas in air to that dissolved in tissue at equilibrium.

$$\text{Partition Coefficient} = \text{Distribution Coefficient}$$

$$= \frac{\text{Volume of gas in medium A}}{\text{Volume of gas in medium B}} \text{ (at equilibrium)} \qquad \text{(eq. 1.38)}$$

In the case of the alveolus, medium A can be thought of as air and medium B as water (representing tissue). Dissolved gas will distribute within the tissue, including the blood, and will partition among aqueous and nonaqueous (e.g., fat or lipid) compartments. Gas dissolved in the blood will distribute throughout the body and deposit in various organs according to their relative blood flows and compositions (e.g., percentages of water and lipid).

If a subject breathes a gas at a fixed concentration in air for a sufficient time, all tissues will effectively reach equilibrium and the rate of movement of molecules into and out of any tissue compartment will balance. This assumes that the gas does not react or chemically bind to any tissue components, simple solution being the only mechanism for uptake.

When the gas concentration in the ambient air drops sufficiently, the process reverses. Tissues with a rich blood supply tend to desorb more rapidly than those with a poor blood supply, and the gas is exhaled until it is completely washed out. Organs such as the brain that have a rich blood supply will tend to wash out more quickly than tissues such as body fat that have poor vascularization. Both brain and fat are rich in lipid and thus can hold large quantities of lipid-soluble gases.

Sophisticated physiologically based pharmacokinetic (PBPK) models are available for investigating inhaled gases and vapors in various species (Dennison et al., 2005).

Expressing Concentration

The concentration of a gas in air can be expressed many ways such as (1) percentage composition, parts per million by volume (ppm), parts per billion by volume (ppb), etc.,

(2) micrograms per cubic meter of air (μg/m^3), and (3) partial pressure in millimeters of mercury (mmHg) or other pressure units. In most toxicological research, gas concentrations are expressed in parts of the gas under consideration to parts of air. Such concentrations may be on a weight or volume basis, but in either case, unless the atmospheric pressure is stated, this means of expressing concentration is problematic. Consider nitrogen as an example. At 79% (790,000 ppm), this component of air is without significant harmful effects. However, at elevated pressures of 4–5 atm, as occur in deep sea diving, 79% nitrogen becomes an anesthetic gas that can produce disorientation and death. Although this is an extreme example, reduced or elevated atmospheric pressures can alter the driving force for uptake of gas molecules by tissues. Thus, volume percent, mass percent, or related units should be used with caution, and extrapolation of toxicity data taken at a given pressure to other pressures should generally be avoided.

Use of the weight of a gas per unit volume of air, for example mg/m^3, to express concentration is reasonable in that this unit is proportional to the number of molecules present regardless of the ambient pressure. Similarly, the partial pressure of a gas relates to the number of molecules present per unit volume of air and to the driving pressure for uptake into tissues. It is frequently useful to convert gas concentrations from the units ppm (volume) to mg/m^3 and vice versa. Assuming that the ideal gas law is sufficiently accurate under the prevailing conditions, the formula in Table 1.9 can be used. For convenience, several conversion factors are listed in the table.

Table 1.9 To Convert mg/m^3 (Mass of Gas/Volume of Air) to ppm (Volume Gas/Volume of Air), Multiply by k.[a]

Gas	Formula	Mol wt	k = 24.45/MW (ppm/[mg/m^3])
Acetylene	HCCH	26.04	0.939
Ammonia	NH_3	17.03	1.44
Carbon dioxide	CO_2	44.01	0.556
Carbon monoxide	CO	28.01	0.873
Chlorine	Cl_2	70.91	0.345
Fluorine	F_2	38.00	0.643
Formaldehyde	HCHO	30.03	0.814
Hydrogen	H_2	2.02	12.1
Hydrogen sulfide	H_2S	34.08	0.717
Methane	CH_4	16.04	1.52
Nitric oxide	NO	30.01	0.815
Nitrogen	N_2	28.02	0.873
Nitrogen dioxide	NO_2	46.01	0.531
Nitrous oxide	N_2O	44.02	0.555
Oxygen	O_2	32.00	0.764
Ozone	O_3	48.00	0.509
Phosgene	Cl_2CO	98.93	0.247
Sulfur dioxide	SO_2	64.07	0.382

[a]Valid for standard temperature and pressure.

$$\text{Formula used: } k = \frac{\text{molar volume (22.4 L/mole)}}{\text{Mol wt}}\left(\frac{T}{273}\right)\left(\frac{760}{P}\right),$$

where T = temperature (K) and P = pressure (mmHg).

Solubility

As previously mentioned, the degree of solubility of a gas in water is an important determinant of where the gas will deposit within the respiratory system. Gases that are highly water soluble are in general effectively removed by the upper airway surfaces and do not reach the deep lung in high concentrations. On the other hand, poorly water-soluble gases are not well absorbed in the upper airways and can better penetrate to the deep lung.

An important exception to these generalizations occurs in the presence of significant amounts of aerosol particles. Liquid droplets can carry dissolved gases into the lung and thus increase exposure of the deep lung to materials that would normally be trapped in the upper airways. Even dry particles can carry significant amounts of gases adsorbed on their surfaces.

2
The Respiratory Tract

INTRODUCTION

It is useful to compare the air breathed per unit of time to the corresponding intakes of water and food. Values for normal intakes of air, water, and food can be found in various references for men, women, and children (Altman and Dittmer, 1974; EPA, 1985; Schleien et al., 1998). As shown in Table 2.1, the *reference man* at rest inhales about 10,800 liters of air per day, and consumes about 1.7 liters of drinking water and about 1.6 liters of food in the same period. Because air at room conditions has a density of about 1.2 g/L, the corresponding daily masses of intake of air, water, and food are about 13,000, 1,700, and 1,600 g, respectively.

Consider a pollutant that is avidly (100%) retained by the body when taken in by any route. In the first case, assume that the pollutant is present in air, water, and food at a level of 1 ppm (weight of pollutant/weight of carrier); that is, each gram of air, water, or food carries 1×10^{-6} g of pollutant. In the second case, assume the pollutant is present at a concentration of 1×10^{-6} g/L of air, water, or food (1 ppm weight of pollutant/volume of air). Table 2.1 gives the doses of pollutant per day for each case. In the first case, pollutant present at 1 μg/g of carrier, the uptake via breathing is about eight times greater than that via drinking or eating. In the second case, pollutant present at 1 μg/L, the dose from air exceeds that from water or food by over 6,000 times. Owing to the great volumes of air breathed continuously, the potential for toxicity via this route can be comparatively overwhelming. This exercise cannot be applied to cases in which uptake is different for various routes of exposure.

Several excellent books that treat in detail the topics presented in this chapter are recommended, including: *Fundamentals of Extrapolation Modeling of Inhaled Toxicants: Ozone and Nitrogen Dioxide* (Miller and Menzel, 1984); *Inhalation Toxicology: The Design and Interpretation of Inhalation Studies and Their Use in Risk Assessment* (Dungworth et al., 1988); *Extrapolation of Dosimetric Relationships for Inhaled Particles and Gases* (Crapo et al., 1989); *Comparative Biology of the Lung* (Parent, 1991); *Respiratory Toxicology and Risk Assessment* (Jenkins et al., 1994); *Concepts in Inhalation Toxicology, 2nd Ed.* (McClellan and Henderson, 1995); *Nasal Toxicity and Dosimetry of Inhaled Xenobiotics: Implications for Human Health* (Miller, 1995); *Aerosol Inhalation: Recent Research Frontiers* (Marijnissen and Gradón, 1996); *Particle Overload in the Rat Lung and Lung Cancer* (Mauderly and McCunney, 1996); *The Lung: Scientific Foundations, 2nd Ed.* (Crystal et al., 1997); *Handbook of Human Toxicology*, (Massaro, 1997);

Table 2.1 Intakes for Air, Water, and Food for an Adult Male (70 kg) and the Dose for Pollutant Exposure by these Routes Assuming 100% Uptake

		Normal rates of intake		Pollutant doses via each route	
Route of exposure	Conditions	Mass/day (g)	Volume/day (L)	Pollutant at 1 ppm (w/w) (mg/day)	Pollutant at 1 μg/L (mg/day)
Air	Resting	12,960	10,800	12.96	10.8
Air	8 h light work per day	20,160	16,800	20.16	16.8
Drinking water	Environmentally not extreme	1,700	1.7	1.70	0.0017
Food	Including water content	1,610	1.61	1.61	0.0016

Source: Adapted from Schleien et al. (1998), Chapter 12.

Particle–Lung Interactions (Gehr and Heyder, 2000); *Pulmonary Biology in Health and Disease* (Bittar, 2002); *The Lung: Development, Aging and the Environment* (Harding et al., 2004); *Toxicology of the Lung, 4th Ed.* (Gardner, 2006); *Inhalation Toxicology, 2nd Ed.* (Salem and Katz, 2006); *Particle Toxicology* (Donaldson and Borm, 2007); and *Inhalation Aerosols: Physical and Biological Basis for Therapy, 2nd Ed.* (Hickey, 2007). New, relevant books are continually being published. A supplement to the *American Review of Respiratory Disease*, on *Comparative Biology of the Lung* (vol, 28, no. 2, August 1983) contains 21 papers that cover a large variety of commonly used species.

POSTNATAL DEVELOPMENT

The birth of a mammal marks the transition from a water environment to breathing air, but it is a minor event in the continuing restructuring of the lung (Burri, 1997; Harding et al., 2004). The nature of lung development is similar across mammalian species, but the timing and mature structures vary considerably (Parent, 1991; Plopper and Pinkerton, 1991; Plopper and Fanucchi, 2004). This variability is also reflected in many functions, including xenobiotic activation and/or detoxification, glandular secretions, aging, and defenses against inhaled materials (Parent, 1991; Harding et al., 2004). Only a brief overview will be presented here.

There are four generally recognized stages in lung development: the *pseudoglandular*, in which there is a transition from a glandular appearance to a tubular branching form; the *canalicular*, characterized by completion of the airway branching, and blood vasculature positioning; the *saccular*, in which epithelial cells differentiate and lung enlargement occurs; and the *alveolar* where the formation of alveoli takes place. The timing of these stages in the mouse is such that at birth (19 days postconception) the lung is in the midsaccular stage, with alveolar formation occurring between postnatal days 5 and 30 (Cardoso, 2004). In contrast, humans are born after completion of the saccular phase and early in the period of alveolar formation (Cardoso, 2004). With respect to various species, Zoetis and Hurtt (2003) provide a brief review for the mouse, rat, rabbit, sheep, and human with references for the guinea pig, hamster, dog, monkey, and human. Of note, the dog is an acceptable species for safety testing of inhaled drugs for pediatric use (Zoetis and

Hurtt, 2003). However, other species are also useful in toxicology investigations, provided the developmental stage at the time of dosing is known and the results are interpreted accordingly. Two references, *Development of the Lung* (Hodson, 1977), and *The Lung: Development, Aging and the Environment* (Harding et al., 2004), provide more details on lung development.

COMPARTMENTS

When an organ system as complex as the respiratory tract is under consideration, it is convenient to simplify by forming conceptual anatomic units or *compartments*. If the process is thorough, the individual compartments can be linked to form a useful description of the complete organ. Compartments also form the basic units of many types of mathematical models, and can be linked by transfer coefficients that describe the rates of transfer of materials or information within a multicompartment system.

The problem of how to compartmentalize the respiratory tract does not have a unique solution. The physician interested in the diagnosis and treatment of diseases may think in terms of four compartments: (1) the large airways (nose, mouth, pharynx, larynx, trachea, and main bronchi); (2) the small airways (smaller bronchi and bronchioles); (3) the acinus (terminal bronchioles, alveolar ducts, and alveoli); and (4) the blood vessels. The histologist, on the other hand, may prefer to use compartments based on cell types; for example, endothelial, epithelial, glandular, cartilaginous, muscular, and neuronal.

Three major compartments, after those proposed by the Task Group on Lung Dynamics (TGLD, 1966) of the International Commission on Radiological Protection (ICRP) have been of great use to inhalation toxicologists. The Task Group divided the respiratory tract into regions based upon anatomical features, and particle deposition and clearance phenomena. The National Council on Radiation Protection and Measurements (NCRP, 1997) modified the regions, as shown in Figure 2.1. Similar models have been proposed (Table 2.2).

Using the NCRP (1997) terminology, the naso-oro-pharyngo-laryngeal (NOPL) compartment begins at the anterior nares and includes the respiratory airways through the of larynx. Particle deposition in this region includes both large particles whose inertia causes impaction at obstructions, and small particles that deposit by diffusing to airway walls (ICRP, 1994; NCRP, 1997). Several pathways describe the clearance of particles which deposit in the nasal compartment. Relatively soluble material moves into the blood,

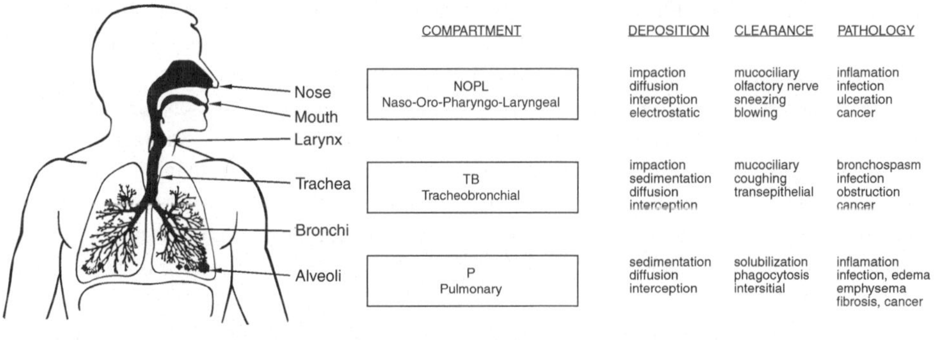

Figure 2.1 Compartmental model of the NCRP (1997).

Table 2.2 Compartments of the Human Respiratory Tract used for Analyzing Particle Inhalation

Region	Anatomic structures	ACGIH®	ICRP	NCRP	TGLD
Head airways	• Nose • Mouth • Nasopharynx • Oropharynx • Larynx	Head airways region (HAR)	Extrathoracic region (ET)	Naso-oro-pharyngo-laryngeal region (NOPL)	Nasopharynx (NP)
Tracheo-bronchial tree	• Trachea • Bronchi • Bronchioles (to terminal bronchioles)	Tracheobronchial region (TBR)	Bronchial region (BB) and bronchiolar region (bb)	Tracheobronchial region (TB)	Tracheo-bronchial region (TB)
Gas exchange	• Respiratory bronchioles • Alveolar ducts • Alveolar sacs • Alveoli	Gas exchange region (GER)	Alveolar-interstitial region (AI)	Pulmonary region (P)	Pulmonary region (P)

Sources: ACGIH®, 1985; ICRP, 1994; NCRP, 1997; TGLD, 1966.

while some persistent material moves by mucociliary transport to the throat for subsequent swallowing. The anterior one-third of the nose clears primarily by blowing, wiping, or other extrinsic means, and effective removal of insoluble particles may require 1–2 days. The posterior portions of the nose have mucociliary clearance, with clearance half-times of 10 min to 30 days, slower clearance being related to tissue damage (NCRP, 1997). Recently, an olfactory clearance pathway toward the brain has been documented for ultra-fine particles (Dorman et al., 2002; Oberdörster et al., 2004). The oral cavity is largely cleared by swallowing or expectoration.

The tracheobronchial (TB) region begins at the base of the larynx and includes the trachea and the ciliated bronchial airways down to and including the terminal bronchioles. A relatively small fraction of all sizes of particles that pass through the NOPL region will deposit in the TB region. The mechanisms of inertial impaction, sedimentation, and Brownian diffusion (for small particles) cause deposition. Interception can be an important deposition mechanism for fibers.

During mouth breathing, as during exercise or speaking, the benefits of the collection of particles in the nose are lost, and particles tend to deposit in the TB region with higher efficiency. The TB region is both ciliated and equipped with mucus-secreting elements so that clearance of deposited particles occurs by mucociliary action followed by swallowing. Relatively soluble material may rapidly enter the blood circulation.

The rate of mucus movement is slowest in the finer airways and increases toward the trachea. Since particles depositing in the TB tree deposit differently with respect to size, with smaller particles tending to deposit deeper in the lung, one expects larger particles to clear more quickly. Clearance of persistent material in the TB compartment cannot be described by a single rate. Studies imply that healthy airways nearly completely clear within 24 h, but recent data indicate that there is a TB compartment which is very slow in clearing (ICRP, 1994; Kreyling and Scheuch, 2000).

The third compartment, the pulmonary (P) region, includes the functional gas exchange sites of the lung, comprising respiratory bronchioles (RBs), alveolar ducts, alveolar sacs, and alveoli. For particles to reach and deposit in this region, they must pass through the NOPL and TB regions, and either by settling, diffusion, or interception come into contact with P surfaces. Since a portion of each breath remains unexhaled, the times available for deposition may be long for some particles. Clearance from the P region is not completely understood, but the Task Group (TGLD, 1966) suggested several mechanisms including: (1) the dissolution of relatively soluble material with absorption into the systemic circulation; (2) direct passage of particles into the blood; (3) phagocytosis of particles by macrophages with translocation to the ciliated airways; and (4) transfer of particles to the lymphatic system.

The usefulness of respiratory tract compartments can be illustrated by considering the deposition of inhaled particles. Deposition phenomena for particles can be complex, especially in such an intricate geometrical structure as the respiratory airways. If one measures the concentration of particles in inhaled and exhaled air and makes a plot versus particle size, a valley-shaped deposition efficiency curve is obtained (Fig. 2.2). The minimum in the total deposition curve at about 0.5 μm occurs, since particles of this diameter are not strongly influenced by either inertial or diffusional forces. By using detailed measurements of regional deposition and performing some mathematical calculations, one is able to break this total deposition curve into three simpler components, one for each compartment. The deposition probabilities in each compartment can sometimes be used to relate various diseases to the sizes of particles that cause these diseases. For example, nasal cancer in machinists can be related to the high nasal collection efficiency for airborne particles above 10 μm in aerodynamic diameter. Figure 2.2B is corrected for *inhalability*, the tendency for large particles to escape entry into the nose or mouth (Soderholm, 1989).

The three-compartment model has been widely used, but it has some important drawbacks. For example, the pattern with which particles deposit within a given compartment is not addressed: The assumption that such deposition is uniform should not be made. For example, bifurcations in the TB compartment can be sites of high regional deposition (Phalen et al., 2006). Also, the model does not adequately separate out the region between the terminal bronchioles and the alveolar ducts and sacs. This region contains RBs that are unique in structure in that they contain mucus-secreting and ciliated cells (like the TB region), and alveoli (like the P region). This intermediate region of RBs should probably be given independent status as a fourth compartment, since it is often the site of airway disease in humans (Bates, 1989).

GROSS ANATOMY

Overview

The gross anatomy of the mammalian respiratory tract includes the nose, oral cavity, pharynges, larynx, tracheobronchial tree, respiratory bronchioles (when present), the parenchyma (alveolar ducts, alveolar sacs, alveoli), and the associated pleura, septa, blood vasculature, lymphatics, and major neural units. As one might expect, these structures are not only anatomically complex, but also vary significantly among species. Several references are available that cover mammalian respiratory tract anatomy in detail (Parent, 1991; ICRP, 1994; NCRP, 1997; Massero, 1997; Bittar, 2002; Harding et al., 2004).

Nose, Nasopharynx, and Larynx

The mammalian nose and its immediately postnasal cavities comprise an elaborate organ that provides for olfaction, detection of airborne irritants, collection of gases and particles,

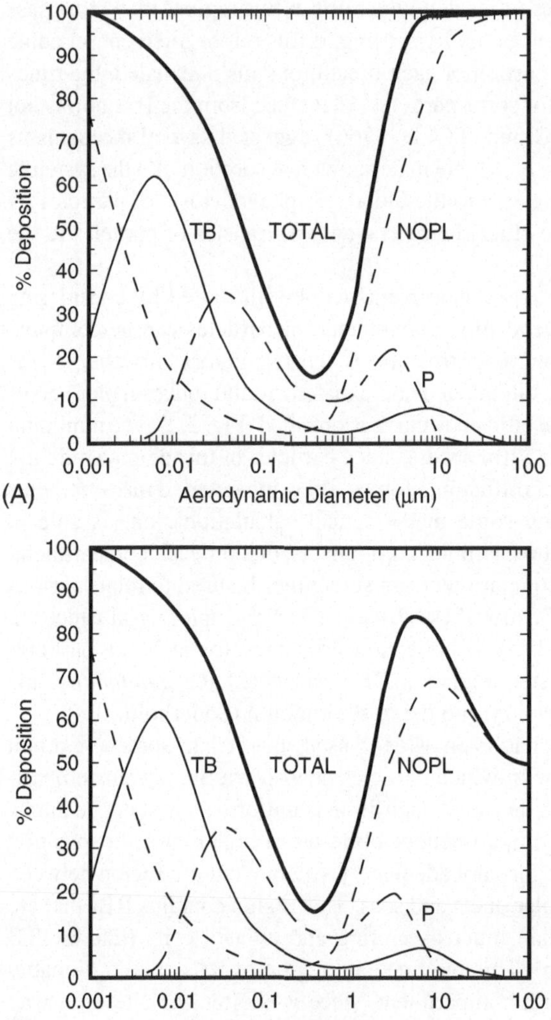

Figure 2.2 (**A**) Particle deposition in the major regions of the human respiratory tract during normal respiration. (**B**) Particle deposition efficiencies from part **A** multiplied by the size-dependent inhalability. *Abbreviations*: See Fig. 2.1. *Source*: From Phalen (2002) *The Particulate Air Pollution Controversy: A Case Study and Lessons Learned*, Kluwer Academic Publishers, Fig. 5.2, p. 60, with kind permission of Springer Science and Business Media. *Note*: The NCRP (1997) model was used to construct the curves.

humidification and temperature adjustment of inspired air, and disposal of fluids that drain from the eyes, sinuses, and inner ears (Fig. 2.3). The critical nature of these functions also makes the nasopharyngeal region an important target for airborne toxicants. It must deal with air pollutants in their raw unfiltered state at ambient concentrations, and failure of any of its functions can lead to serious, even life-threatening, conditions.

In humans, the nose contains two channel-shaped cavities separated by a cartilaginous bony central septum. The average adult male's nose has an air volume of about 17 mL (Gross and Morgan, 1991). Each nasal cavity is entered through a naris (nostril) having a cross-sectional area of about 0.7 cm. The nasal cavity is supported by walls consisting

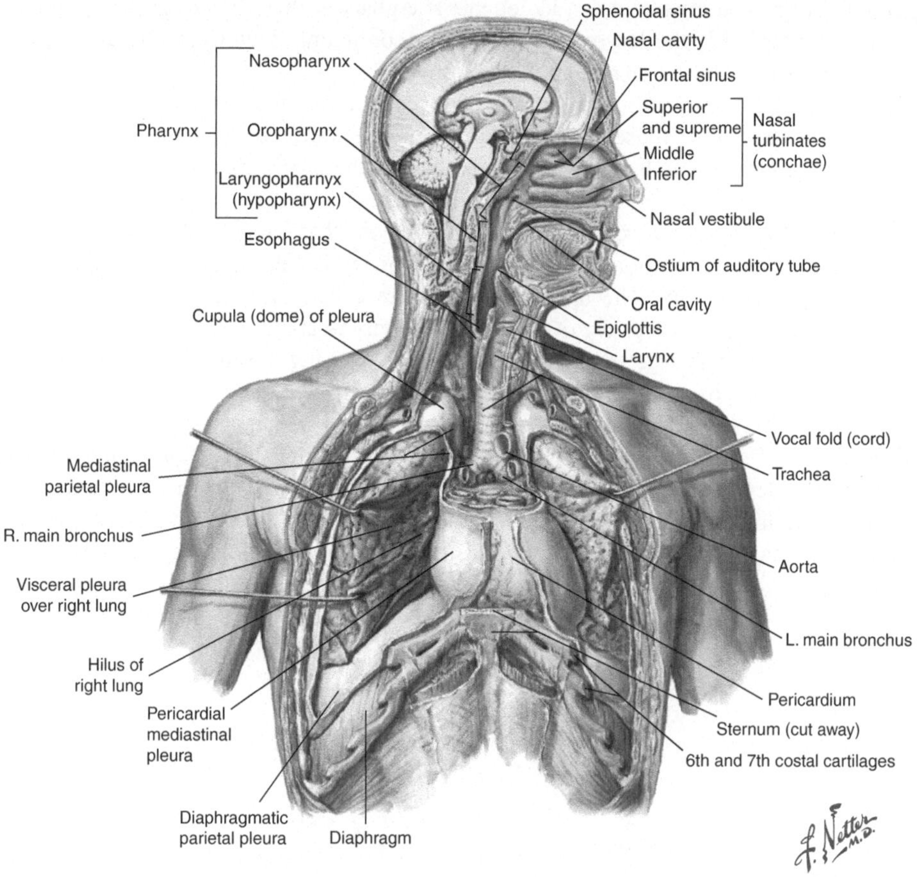

Figure 2.3 The human respiratory tract. *Abbreviations*: L, left; R, right. *Source*: Netter medical illustration used with permission of Elsevier.

of bone, cartilage, and connective tissue that provide sufficient rigidity to prevent collapse during breathing. The anterior one-third of the nasal cavity is covered with skin much like that on the face, and does not have an effective coating of mucus. The posterior two-thirds of the cavity are covered with mucus that moves rearward driven by cilia at a velocity of about 1 cm/min to a point where it is swallowed. This mucus, produced by goblet cells and glands, is mixed with fluids, including tears, that drain into the nasal cavity from the eyes and sinus cavities of the facial bones. The anterior portion of the nasal cavity is partially covered with hair that traps large inhaled bodies and warns of their presence via nerves at the base of the hair follicles.

In the central and rear portion of each human nasal cavity three turbinates (conchae), shelf-like plates that curve downward, stratify the inspired air, and increase the moist surface area that humidifies and warms or cools the inhaled air. Ducts from the eyes and sinus cavities open into the spaces between the turbinates and nasal cavity walls. The olfactory area is located above the topmost turbinates.

At its rear, the nasal cavity narrows and turns sharply downward. This area, the nasopharynx, is a region of collection of some particles that eluded previous capture. The nasopharynx, roughly tubular in shape, is joined by the oral pharynx (rear portion of the mouth) a few centimeters down its length. The pharynx then divides at the epiglottis to turn and enter either the larynx and trachea, or continuing downward, the esophagus. The pharynx is lined with mucociliary epithelium.

The epiglottis is a muscular flap that moves to cover the entrance to the larynx during swallowing. Other muscular action also prevents swallowed material from entering the trachea because people having surgical removal of the epiglottis can swallow without choking.

The larynx, or voicebox, is a short tubular cavity that has a slit-like variable-size narrowing in its central portion (Fig. 2.4). The narrowing is caused by two pairs of folds in the walls of the larynx. The uppermost folds are called the *false vocal chords* and the lower folds the *true vocal chords*. The adult larynx is about 3–6 cm long and has a variable cross-section that depends on the airflow rate passing through it (Stanescu et al., 1972). The dimensions of the average adult female larynx are about 78% those of adult males (Gross and Morgan, 1991). The larynx represents a major resistive element to airflow and also forms an inspiratory air jet that leads to particle impaction on the wall of the trachea (Schlesinger and Lippmann, 1976). The larynx is encased by muscle, bone, and cartilage and is lined by a mucus-covered membrane very similar to that found in the rear portion of the nasal cavities and pharynx. In the larynx, mucus is propelled upward for swallowing.

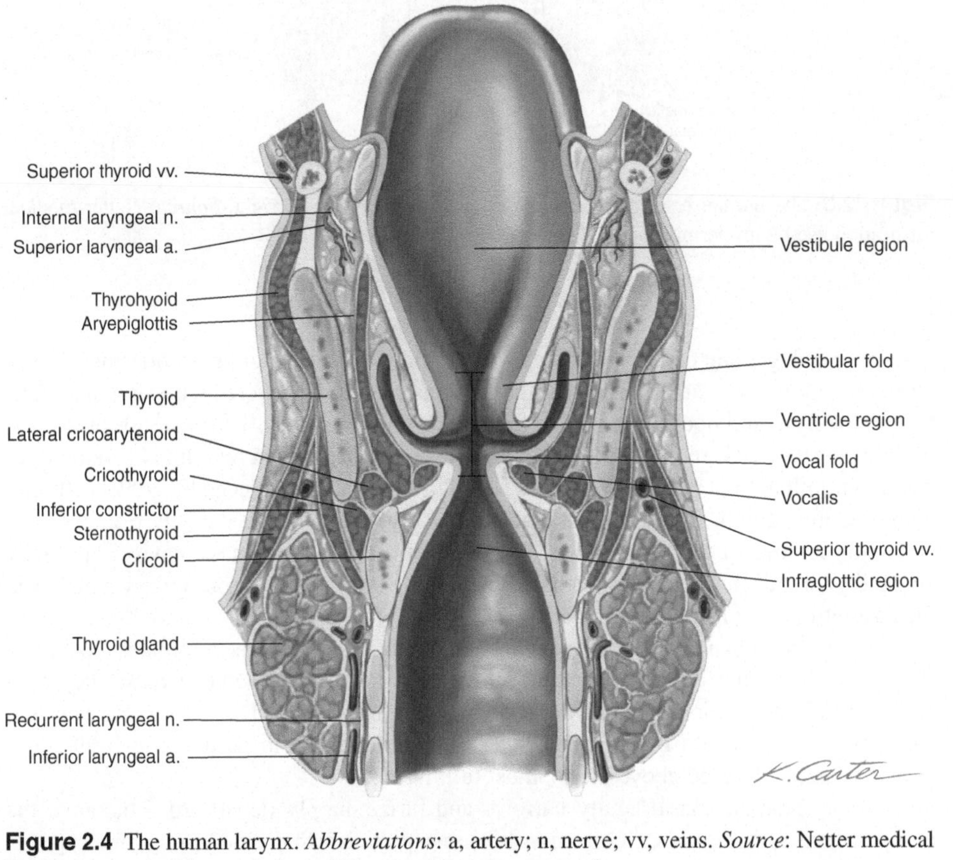

Figure 2.4 The human larynx. *Abbreviations*: a, artery; n, nerve; vv, veins. *Source*: Netter medical illustration used with permission of Elsevier.

Although the anatomical descriptions above strictly apply only to humans, mammals in general have similar structures. The dimensions and shapes of structures vary considerably from one species to another, and even among varieties within a given species (Gross and Morgan, 1991). Imagine, for example, differences in the nasal cavities of the shrew, the dolphin, and the elephant. A common laboratory animal, the dog, differs from the human in several important ways. The nasal cavity of the dog can be several times longer than that of man and contain over two dozen turbinates (Schreider and Raabe, 1981) (Fig. 9.7, Chapter 9).

Tracheobronchial Tree

Trachea

At its entry, the trachea is continuous with the larynx. It is a flexible tube that divides into primary bronchi (Fig. 2.5). In humans, about 20 roughly, C-shaped cartilages in its wall prevent its collapse. The gap between the ends of the cartilaginous rings is filled with a flexible muscular sheet of tissue. Thus, in cross-section, the trachea tends to have a D- or O-shape depending upon the internal air pressure. The inner walls of the trachea are covered with mucus supplied by goblet cells and mucous glands. The mucus is moved toward the larynx by ciliated cells.

During breathing, both the diameter and length of the trachea vary, the trachea elongating on inspiration. Tracheal dimensions vary greatly from species to species as described in Chapter 9.

Some species have a bronchus that branches off of the trachea near its midregion and enters a lobe of the lungs. This tracheal bronchus is present in the pig and goat, but only rarely in the human and most other mammals.

Bronchi and Bronchioles

In mammals, the trachea divides into two main branches called bronchi. These bronchi enter the right and left lungs and continue to divide for several generations, about 16 in humans, before alveoli begin to appear in the bronchiolar walls (Fig. 2.6). This appearance of alveoli marks the end of the TB tree and the beginning of the respiratory RB region.

Bronchi and bronchioles are roughly circular in cross-section, and smooth muscle completely encircles bronchial airways. The C-shaped cartilages of the trachea are replaced in the bronchial walls by irregularly shaped cartilage plates situated outside of the smooth muscle. Further down the TB tree where the tube diameters are about 1 mm or less, the cartilage disappears. These tubes, called bronchioles, have mucus-secreting goblet cells but do not have mucous glands in their walls. The outermost layer of the bronchi consists of a mixture of tough connective tissue and elastic fibers.

The inner lining of the bronchi is pseudostratified columnar epithelium having ciliated cells, mucus-secreting goblet cells, and underlying mucus-secreting glands (Fig. 2.7). Thus, the TB tree possesses an active clearance mechanism due to the propulsion of mucus toward the pharynx to a point where it is swallowed. The bronchioles are lined with ciliated columnar epithelium that is not pseudostratified.

Mammalian TB trees have two basic forms of branching, *monopodial* and *regular dichotomous* (Schlesinger and McFadden, 1981). These terms strictly relate to the manner of formation of branches before birth, but are also convenient in describing mature lungs. The monopodial division, which is common in dogs, cats, rabbits, sheep, monkeys, pigs, rats, mice, hamsters, and several other species, is characterized by long, tapering airways

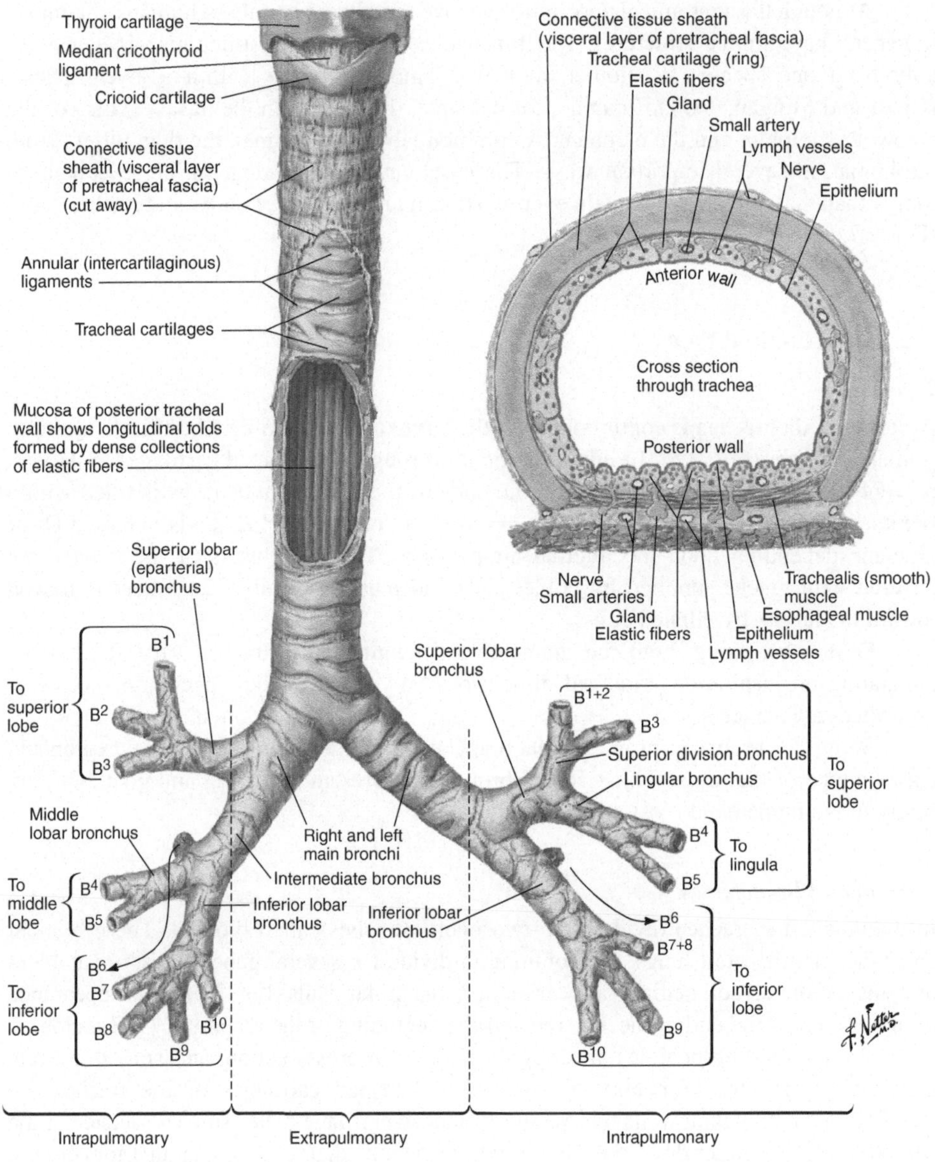

Figure 2.5 The human trachea and major bronchi. *Source*: Netter medical illustration used with permission of Elsevier.

with small lateral branches which usually take off of the main tube at an angle of about 60° (Phalen et al., 1978a). Regular dichotomous (symmetric) branching, which is typical of the human lung, involves the division of a tube into two daughters having nearly equal diameters and nearly equal angles of branching with respect to their parent tube. However, branching asymmetry does exist in the human lung (Phillips and Kaye, 1997). Figure 9.4 (Chapter 9) illustrates the two types. The mode of branching is a function of generation number, in that branching becomes increasingly symmetrical when progressing deeper into the lung.

Data taken from measurements on silicone rubber TB replica casts from human, dog, rat, and hamster have been published by the author and colleagues (Phalen et al., 1973;

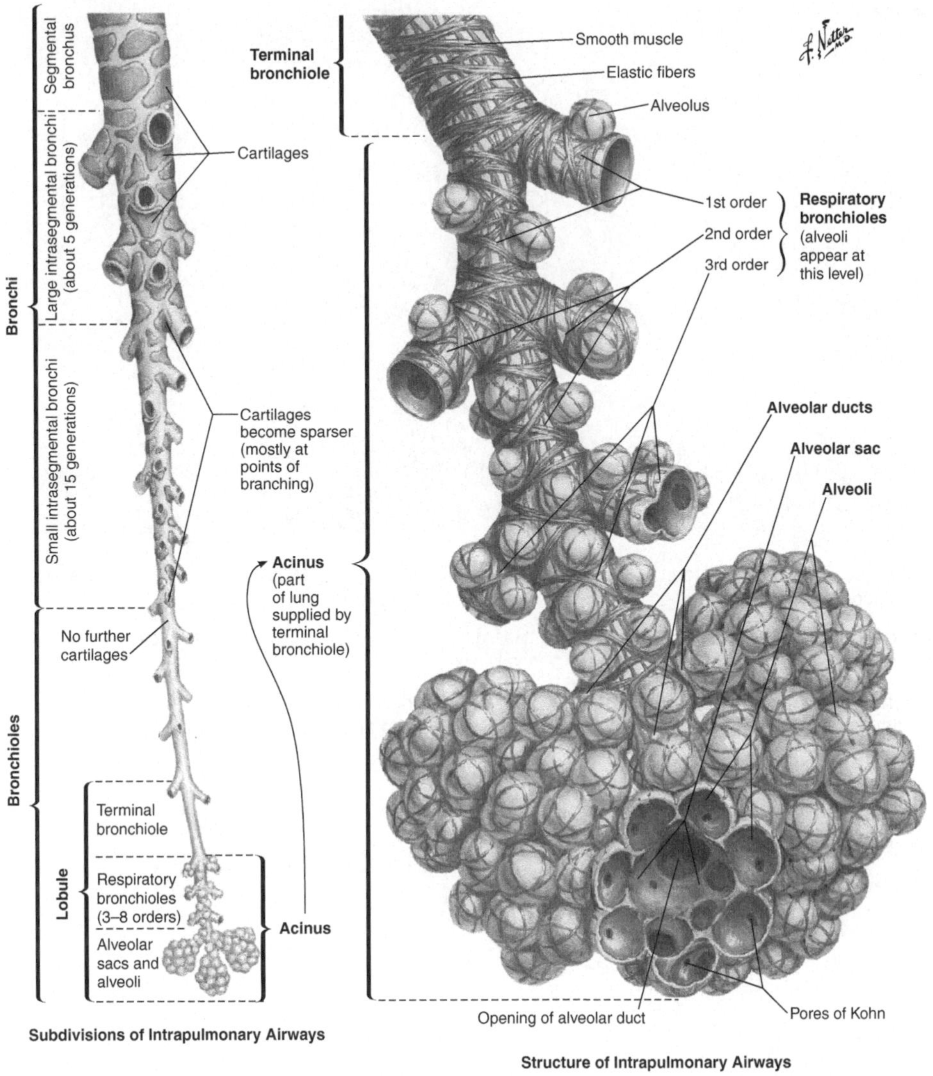

Figure 2.6 Bronchi, bronchioles, alveolar ducts, alveolar sacs, and alveoli. *Source*: Netter medical illustration used with permission of Elsevier. All rights reserved.

Phalen and Oldham, 1983). The replica casts were made and cured in the thorax, the tissue digested away, alveoli trimmed away, and measurements made. An idealized model of an airway branch (Fig. 2.8) was used for defining parameters that were measured. The length of a segment was defined by points a and b which fall midway between the intersections of the axes of the two daughter segments and the central axis of the parent segment. The total path length followed by an inhaled particle is represented by the summation of individual lengths of tubes it passes through. Two or three measurements of diameter along each tube were averaged to give the diameter, D. The branching angle (Θ) of each daughter segment was defined as the change in direction of the bulk airflow, moving from the parent segment into the daughter segment. Two populations of daughter branches were recognized: the major daughters and the minor daughters. (The daughter having the larger

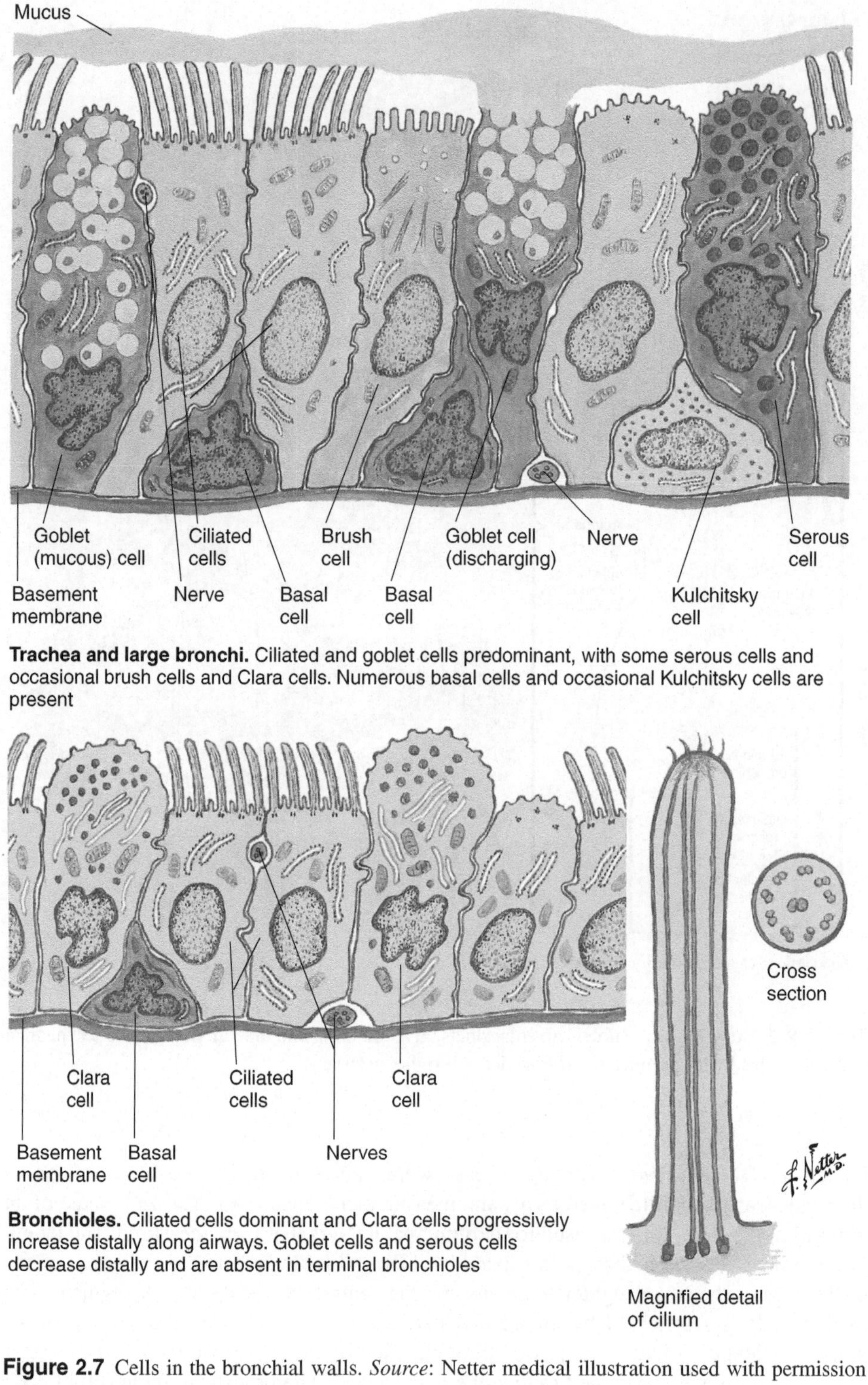

Figure 2.7 Cells in the bronchial walls. *Source*: Netter medical illustration used with permission of Elsevier.

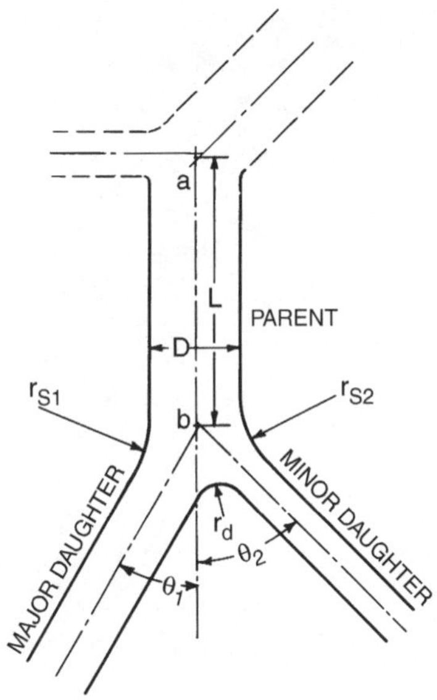

Figure 2.8 Idealized model of the airway branch, the basic unit of structure in the tracheobronchial airway. Diameters (D), lengths (a–b, L), radii (r), and branch angles (θ) are defined by this model.

diameter was the major, and the daughter with the smaller diameter, the minor.) It was found that major and minor daughters differed from one another in several respects. Major daughters had smaller branch angles and were usually, though not always, shorter in relation to their lengths than were minor daughters.

The symmetrical TB tree model for the human, described by Weibel (1963a), is widely used and contains information on airway lengths, diameters, and number. More recent TB models exist for the human (ICRP, 1994; NCRP, 1997), and for the rat (Yeh et al., 1979; Yeh and Schum, 1980). Differences among such models may relate to differences in technique, and/or individual differences in anatomy. TB models for humans, dogs, and rats have been reviewed (McBride, 1991; Phillips and Kaye, 1995, 1997; Brown et al., 2005). Figure 2.9 shows a human airway cast with trimmings representing bronchioles and alveoli.

Why is the human TB tree more symmetrical than those of other studied mammals? The difference may relate to the human's upright posture and more spherical thoracic cavity, as opposed to the horizontal posture and more elongated thorax of most other mammals. Also the lack of fur on the human reduces the requirement for body-cooling through evaporation from the respiratory system. Thus, fur-covered animals may be able to pant effectively by using the long monopodial airways to move air rapidly over moist surfaces without overventilation.

Respiratory Bronchioles

In humans, the terminal bronchioles of the tracheobronchial tree, that is those with diameters of about 0.6 mm, branch to form the first-order RBs. These RBs continue to divide and branch to give a total of about 2–5 orders of RB.

As they branch, RBs exhibit an increasing number of alveoli opening into their lumens (Fig. 2.6). These alveoli are thin walled, surrounded by blood capillaries, and presumably

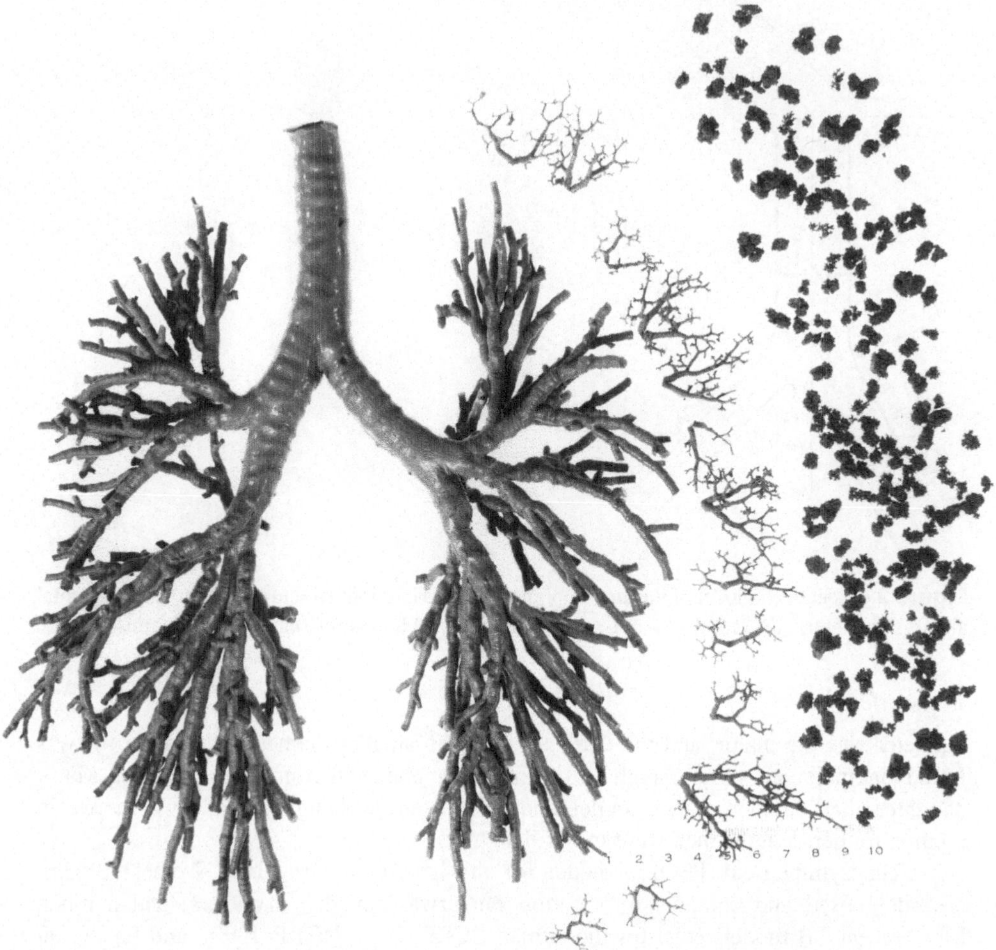

Figure 2.9 Replica cast of the human tracheobronchial airway with trimmings, made in the thoracic cavity to preserve branch angles. The largest portion represents bronchi, the finer branches represent nonalveolarized bronchioles, and the small pieces acini.

participate in the gas exchange function of the lung. Between alveoli, the surface of the RB is ciliated epithelium. Within alveoli, ciliated cells are not found.

Three important points must be made with respect to RBs. First, there is enormous species variability in these structures (Phalen and Oldham, 1983; Tyler and Julian, 1991). Some mammals, for example rats and mice, even appear to be totally devoid of RBs. Second, the RB has been acknowledged as an important site for disease in humans (Bates, 1973, 1989). And third, these structures form part of the *silent zone* of the lung, a region in which respiratory disease is difficult to detect by conventional pulmonary function testing. In fact, about 90% of the airway resistance to airflow through the respiratory system is normally produced in the large airways. Therefore, a narrowing or partial occlusion of small airways may not be detectable in clinical measurements of airway resistance. Also, since RBs are essentially uniformly distributed throughout the lung and are outnumbered by alveolar ducts and alveolar sacs, abnormalities in their structure are not easily seen in radiographs. For these reasons, detection of disease in small airways requires specialized techniques (Bates, 1989; Garay, 1992).

Parenchyma or Pulmonary Region

The term *parenchyma* relates to the primary functional tissue of an organ as distinguished from its supporting framework or secondary tissues. When applied to the lung, the parenchyma relates to the alveoli, and does not include the trachea and bronchial tree which are often viewed as having the purpose of delivering (or conducting) air to and from the alveoli. The major structural elements of the parenchyma of the lung are alveolar ducts, alveolar sacs, alveoli, alveolar capillaries, and the pulmonary lymphatics. These units are shown in Figures 2.6 and 2.10. Note that the *acinus* is commonly defined as the gas-exchange unit supplied by a terminal bronchiole. The human acinus contains about 10,000 alveoli (Mercer and Crapo, 1991).

The alveolar duct is a tubular structure whose walls are completely covered with alveoli. The alveolar duct usually branches to either two other alveolar ducts or two blind-ended tubes called alveolar sacs.

Although it is often depicted as spherical, the alveolus (or air sac) more closely resembles an incomplete polyhedron (Fig. 2.10). The open face of the alveolus is exposed to the air in either a respiratory bronchiole, an alevolar duct, or an alveolar sac, the closed portions being surrounded by a network of fine blood capillaries. Thus, in the alveolus, the atmosphere and the blood are brought into intimate contact where equilibration of CO_2 and O_2 can take place. In addition to the surrounding capillary net, alveoli are partially surrounded by elastic and nonelastic fibers that provide mechanical support. The size and total number of alveoli vary considerably from species to species, with age in a given species, and even from individual to individual. The average diameter of the adult human alveolus is about 150–300 μm. Alveolar number depends on body size in humans, varying from about 275 to 800 million (Ochs et al., 2004).

Subgross Lung Types

McLaughlin and associates (McLaughlin et al., 1961a, b) examined sections of mammalian lungs which were injected with colored latex. As a result of these studies they defined three basic lung types. Within a given lung type they found similarities in the degree to which lobes are separate, the characteristics of the pleura (membrane that encloses the lungs), the presence or absence of terminal and respiratory bronchioles, and the nature of the blood supply to the lung. These parameters were selected for their importance in the development of lung disease and its possible spread throughout the organ (Tyler and Julian, 1991). Thus, one might expect similarities in the types of lung disease and the extent of these diseases within mammals that exhibit a given lung type. Lung type I, which includes the cow, sheep, and pig, has a thick pleura, well-developed secondary lobulation, and marked interlobular septa. The dog, cat, and rhesus monkey, examples of type II, have very thin pleurae, absent secondary lobulation, and ill-defined interlobular septa. Type III, which includes the horse and human, is an intermediate type having thick pleurae, incompletely developed lobules, and haphazard interlobular septa. With respect to the anatomy of the distal airways, the cow, rat, and guinea pig had very little alveolarization of RBs. At the other extreme were the dog, cat, and rhesus monkey with heavily alveolarized RBs. Man, horse, and rabbit appeared to fall between the extremes, showing an intermediate degree of alveolarization of the RBs. These findings appear to represent careful work and should

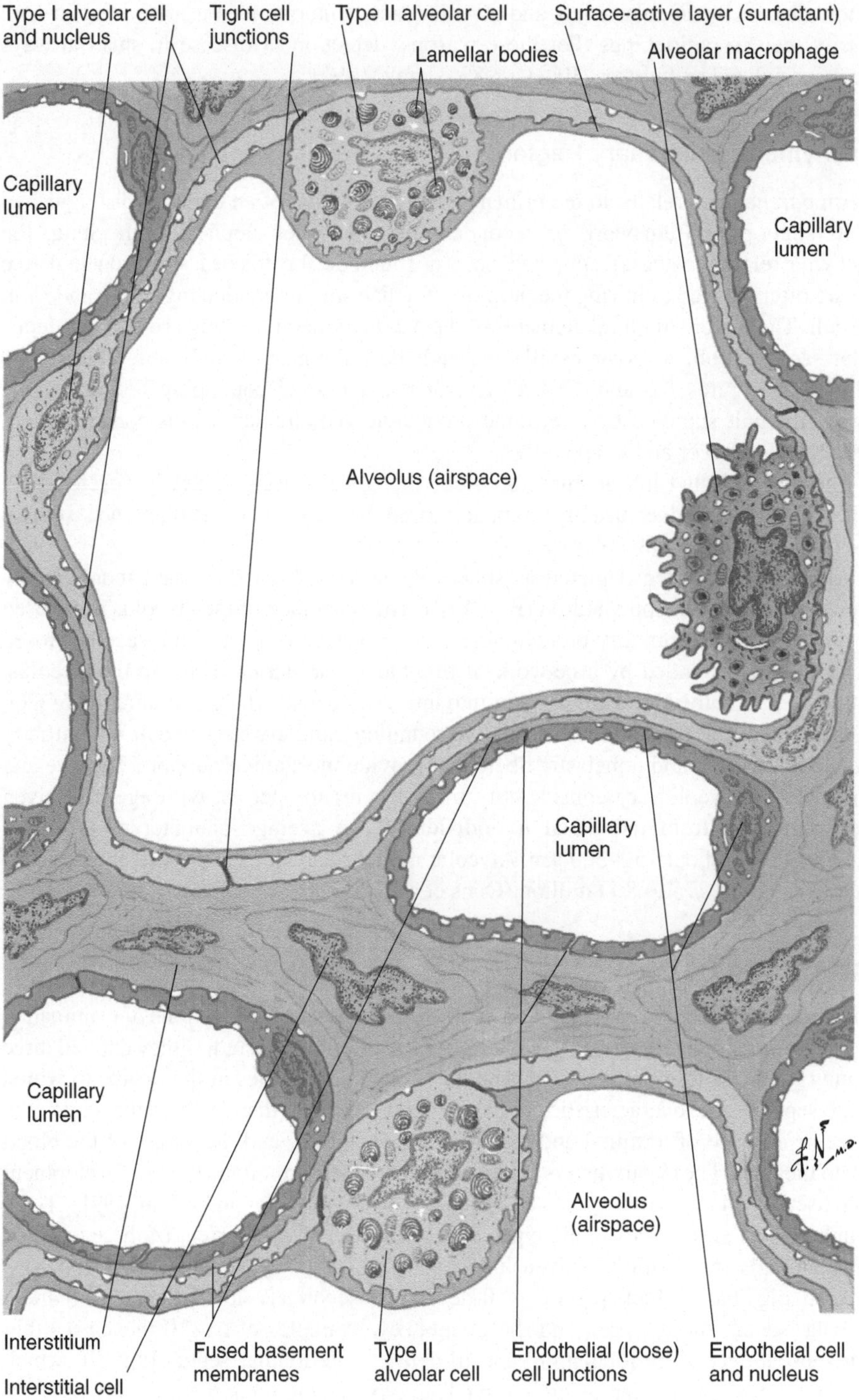

Figure 2.10 The pulmonary alveolus. *Source*: Netter medical illustration used with permission of Elsevier.

be borne in mind when selecting animals for studies, or when extrapolating results across species.

CELLS AND TISSUES

Ciliated Mucosa

The tissue that lines the rear portions of the nose, the larynx, the trachea, the bronchi, and the bronchioles is described as *ciliated mucosa*, or *mucociliary epithelium* (Fig. 2.7). As the name implies, such tissue is characterized by the presence of cells with numerous hair-like projections (cilia), and by the presence of individual cells and glands that secrete the components that make up mucus, a sticky, viscoelastic fluid. The cilia, which are surrounded by low-viscosity fluid, are motile and beat in a coordinated fashion, resulting in movement of the overlying mucus, usually toward the throat where it is swallowed. Although it is commonly assumed that the overlying mucus is continuous and unbroken, Salathé et al. (1997) argue that it is likely to be discontinuous.

The ciliated cells of the human respiratory system have cell nuclei and are columnar in shape, being about 10–15 μm in diameter and 20–40 μm in height. The ciliated cells are attached to a basement membrane, and replacement cells appear to form beneath the mature cells to replace lost cells. At the top surface protruding into the lumen of the airway, there are 15 to perhaps 150 or more filamentous cilia that are 5–15 μm long and about 0.3 μm in diameter. The cilia bend and then lash forward at rates up to several hundred cycles per minute. It is the coordinated beating of cilia on adjacent cells that propels the mucus at rates in the order of millimeters per minute.

Interspersed among the ciliated cells are columnar goblet cells similar in size to the ciliated cells, but lacking cilia and having a narrow base, and thus a goblet shape (Fig. 2.7). These cells, also attached to the basement membrane, manufacture mucus and open at the top to discharge their contents onto the airway surface. Beneath the basement membrane, there are mucous glands consisting of clusters of cells which secrete into a duct that leads to the epithelial surface (Fig. 2.11). The total amount of mucus secreted onto the surface of the human airway is often stated to be about 100 mL/day, possibly 10 mL/day being swallowed.

The action of the ciliated mucus-secreting tissues of the NOPL and TB compartments is responsible for sweeping particles from the surfaces of the airways. This function depends upon the quality and quantity of mucus and the quantity and synchronization of cilia. Viral and bacterial infections as well as several lung diseases and toxicants can lead to over- or undersecretion of mucus and to loss or paralysis of cilia (Salathé et al., 1997). During such states, sneezing and coughing become major clearance mechanisms that serve to clear the mucociliary epithelium. Often thought of as an annoying symptom, coughing is a health-preserving mechanism for removing toxicants and infectious organisms from the respiratory tract.

The Alveolus

As was previously mentioned, the human alveolus is a polyhedral structure, about 150–300 μm in diameter, having one face open to the atmosphere (Fig. 2.10). The walls are mostly formed by very thin, alveolar epithelial cells whose nuclei bulge into the alveolar airspace. At its thinnest portions, the type I alveolar epithelial cell is about 0.1 μm or slightly less in thickness. These cells appear to have relatively smooth surfaces and lie on top of

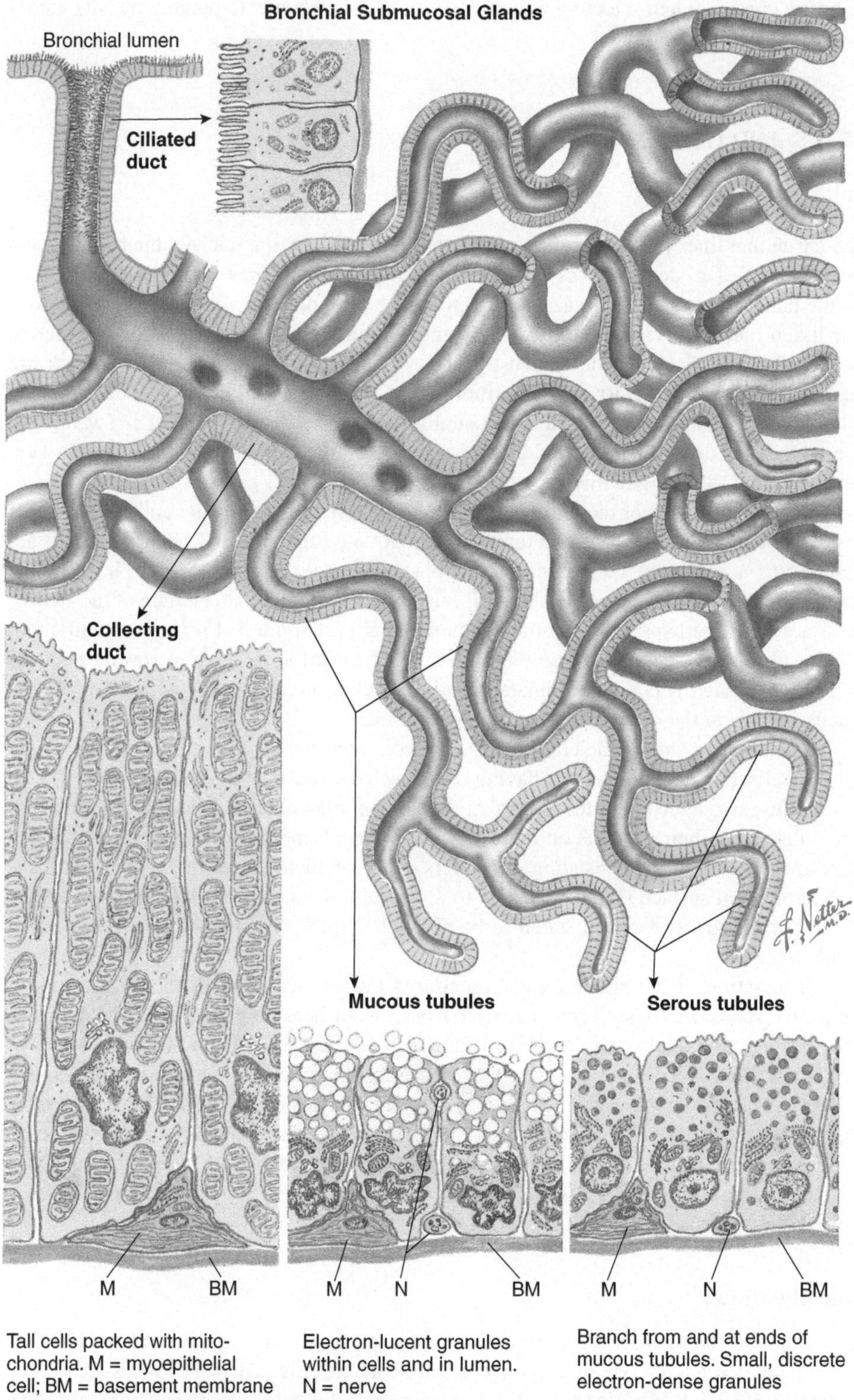

Figure 2.11 Glands of the bronchial walls. *Source*: Netter medical illustration used with permission of Elsevier. All rights reserved.

a basement membrane that is about 0.02–0.04 μm thick. Another basement membrane supports the blood capillary endothelial cells. These cells join to form the capillary wall and are similar in size and shape to the thin alveolar cell. The total thickness of the air-to-blood interface has been measured by H. Meesen and reported by Weibel (1964) and Weibel and Gil (1977) (Table 2.3). The interface thickness is greater in larger animals (Pinkerton et al., 1991).

A thicker, roughly cube-shaped cell, the type II alveolar epithelial cell, has a surface covered with small protrusions (microvilli). These microvilli on the airspace side greatly increase the surface area of this cell and imply, along with the presence of inclusions within the cell body, that this cell manufactures and secretes substances onto the surface of the alveolus. This cell is apparently involved in the manufacture and secretion of surfactant, a surface-tension–lowering agent that reduces the tendency for collapsing that alveoli have (Clements et al., 1958; Pattle, 1965). Abnormalities in lung surfactant can be related to disease states, including *edema*, *infant respiratory distress syndrome*, *hyaline membrane disease*, and *alveolar collapse* (Ganong, 1999).

The types I and II alveolar epithelial cells appear to be similar in structure and function for various mammals, reptiles, and amphibians. In several birds, the type II (secretory) alveolar epithelial cell is not found in the alveolus, but is located further up the TB tree in the bronchioles. The type II cell may divide and produce type I cells during lung growth and repair.

Other cells present in the alveolar region include the macrophage, alveolar brush cells, and interstitial cells. The macrophage will be described in detail later. The alveolar brush cell, which is described as roughly pyramidal, sits on the alveolar basement membrane and protrudes into the alveolar air space. It has large microvilli on its air-exposed side but its functions are unknown (Mariassy, 1991).

In some areas, the basement membranes of the alveolus and capillaries are separated by a space called the interalveolar septum or interstitium. This interstitium contains both elastic and inelastic fibers and fibroblasts. Fibroblasts are irregularly shaped cells that are involved in the formation of connective tissue. In pathologic conditions, such as edema and infections, the interstitial space may become enlarged due to the presence of excess fluid and cells such as blood leukocytes (white cells).

Alveolar walls in fixed tissue are frequently observed to have pores that appear to connect the airspaces of adjacent alveoli. These pores, called *pores of Kohn* (Fig. 2.6), were discovered by A. Adriani in 1847, according to Miller (1947). They have been reported by Loosli (1937) and Port et al. (1977) in a wide range of mammals including

Table 2.3 Thickness of the Air–Blood Barrier in Thin Areas for Various Species

Species	Capillary endothelium (μm)	Basement membrane (μm)	Alveolar epithelium (μm)	Total barrier (μm)
Human	0.02–0.4	0.11–0.16	0.04–0.065	0.36–2.5
Dog	0.25	0.12–0.18	0.03–0.08	0.4–0.65
Rabbit	0.015–0.075	0.06	0.1	0.17–0.24
Rat	0.018–0.1	0.065	0.05–0.1	0.13–0.26
Chicken	0.15	0.05–0.057	0.02–0.12	0.2–0.3
Pigeon	0.0385–0.075	0.0295–0.047	0.0145–0.0175	0.1–0.14

Source: Data adapted from Meessen (1960).

mouse, rat, guinea pig, cat, rabbit, monkey, dog, pig, baboon, horse, and humans (see Table 9.2). Martin (1963), in studies of the effect of aging on alveolar pores in the dog lung, found from 3 to 8 pores per alveolus and determined their diameters to be between 2 and 10 μm (mean 8 μm). He concluded that newborn dogs do not have alveolar pores but in older dogs these pores are lined with alveolar epithelium and therefore not artifacts of fixation. Environmental stress may be an impetus for development of pores (Pinkerton and Green, 1991). Mitzner (1997) reviewed collateral ventilation in the lung and provided evidence that such pores are probably not open in vivo. Mitzner also challenged the practice of calling alveolar pores "pores of Kohn". Alveolar pores were seen on casts of lungs by Roosevelt (1890).

The Macrophage

Alveolar macrophages are relatively large, nucleated cells which possess the ability to engulf foreign materials (Fig. 2.10). Mamallian macrophage diameters vary; being about 13 μm in rats and 21 μm in humans (Kromback et al., 1997). Macrophages have many functions, including foreign-body sequestration and secretory activities (Valberg and Blanchard, 1991). Over 100 secretory products with a broad range of actions are known (Nathan, 1987). Macrophages can change shape, presumably by liquefaction of their cell membrane, subsequent flowing of the cell contents, and reformation of their membrane. Thus, the cells are mobile and can engulf liquids and small bodies.

Phagocytosis and *pinocytosis* are two aspects of *endocytosis*, a term used to describe the engulfment of substances by cells, such as macrophages. Phagocytosis refers to the incorporation of solid materials, pinocytosis refers to the incorporation of liquid droplets, and endocytosis includes both processes.

Macrophages move freely on the surfaces of the alveoli, where they are credited with maintaining the sterility of the lung by virtue of their ability to engulf and kill infectious microorganisms. Macrophages also engulf other particles that deposit in the deep lung upon inhalation. It appears that *pulmonary alveolar macrophages* (PAMs) find their targets by chance, and/or by chemotaxis (movement in response to chemical stimuli). Chemotaxis may be positive, toward the debris, or negative, away from the debris. There is a need for data on the chemotactic potential of inhaled particles.

The process of phagocytosis has been decribed as occurring in sequential steps (Stossel, 1976; Valberg and Blanchard, 1991). The steps are (1) target recognition, (2) reception of the message to initiate phagocytosis, (3) transmission of the message to an effector, (4) attachment of the macrophage membrane to the target, (5) formation of pseudopodia, (6) engulfment by the pseudopodia, and (7) fusion of the pseudopodia with the macrophage cell body. Failure of any of these subprocesses could inactivate the macrophage's defense of the lung.

Hocking and Golde (1979) reviewed the research on antimicrobial mechanisms of the macrophage and listed several chemicals present within such cells that have antimicrobial properties, including hydrogen peroxide, catalase, superoxide anion, and lysosomal cationic proteins. With respect to nonviable materials deposited in the deep lung, Brain and Corkery (1977) have suggested that a major role of the macrophage is the prevention of incorporation of insoluble materials in tissues of the lung that have slow rates of clearance. Thus, if phagocytosis occurs, potentially toxic particles are prevented from entering a tissue in which they will be in contact with vulnerable cells for protracted periods.

Macrophages reside in locations other than the lung, for example, bone marrow (where they originate), connective tissue, and the abdominal cavity. There appear to be important differences in the actions of alveolar and peritoneal macrophages. For example, Lehrer et al. (1980) found that rabbit alveolar macrophages were nearly twice as efficient in killing *Candida albicans* (a yeast) than were unstimulated peritoneal macrophages.

Macrophages have an amazing efficiency in engulfing particles. Within minutes of deposition of an inhaled particle, the PAM is seen to have begun ingestion. Also, these cells appear to be able to phagocytize even when packed nearly full of debris. Data generated by Steve McAllen and Sonny Chiu (McAllen et al., 1981) at the author's laboratory show that PAMs recovered by lavage (rinsing of the lung with saline) can carry more than 10 times their weight in colloidal gold particles without appreciable loss of mobility or phagocytic ability. The term macrophage, meaning "big eater", is an appropriate name indeed. However, Stöber et al. (1994) proposed that overloaded macrophages lose their mobility, which is consistent with the observed clearance overload phenomenon observed in heavily exposed lungs (Morrow, 1988; Mauderly and McCunney, 1996). On the other hand, certain dusts are clearly toxic to the macrophage and result in its death or debilitation. Examples of such cytotoxic particles include cadmium, nickel, manganese, chromium, silica, asbestos, vanadium, and coal dust (Brain and Corkery, 1977; Hocking and Golde, 1979).

Much of the early work on cell phagocytosis and efficiency in relation to particle size was performed by W. O. Fenn (1921, 1923). Holma (1969) suggested 1.5 μm as optimal particle size for maximally efficient uptake by macrophages. He found that phagocytic uptake had an upper limit on particle diameter of 8 μm. However, it is unlikely that there is a lowest particle size for phagocytosis (Valberg and Blanchard, 1991), and engulfment of 15-μm diameter particles has been reported (Cannon and Swanson, 1992).

Mucus-Secreting Glands

The mucus-secreting glands are present in greater numbers in large airways and become more sparse toward the periphery, disappearing in humans at the level of the bronchiole. Along with goblet cells, these branched submucosal glands produce the mucus that covers the ciliated portions of respiratory tract (Fig. 2.11). The tubes into which mucus is secreted join a collecting duct that is ciliated just before it enters the bronchial airspace. These ciliated ducts may appear as pinholes on the surfaces of bronchi, having a maximum surface concentration of about one opening per square millimeter in the trachea (Netter, 1979). Two types of cells, mucous and serous, rest on a basement membrane and line the tubules. Serous cells are found lining the blind ends of the tubules, and mucous cells line the more proximal (upper) portions. The secretions of these cells form mucus, which is primarily an acidic glycoprotein with viscous, elastic, lubricant, adhesive, and wetting properties (King et al., 1989; Girod et al., 1992).

Its elastic property allows mucus to store energy, like a spring, and return toward its original dimensions after being stretched or compressed. This property allows the cilia to effectively propel the mucous layer. Using a frog palate, which is covered with mucociliary epithelium, Gilboa and Silberberg (1976) found that synthetic mucus consisting of cross-linked gels was effectively transported. If the simulants were either too heavily crosslinked, that is too solid, or too lightly crosslinked, that is too fluid, transport by cilia did not occur. In similar studies, Giordano et al. (1978), using dog tracheas, demonstrated that the rate of mucus movement increased as the elastic modulus decreased. The elastic modulus, or ratio of stress (force per unit area) to strain (deformation), is greater for stiffer materials. Thus, there appears to be a range of mucus elasticity within which cilia can effect transport. Abnormal mucus depth and/or physical properties are seen in bronchitis, asthma, cystic fibrosis, and other diseases. The roles of mucus in disease has been described by Jeffery (1994) and Kim (1997). The effect of inhaled materials on the properties of mucus is of interest in inhalation toxicology.

Innervation of the Respiratory System

The nervous system receives, generates, conveys, stores, and processes information. Portions of the nervous system, found in nearly every tissue of the body, play an important part in the voluntary and involuntary control and coordination of muscles, organs, glands, tissues, and cells. In the respiratory system, nerves are responsible for: (1) control of muscles for breathing, adjustment of the size of bronchial airways, and the cough, sneeze, and gag reflexes; (2) initiation and control of protective breathing patterns; (3) control of secretions; (4) adjustment of the distribution of blood flow; and (5) provision of sensory information on odor, irritancy, and the composition of lung tissue fluids and blood (Fig. 2.12). As for the body in general, most of the information carried by the nervous systems of the respiratory tract is not noticed at the conscious level. Table 2.4 covers innervation of the main muscle groups of the respiratory tract, which is by no means a complete description of the innervation of the respiratory system.

Especially important in toxicological studies are nerves that trigger the cough reflex, nerves that lead from pressure, stretch, and chemical receptors, and nerves involved in bronchial muscle constriction, protective breathing patterns, and mucous gland secretion (Zorychta and Richardson, 1991). Innervation of the respiratory tract is present in nearly every region from the nose down to the alveoli (Alarie, 1973). The interaction of inhaled substances with this system is a fruitful research area.

Bronchial Musculature

The smooth muscle that runs along the rear wall of the trachea and forms a spiral wrap around bronchi and bronchioles can greatly reduce the airway diameter via constriction (Figs. 2.5 and 2.6). Constriction of bronchial muscles is a major precipitating factor, along with excess mucus secretion, in an asthma attack. Such an attack can be caused by inhaled particles or gases, infection, ingested foods or drugs, and probably psychological factors. Bronchial constriction can be due to direct chemical action on smooth muscle, via a nerve reflex or by liberation of bioactive substances, including histamine.

VENTILATION

Normal Breathing

The average, resting, adult male inhales about 500 mL of air per breath (the *tidal volume*). This volume is small in comparison to a total lung capacity of about 6 litres. The inspired air fills the upper airways and TB tree which have a cumulative total volume of about 200 mL. Alveolar ventilation, about 300 mL, mixes with the air already residing in the alveoli. Upon normal exhalation, air that was in the TB tree from the previous inspiration is essentially completely expelled from the body along with approximately 300 mL of alveolar air. This exhaled alveolar air is not the same 300 mL that was just inhaled, rather it is mixed alveolar air. Thus pollutant gases or particles that are inhaled into the alveolar region may be washed out over several successive breaths. The times available for deposition and uptake of inhaled materials in the bronchial tree are short (of the order of

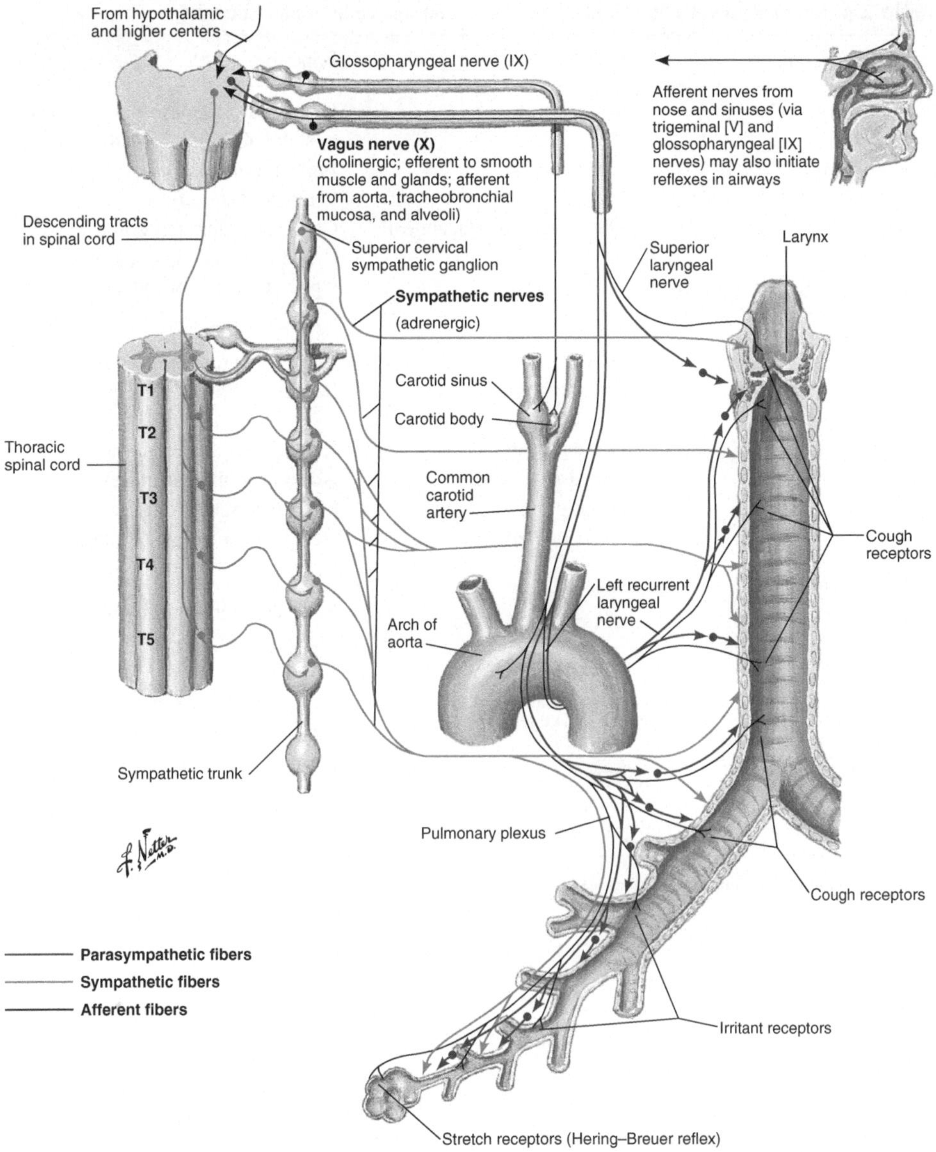

Figure 2.12 Innervation of the human respiratory airways. *Source*: Netter medical illustration used with permission of Elsevier. All rights reserved.

a few seconds), while residence times for deposition in the alveolar region can be of the order of minutes.

Breathing can be better understood by reference to Figure 2.13 and Table 2.5, which show various volumes and capacities of interest to the respiratory scientist. During quiescent breathing (*eupnea*), inhalation begins about every 6 s and brings the volume of gas in the lungs up to about half of the maximum, or *total lung capacity*. Exhalation reduces the volume of gas in the lung by about 500 mL, leaving the lung at its resting expiratory level, or *functional residual capacity*, with a gas volume of about 2,500 mL. At an elapsed time of 24 s (Fig. 2.13), the person makes a maximal inspiratory effort

Table 2.4 Motor Pathways that Control and Coordinate Breathing and Speaking

Nuclei and segments	Nerves	Function
Pons—motor nucleus of V	Mandibular	Open and close jaw, open airway, stabilize thyroid cartilage
Pons—motor nucleus of VII	Facial	Prevent collapse of soft tissue during inspiration
Medulla—nucleus ambiguous	Pharyngeal and laryngeal	Close nasopharynx, control speech and larynx as in coughing
Medulla—hypoglossal nucleus	Hypoglossal	Open and close oropharynx
Cervical cord		
1	Hypoglossal	Control suprahyphoid muscles
1–3	Ansa hypoglossi and descendents cervicalis	Stabilize thyroid cartilage
1–8	Posterior rami	Extend neck, open airway
2–4	Spinal root of XI	Accessory inspiratory muscles
3–5	Phrenic	Main inspiratory muscles
4–8	Anterior rami	Accessory inspiratory muscles
Thoracic cord		
1–11	Intercostal	Main inspiratory and possibly expiratory muscles
1–12	Posterior rami	Extension of spine
7–12	Intercostal and subcostal	Main expiratory muscles
Lumbar cord		
1 and 1–5	Ilioinguinal, iliohypogastric, and posterior rami	Expiratory muscles and possibly accessory inspiratory muscles
Sacral cord	Sacral and pudental	Pelvic and sphincter muscles

Source: Adapted from Campbell (1964), Chapter 21.

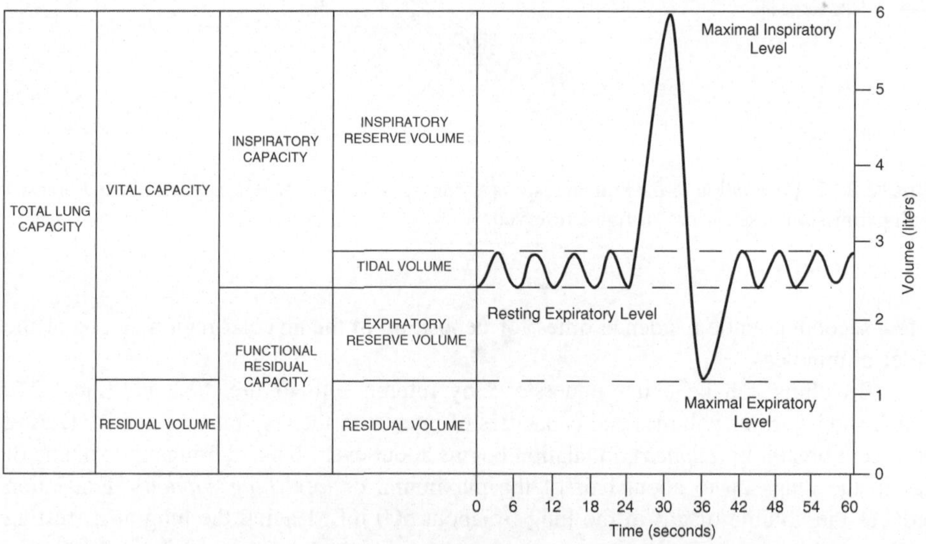

Figure 2.13 Capacities and volumes of the human lung as measured by spirometry.

Table 2.5 Lung Volumes and Capacities

1. *Tidal Volume*: The volume of gas inspired or expired during each respiratory cycle. Also the depth of breathing.
2. *Inspiratory Reserve Volume*: The maximum volume of gas that can be inspired from the end-inspiratory level.
3. *Expiratory Reserve Volume*: The maximum volume of gas that can be expired from the end-expiratory level.
4. *Total Lung Capacity*: The volume of gas in the lungs at the end of a maximum inspiration (the maximal inspiratory level).
5. *Vital Capacity*: The maximum volume of gas that can be expelled from the lungs by forceful effort following a maximum inspiration.
6. *Functional Residual Capacity*: The volume of gas remaining in the lungs at the resting expiratory level.

inspiring about 3,500 mL of air and bringing the lung volume to a *maximal inspiratory level* of 6,000 mL. This is followed by a maximal expiration that expels the vital capacity of 4,800 mL, leaving the lung at a *maximal expiratory level* having the residual volume of 1,200 mL of gas within. The *expiratory reserve volume*, about 1,300 mL, is the maximal volume that can be exhaled from the resting expiratory level. The normal values for these lung volumes vary with age, sex, body size, species, and other factors (Newton, 1995; Sahebjami, 1991; Lai, 1991).

Values for lung volumes for nonhuman mammals are more difficult to obtain since cooperation is required for the maximal efforts. Also, such measurements usually require the use of a mask or mouthpiece, which may alter the breathing pattern. Anesthesia will also usually alter the breathing patterns. Species such as dogs, horses, sheep, and some rodents have, however, been trained to wear masks, so relatively normal lung volumes are obtained. Examples of such data are discussed in Chapter 9. Pulmonary function testing in small mammals has been reviewed by Mauderly (1988) and Costa et al. (1991).

Exercise

The effects of exercise on the response to inhaled toxicants are of critical importance in inhalation studies. During exercise, several events occur that may change the response to a toxicant (Mautz et al., 1988; Mautz, 1997). The increased volume of air breathed in leads to greater exposure. A person walking may inhale 2–3 times the air volume per minute than while sitting. During maximal exertion, the minute volume can be increased more than ten-fold in humans and other mammals (Newton, 1995). An offsetting factor to the greater rate of delivery of air during exercise is the fact that the functional residual capacity may be increased in the exercise state. Thus, the freshly inhaled material is diluted by a larger volume of residual air. Usually, the percentage increase in functional residual capacity will be small with respect to the percentage increase in minute ventilation. In exercise, there is often a shift from nasal to oronasal (combined nose and mouth) breathing. This shift, which occurs at different workloads in different people (Saibene et al., 1978; Bennett et al., 2003), has the effect of decreasing the resistance of the upper airways to airflow. The fraction of air that enters the mouth does not receive the benefit of nasal scrubbing of larger particles and pollutant gases, which can increase the biological impact of many inhaled substances.

Other factors in exercise that may modify a person's response to an air pollutant are: (1) the potential overriding of a protective reflex breathing pattern (e.g., a shift to more shallow breathing); (2) the widening of the larynx in response to exercise; and (3) changes in tissue metabolism which may alter local tissue sensitivity.

Muscular exercise leads to the production of internal heat which must be dissipated. If the skin temperature is not low enough to allow sufficient loss of heat, other physiological mechanisms will come into play. In humans, evaporation of perspiration is an effective cooling mechanism. The combination of muscular work and high environmental temperature and humidity can lead to the production of sweat at rates that exceed 1.5 L/h (Ganong, 1999). In animals that are covered with fur, such as the dog, evaporation of moisture on the tongue and in the oral cavity can lead to cooling of blood that passes through a vascular plexus at the rear of the oral cavity. In rats, increased blood flow through the tail dissipates heat. Allowing airflow over rats' tails during exposure in tubes can facilitate thermal regulation. Caution must be applied when laboratory animals are used as models for exercising humans. Stavert et al. (1982a,b) developed a refrigerated treadmill controlled by an algorithm based on skin and rectal temperatures that permits one to expose the exercising dog to air pollutants and simultaneously prevent panting.

Reflex Responses to Inhaled Irritants

The innervation of the mammalian respiratory tract leads to an important category of interaction with inhaled pollutants; viz. reflex changes in breathing pattern due to sensory irritation. This irritation may or may not be associated with pain, smell, or other perception, depending on which nervous structures are stimulated. In general, the reflex actions appear to be protective, in that the changes in breathing pattern serve to limit or prevent exposure of the respiratory tract. Familiar reflex actions include coughing and sneezing; less familiar are bronchial constriction, reductions in the volumes and rates of ventilation, and excessive secretion of mucus. The papers of Alarie (1973), Amdur et al. (1978a) and Kane et al. (1979) demonstrate the importance of studies of the effect on inhaled materials on sensory receptors. Alarie's classification of inhaled materials, based on their irritancy and effects on various receptors in the respiratory system, is presented in Table 2.6. An alternate approach for measuring the effects of irritation has been described by Phipps and Richardson (1976). In their method, a tracheal cannula is used to collect mucous from the anesthetized cat.

Inhaled materials can also present as discriminable events such as intoxication, olfaction, or pain. Such events, whether pleasurable, neutral, or unpleasant can modify behavior, producing avoidance, or in extreme cases, addiction (Bowen et al., 2006). A brief review of these aspects of inhaled materials, including methods for laboratory studies, has been published by Wood (1978). The importance of the foregoing to the inhalation toxicologist is twofold. First, sensory irritation represents a potentially important biological endpoint. Second, the responses to substances with stimulus properties can modify the behavior of subjects and produce a greater or lesser degree of exposure than would otherwise occur.

Due to differences in innervation, different species may react differently to inhaled irritants (Zorychta and Richardson, 1991). The importance of species differences in this regard was also pointed out by a National Academy of Sciences Committee (Committee on Sulfur Oxides, 1978). The Committee reported: "Mice are said to be virtually incapable of bronchoconstriction,... Other common species of laboratory animals—including guinea pigs, cats, dogs and monkeys—respond to sulfur dioxide inhalation principally with narrowing of airways. The narrowing is reflected in an increase in flow resistance... In cats and rabbits irritation of the nasal mucosa is associated with bronchodilitation rather than bronchoconstriction." In the same review, humans are said to sometimes respond to SO_2 inhalation by bronchoconstriction.

Table 2.6 Classification of Airborne Chemicals that Stimulate Respiratory Tract Nerve Endings

A. Sensory irritant	**C. Bronchoconstrictor**
Definition—when inhaled via nose will stimulate trigeminal nerve endings, evoke burning sensation in nose, and inhibit respiration. Most will induce coughing from laryngeal stimulation **Other characteristics**—stimulate trigeminal endings in cornea and induce tearing, induce burning sensation on face, may induce bronchoconstriction **Examples**—Chloracetophenone, SO_2, ammonia, acrolein, inert dust	**Definition**—when inhaled will induce increased resistance to airflow. Action can be directly on smooth muscle, via axonal reflex or by liberation of histamine **Other characteristics**—most produce pain via action on bronchial mucosa **Examples**—SO_2, ammonia, inert particles, allergens
B. Pulmonary irritant	**D. Perspiratory irritant**
Definition—when inhaled will stimulate sensory receptors in lung and induce rapid shallow breathing. Cause sensation of dyspnea and breathlessness rather than pain **Other characteristics**—can induce pulmonary edema and then painful breathing **Examples**—phosgene, NO_2, O_3, sulfuric acid mist, sulfur and nitrogen mustard, sulfur pentafluoride	**Definition**—when inhaled acts as sensory irritant, bronchoconstrictor, and pulmonary irritant. **Other characteristics**—similar to sensory and pulmonary irritants **Examples**—chlorine, ketone, dichloromethyl ether, chlorine pentafluoride

Source: Adapted from Alarie (1973).

DEPOSITION OF INHALED PARTICLES

Introduction

When an aerosol is inhaled, some particles will be exhaled and the rest will either deposit on respiratory tract surfaces or remain suspended in unexhaled air. The deposition of particles is the result of their deviating from the airflow and touching an airway surface (the sticking coefficient is assumed to be 1, that is, any particle touching the surface will deposit). Except for very large or very small particles, the total deposition efficiency will be less than 100%. Table 2.7 shows a partial list of the factors known or suspected of affecting inhaled particles deposition. It is clear that modeling aerosol deposition is a difficult task.

Aerosol Deposition Models

Three basic types of models are used to estimate inhaled aerosol deposition: *empirical* (based on experimental data); *mechanistic* (based on theoretical models of particle deposition in tubes and other simplified structures); and *computational fluid dynamic* (CFD) (based on solving airflows in boundary conditions designed to simulate airways followed by placing particles in the flow and counting all particles that have trajectories that deposit by intersecting the boundary). Each type of model has its strengths and weaknesses (Table 2.8). Empirical and mechanistic models, which are widely used in risk assessment, are discussed by the National Council on Radiation Protection and Measurements (NCRP, 1997) and the International Commission on Radiological

Table 2.7 Some Factors that are Either Known or Believed to Influence the Deposition of Inhaled Particles

Environmental characteristics: gravitational force constant; magnetic field strength; electrical field strength; electrical ions; temperature; relative humidity; wind velocity; composition of air; barometric pressure; illumination intensity
Particle characteristics: geometrical size; shape; density; hygroscopicity; surface area; surface composition; electrical charge; electrical conductivity; state of agglomeration; number of particles per unit volume; temperature; irritancy
Respiratory tract characteristics: nasal, oral, and pharyngeal anatomy; nasal hairs; electrical charge on body, nose, or hairs; size and shape of laryngeal opening; tracheal anatomy; bronchial anatomy; mucus distribution; alveolar anatomy; surface temperature; surface composition
Breathing pattern characteristics: tidal volume; air velocities; respiratory rate; functional residual capacity; air distribution among and within lobes; air-mixing characteristics; breath-holding

Protection (ICRP, 1994). The NCRP and ICRP reports are excellent, well-referenced documents with descriptions of the anatomical, physiological, and physicochemical aspects of inhaled aerosol deposition, whether or not the aerosol is radioactive. Also, both reports describe associated software for predicting particle deposition and clearance in men, women, and children at rest and during physical activity. The mechanistic approach to particle deposition modeling dates back to Findeisen (1935).

Empirical, mechanistic, and CFD models must consider aerosol particle *inhalability*, which is the efficiency with which particles in the inspired air actually accompany the inhaled gas and enter the nose and/or mouth. Large particles will be undersampled (inhalability <100%), because their settling velocities are greater than the change in velocity of air as it enters the nose and/or mouth. Inhalability depends on the particle aerodynamic diameter, environmental wind speed, upper body anatomy, inhalation rate, body orientation to the wind, and other factors. Thus there is not a single particle-size dependent inhalability curve for humans or any other species. The American Conference of Industrial Hygenists (ACGIH®, 1985, 2007), the International Standards Organization

Table 2.8 Strengths and Weaknesses of Models Used for Predicting Inhaled Particle Deposition

Model	Strengths	Weaknesses
Empirical	Fit laboratory data well	Apply only to the cases that were measured
	Self-validating	Based on costly experiments
Mechanistic	Do not require much computer power	Airway geometry must be simplified
	The major mechanisms are easy to model	Some mechanisms are difficult to model
	Can apply to many cases	Don't provide much detail
Computational Fluid Dynamic	Provide aerosol deposition pattern in detail	Require large computational effort
	Can use real airway anatomy	Flow field solution can be ambiguous
		Interactions of particles with flow field may not be realistic
		Computer codes used are difficult to understand and/or modify

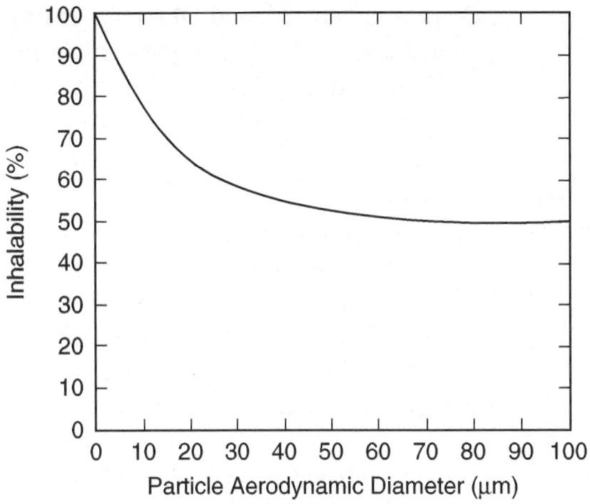

Figure 2.14 Aerosol inhalability, which is the wind speed and orientation-averaged efficiency with which the human head samples aerosol particles from the ambient air. *Source*: From Phalen (2002) *The Particulate Air Pollution Controversy*, Kluwer Academic Publishers, Fig. 5.1, p. 56, with kind permission of Springer Science and Business Media.

(Soderholm, 1989), and the European Committee of Normalization (CEN) (Lidén and Harper, 2006) have accepted an inhalability curve for particles up to 100 μm aerodynamic diameter (Fig. 2.14) for workers in low wind speeds, averaged over all orientations of the face to the wind. Inhalability has been addressed for larger particles and high, low, and still wind speeds (Breysse and Swift, 1990; Vincent et al., 1990; Aitken et al., 1999; Brown, 2005; Dai et al., 2006), and for some laboratory animals (Ménache et al., 1995; Asgharian et al., 2003; Jarabek, et al., 2005). Inhalability data are needed for children (Phalen et al., 1992), other mammals, and various exposure conditions. Note that the inhalability in laboratory exposures is affected by the specific exposure system used.

Particle deposition in the airways of the head (nose, mouth, pharynges, and larynx) is generally calculated using equations that are fitted to empirical data (ICRP, 1994; NCRP, 1997). Although the fitted equations imply precision, the clinical data they represent vary greatly. For example, the average resting inspiratory nasal deposition from seven studies (for 3.3-μm aerodynamic diameter particles) is 0.31 (31%). However, the actual deposition data under these conditions range from 0% to 90% in human subjects. Thus, use of the 31% average deposition will greatly overestimate the deposition for some people, and greatly underestimate it for others. The empirical equations for nasal and oral particle deposition, during inhalation and exhalation along with the data scatter are found in Chapter 5 of NCRP (1997).

Total deposition (the probability of particles depositing anywhere in the respiratory tract) can be directly measured, so the values are well known for adult humans. In order to estimate particle deposition in the TB and P airways, mathematical adjustments are required. The basic model for constructing regional deposition curves is:

$$DE_{TOTAL} = DE_{NP} + DE_{TB} + DE_{P} \qquad \text{(eq. 2.1)}$$

where DE_{TOTAL} = total deposition efficiency, DE_{NP} = nasopharyngeal deposition efficiency, DE_{TB} = tracheobronchial deposition efficiency, and DE_{P} = pulmonary deposition efficiency.

DE_{TOTAL} and DE_{NP} are obtained experimentally, DE_{TB} is calculated, and DE_{P} obtained by subtraction of DE_{NP} and DE_{TB} from DE_{TOTAL}. One must appreciate the many assumptions, and the fact that because averaged data are used: any individual's deposition may not be accurately estimated. The results of the foregoing approach were presented in Figure 2.2, considering the case where inhalability is, or is not, included. The curves apply to an average resting adult breathing uncharged nonhygroscopic particles.

Software is available for generating curves similar to those in Figure 2.2 for adults, children, for various states of exercise, for rats, and for particle-size distributions (Birchall et al., 1991; ICRP, 1994; Brown et al., 2005). The software typically considers three deposition mechanisms (impaction, sedimentation, and diffusion), or sometimes four (plus interception, which allows for particles of large particle dimension, e.g., long fibers). Although not usually modeled, other mechanisms such as electrical, thermal, and transpirational can also modify particle deposition.

Computational fluid dynamic (CFD) modeling of inhaled particle deposition was introduced around 1990 (Gradón and Orlicki, 1990; Yung et al., 1990; Lee and Goo, 1992; Balásházy and Hofmann, 1993; Kinsara et al., 1993). Although CFD modeling provides very detailed maps of particle deposition, experiments to verify their accuracy are difficult (Oldham et al., 2000). However, the CFD prediction of heavy particle deposition at bifurcation points in the bronchial tree (Balásházy et al., 1999; Oldham et al., 2000; Broday, 2004; Longest and Oldham, 2006) has strong experimental support (Phalen et al., 2006).

Progress has also been made on calculating particle doses for special populations including: infants, children, and adolescents (Yu and Xu, 1987; Schiller-Scotland et al., 1992; Martonen and Zhang, 1993; Bennett et al., 1996, 2003; Phalen and Oldham, 2001; Asgharian, et al., 2004; Kreyling et al., 2007; Ménache et al., 2008); women (Prichard et al., 1986; Kim and Hu, 1998); laboratory animals (Schlesinger, 1985; Raabe et al., 1988; Martonen et al., 1992; Newton, 1995; Brown et al., 2005; Jarabek et al., 2005; Tran and Kuempel, 2007); and for people with some lung diseases (Sweeney et al., 1995; Bennett et al., 1997; Kim and Kang, 1997; Segal et al., 2002; Chalupa et al., 2004; Kreyling et al., 2007).

UPTAKE OF INHALED GASES

Many gases react with lung fluids and tissue elements. The consideration of such reacting gases is beyond the scope of this book, but an introduction by Kimbell and Miller (1999) is available. The behavior of nonreacting gases will be introduced in order to present the major physical factors that lead to the uptake and distribution of inhaled gases. These factors include properties intrinsic to the gas molecules as well as physiological factors.

The movement of a gas from air into the body tissues can be described in steps. Breathing pulls the gas through the upper airways, TB tree, and into the alveoli. During this passage some of the gas, a small fraction, or perhaps essentially all of it, will dissolve in the aqueous lining of the air passages. Of that portion reaching the proximal alveoli, some will diffuse further down into the deeper alveolar regions.

Inspired gas moves into tissues by diffusion, the net movement of molecules from regions of higher concentration into regions of lower concentration. Such net movement will continue until equilibrium (equal partial pressures) is reached. For oxygen, the normal partial pressure is 152 mmHg in the atmosphere, 105 mmHg in the alveoli, 90 mmHg in the blood, and 0 mmHg in the tissues. This gradient, produced by oxygen consumption in tissues, produces a net inward flow from air into the blood and body tissues that is aided by breathing. The rate (R) of diffusion of a gas into a liquid is:

$$R = \text{Constant [solubility in the liquid / (molecular weight)}^{1/2}] \qquad \text{(eq. 2.2)}$$

As an example, the diffusion coefficient through epithelium into the blood (in units of cc/min per unit of driving pressure) is about 35 for oxygen and about 500 for carbon dioxide. This great difference is primarily due to the greater water solubility of carbon dioxide.

Once in the blood vessels, gases are transported either in simple solution or bound to constituent molecules in the blood. For example, 90% of the carried oxygen is bound in hemoglobin molecules and 10% is in solution. Nitrogen and most other nonreacting gases are carried totally in solution. The solution may be in plasma water or in lipids which are found in high concentrations in red blood cells. Ether is primarily dissolved in plasma and chloroform is primarily dissolved in the lipids of red blood cells; neither is chemically bound. The concept of solubility can be more sharply defined when an air and blood interface (or any other interface) is considered. The distribution coefficient in two media is defined as the equilibrium ratio (ER):

$$ER = \text{concentration in medium A / concentration in medium B} \qquad \text{(eq. 2.3)}$$

Table 2.9 shows the equilibrium distribution coefficients of several gases in water/air and in oil/water; water being a surrogate for plasma or for tissue fluids, and oil a surrogate for lipids. Gases that have a small water/air distribution coefficient can rapidly saturate the water phase when brought into contact by the air, and the air concentration may not be greatly depleted during such contact. Ethylene in air, for example, can rapidly equilibrate and, when fresh air is presented, desorb again within minutes.

The uptake by an organ of a gas dissolved in blood will depend upon the composition of the organ in terms of its water and its lipid content. The brain, which is high in lipid content and has a rich blood supply, is the site of action of many anesthetics. The brain soaks up lipid-soluble anesthetics avidly in comparison to body fat deposits which have a relatively poor blood supply. The rapid rate of blood perfusion through the brain (50 cc of blood per 100 g of tissue per minute) also causes the brain to desorb gases rapidly when the blood levels drop after the external supply has been discontinued. As long as the air/tissue gradient of a gas is positive, the gas will move into the tissues. When the supply in air is discontinued, the gradient reverses and gas is eliminated from organs and blood, and is exhaled.

Table 2.9 Equilibrium Distribution Coefficients for Selected Gases in Air and Tissue Components[a]

Gas	Water/air	Oil/water
Helium	0.0097	1.7
Nitrogen	0.0144	5.2
Ethylene	0.089	14.4
Chloroform	4.6	110
Ethyl ether	15.5	3.2

[a]*Note*: Water is used to approximate serum and other tissue fluids, and oil is used to approximate cellular and tissue lipids.

An interesting example of the application of the foregoing principles in gas toxicity is found in deep-sea diving. At ambient pressures of 4–5 atm, divers frequently become disoriented due to the narcotic effect of high levels of nitrogen dissolved in the brain. At lower pressures, equilibrium levels of nitrogen in the brain are not narcotic. Note that nitrogen is about five times more soluble in lipids than in blood, while helium is the least water-soluble of the elemental gases and has an oil/water distribution coefficient of only about 1.7. Thus, the substitution of helium for nitrogen in compressed diving gas solved the narcotic problems seen in deep-sea diving. Incidentally, helium is less dense than nitrogen and a mixture of 80% helium and 20% oxygen has one-third the density of air, reducing the effort expended in breathing.

Some gases are highly soluble in aqueous media. For example, at 20°C, 1 cm^3 of liquid water will hold, at saturation, 0.87 cm^3 (at standard temperature and pressure, STP) of CO_2, 39 cm^3 of SO_2, and 715 cm^3 of NH_3. Gases of low solubility in water under the same conditions include N_2 (0.016 cm^3/cm^3 of water), O_2 (0.031 cm^3/cm^3 of water), CO (0.023 cm^3/cm^3 of water), and NO (0.071 cm^3/cm^3 of water).

The relationship between the solubility of a gas in aqueous media and its depth of penetration into the respiratory airways is evident in a set of experiments performed by Ichioka (1972) in which SO_2 was passed through a wet model airway. The model airway consisted of a segmented glass tube, 9 mm inside diameter and 60 cm length, which was lined with filter paper made wet with various fluids including distilled water, physiological saline, or a 0.1% solution of bovine albumin in saline. The fraction of SO_2 penetrating the tube was measured at airflow rates ranging from 0.5 to 3.0 L/min. The concentration of SO_2 entering the tube was 4–10 ppm in nitrogen gas. As an experimental variation, gases were sometimes mixed with a polydisperse sodium chloride aerosol with a count median diameter of 0.95 μm and an unspecified concentration. For SO_2 without particles, most of the gas was absorbed in the first segment (of six total) regardless of the wetting fluid and flow rate. The absorption by bovine albumin was slightly greater than that seen with water or saline. The penetration of SO_2 was less than 4% under all conditions of fluids and flow rates. When the mixture of SO_2, and aerosol was passed through the tube, the penetrating percentage of SO_2 typically increased to between 11% and 17%.

From Ichioka's work one might predict that inhaled SO_2 would primarily affect the upper airways rather than the deep lung, and that the presence of an aerosol might alter the effect in some way. Several biological experiments have indeed been consistent with this expectation. Frank et al. (1967) found that the uptake of radiolabeled $^{35}SO_2$ in the upper airway of dogs was greater than 95%. Amdur (1957) reported that the bronchial constriction seen in guinea pigs exposed to SO_2 plus submicron sodium chloride aerosol was markedly increased over that seen with SO_2 alone; sodium chloride alone being relatively inert with respect to bronchial constriction. In this experiment, the bronchial constriction produced by SO_2 alone was believed to be produced by nasal irritation. Amdur's findings were clarified by McJilton et a1. (1976, 1973) who also measured airway resistance in guinea pigs exposed to SO_2 with and without sodium chloride aerosol. McJilton and associates studied the combination at low relative humidity in which the aerosol was dry and at high humidity in which the aerosol was in the form of wet droplets. They found that intensification of bronchial constriction over that for SO_2 alone occurred only when the combination was inhaled at high relative humidity and concluded that the effect was due to absorption of the highly water-soluble SO_2 into the droplets prior to inhalation.

As an example of a gas with low water solubility, consider ozone. Although ozone can produce damage in upper airways, its most serious effects are deep within the lung. These effects include killing of type I alveolar cells, the production of increases in alveolar wall thickness and, at sufficiently high concentrations, edema. Mathematical modeling of ozone uptake by Miller et al. (1978) indicates that in humans and a variety of laboratory animals, ozone doses can be expected to be high in the proximal alveolar regions.

Useful approaches to modeling pollutant gas uptake based upon biological, physical, and chemical considerations are available (Morgan and Frank, 1977; ICRP, 1994; Martonen et al., 1995; Medinsky et al., 1999; Andersen, 2003). Such models can be instrumental both in comparing experimental data taken under a variety of conditions and in performing extrapolations from one species to another.

DEFENSES

Introduction

Because the respiratory tract is continually exposed to large numbers of particulate toxicants, infectious agents, allergens, and contaminant gases and vapors, a variety of active defenses is critical for sustaining life. Respiratory tract defenses include behavioral (avoidance of exposure, and reflex changes in breathing), physical (cough, sneeze, and mucus secretion), chemical (detoxification, and oxidation), mucociliary transport, immunological, cell population changes, and enzyme induction. In addition to medical texts, several good references cover this enormous topic (Parent, 1991; ICRP, 1994; McClellan and Henderson, 1995; NCRP, 1997; Gehr and Heyder, 2000; Harding et al., 2004; Hickey, 2007). Here, particle clearance will be covered in more detail.

Proximal Airways' Clearance Mechanisms

The anterior portion of the nose is cleared of deposited, slowly-dissolving particles by blowing, wiping, or other similar extrinsic means; rapidly-dissolving particles may be absorbed within tissues or carried away in the blood or lymph.

The posterior portions of the nose, where mucociliary clearance occurs, show great variability in the rates of particulate clearance. The variability is due to the presence or absence of disease, to the specific location of deposition, and to differences among individuals. The ICRP (1994) recommended a multicompartment model for the clearance of particles from the nose, with clearance pathways from the nose to the gastrointestinal tract, to the environment, and to the lymphatics. Half-times for clearance ranged from 10 min (to the GI tract), to 17 h (to the environment), to 700 days (to lymph nodes).

Mechanisms that operate to rapidly clear large quantities of mucus and entrapped contaminants include sneezing and coughing. These mechanisms, triggered by physical or chemical irritation, act sporadically and serve as an important backup when mucociliary clearance fails. Failure of the mucociliary clearance mechanism occurs under several conditions such as excessive thickening or drying of mucus, loss of cilia (as occurs during respiratory infections), and inactivation of cilia during inhalation of toxic gases or smokes. Mossberg (1980) described the role of cough in clearing particles deposited in the TB tree. He pointed out that healthy humans with normal mucociliary transport were inefficient in clearing 6 μm diameter particles via coughing, but that mucociliary impaired subjects were usually quite efficient in clearing by cough. His conclusion was that excess quantities of mucus were necessary for effective particle clearance during coughing. He also indicated that cough is most effective in clearing larger airways, but that during rapid successive coughs without intervening inspiration, clearance of smaller airways occurs.

A clearance pathway from the nose to the brain has been documented for small, slowly-dissolving particles (Dorman et al., 2002; Oberdörster, et al., 2004). The particles travel along the olfactory nerve. The toxicological significance is, as yet, unclear.

Mucociliary Clearance

In healthy mammals, the TB tree is cleared of debris by a moving layer of mucus that is driven toward the oral pharynx and subsequently swallowed. It is generally (although not universally) accepted that the mucus lining is continuous and constantly moving due to the action of ciliated cells, and that particles deposited in the healthy TB airways are cleared within about 1 day after deposition (Wolf, 1991; NCRP, 1997; Salathe, 1997). The rapidity of TB clearance has been challenged by Patrick and Stirling (1977) and by Stahlhofen et al. (1995). These investigators concluded that particles may remain uncleared for more than 30 days in both rats and humans. A slow-cleared fraction has been incorporated into the ICRP (1994) lung model. In humans, the slow-cleared fraction (after bolus inhalation) is reported to linearly decrease from about 70% for 1-μm geometric diameter particles to about 5% for 6-μm particles (Kreyling and Scheuch, 2000).

The rate of mucus-mediated clearance appears to be faster in larger airways than in smaller ones (Morrow, 1977). The use of discrete clearance half-times to model clearance was not meant to imply that mucus velocities are discontinuous. Apparently there is a continuous gradient of mucus velocity increasing from the terminal bronchioles to the trachea (NCRP, 1997), with tracheal mucus velocities being faster in larger mammalian species (Wolff, 1991).

Mechanisms for clearance (mucociliary transport, macrophage action, dissolution in lung fluids, etc.) appear to be similar for most mammals. Therefore, one expects similarity in clearance phenomena. This expectation does not appear to be substantiated with respect to normal early clearance. Interspecies differences exist in short-term clearance rates for particles with similar properties deposited under similar conditions. Rats and mice appear to have a relatively sustained, rapid clearance phase at times after the disappearance of rapid clearance in human and dogs (Snipes et al., 1989; Wolff, 1991).

The effect of a variety of factors on particle clearance rates was reviewed by Pavia et al. (1980). Their conclusions are summarized in Table 2.10. More recent research has found no gender-related effects (Hasani et al., 1994) that brisk exercise accelerates clearance (Wolff, 1991), and that some lung diseases and genetic defects impair clearance (Pavia, 1987; Smaldone et al., 1993; Abu-Musa et al., 2008).

For ozone and many other agents, the effects on clearance of particles appear to be similar for various mammals (Newton, 1995). Though one cannot assume *a priori* that related agents will always influence clearance phenomena similarly in all mammals, the data do support the use of animals in toxicology studies in which disturbances in clearance are examined.

Two conclusions currently appear to be valid with respect to comparative mammalian early clearance: the rates of clearance are species dependent; and the effects of inhaled agents on clearance rates are qualitatively similar.

Alveolar Clearance

The alveolar region of the lung does not possess a mucociliary clearance apparatus. Deposited particles are cleared by several probable mechanisms including (1) dissolution and uptake by the systemic blood, (2) intact passage into the blood, (3) phagocytosis by macrophages followed by transport to ciliated airways, lymphatics, and possibly blood,

Table 2.10 The Probable Effects of Various Factors on Particle Clearance from the Human Respiratory Tract

Gender	Males and females probably have similar clearance rates
Age	Increasing age is associated with slower clearance
Posture	No apparent effect
Exercise	Brisk exercise may accelerate clearance
Sleep	May decrease clearance
Smoking	Chronic cigarette smoking impairs clearance. Acute exposure to cigarette smoke appears to speed clearance from deep airways and inhibit clearance from large airways. Effects vary with dose
Sulfur dioxide	Variable effects seen by different investigators
Freon propellants	No effect
Hair spray	Inhibition of clearance
Chronic bronchitis	Decrease in clearance rate when cough is absent. However, more proximal deposition of inhaled particles may mask this effect
Emphysema	Normal or possibly increased rate of clearance
Bronchial asthma	Decreased rate of clearance that can, in some cases, be overcome by medications
Influenza	Decreased rate of clearance with impairment lasting about 2 to 3 months
Pneumonia	Decreased rate of clearance with impairment lasting up to 1 year
Bronchogenic carcinoma	Probably no effect
Cystic fibrosis	Impaired clearance
Kartagener's syndrome (immotile spermatozoa)	Greatly impaired clearance
Asbestosis	Probably no effect

Source: Adapted from Pavia et al. (1980).

and (4) movement of particles into lymphatics either in macrophages or bare (ICRP, 1994; Stöber et al., 1994; NCRP, 1997; Kreyling et al., 2007). In addition, particles or their components may remain in lung tissues for long periods. Inactivation of microorganisms, with or without clearance, by alveolar macrophages is another important aspect of alveolar defense (Pavia, 1987; Murray and Driscoll, 1991).

The fate of particles deposited in the P compartment is strongly dependent on the mechanical stability of the particles. Particles that undergo significant dissolution in the fluids found in the lung may dissolve while still within the airspace, inside lung cells, or while in interstitial spaces (ICRP, 1994; NCRP, 1997).

Studies of the effects of disease states, or of inhaled pollutants, on deep lung clearance are of great importance. Cigarette smoke exposures have been shown to increase the survival of inhaled viable bacteria in hamsters; excess deaths due to bacterial infections were seen in animals exposed for 2 h at a concentration of 3% v/v (Henry et al. 1970). Preexisting influenza infection has been shown to impair both upper and lower respiratory tract clearance. Studies by Green (1965) with virus-infected mice that were exposed to viable staphylococcus bacteria showed that infected animals did not effectively kill the bacteria. Similarly, Creasia et al. (1973) found that virus-infected mice had drastically impaired clearance of radioactive, insoluble particles. Goldstein et a1. (1971) reported work in which mice were challenged with radiolabeled viable staphylococcus both before and after exposures to ozone and NO_2. Prior exposure to ozone (0.6–2 ppm for 17 h) or ozone plus NO_2 (0.1–0.2 ppm and 1.5–4.2 ppm for 17 h) led to decreased overall deposition of bacteria, and impaired killing of deposited bacteria. In the same series of studies, exposures to ozone plus NO_2 (0.4 ppm and 4–6.8 ppm for 4 h) after inhalation of bacteria led to increased survival and increased clearance of the inhaled bacteria. Sulfur dioxide (1 ppm, 7 h/day for 5–25 days) was shown by Ferin and Leach (1973) to diminish the clearance of inert particles in both deep lung and TB tree in the rat. A similar effect on bronchial clearance in donkeys was seen after brief exposure (300 ppm SO_2 for 30 min) by Spiegelman et al. (1968).

Studies of deep lung clearance by the author and associates at the University of California at Irvine utilized rats whose lungs were labeled by nose-only inhalation of insoluble radioactive particles. Following this labeling step, animals were divided into groups for chamber exposures to either clean or polluted air; the effects on clearance were followed by serial counts of the whole body for about 2 weeks. Rats were killed at 30 days postexposure for determinations of remaining radioactivity. In these studies, ozone (4 h at 0.8 ppm or more) appeared to stimulate removal of particles from the lung (Phalen et al., 1980; Kenoyer et al., 1981). SO_2 plus 1.5 mg/m^3 sulfate aerosol for 4 h was not found to alter lung clearance by Mannix et al. (1982).

A limitation on the use particle clearance evaluations as a tests of toxicity in rats is that excessive particulate pollutant exposures can lead to clearance slowing or stasis, that is, particle overload (Morrow, 1988; Mauderly, 1996a; Mauderly and McCunney, 1996). This condition occurs when particle deposition rates exceed the capacity of lung macrophages to clear them. Of importance to toxicologists is the fact that the use of particle exposures at high concentration in rodents can produce clearance stasis, which may invalidate the applicability to human environmental (or industrial) exposures.

3

Establishing and Controlling Exposures

INTRODUCTION

Modern inhalation studies must conform to rigorous criteria with respect to generation and control of the exposure environment. Pure air, having precisely adjusted temperature and relative humidity, is intentionally polluted with substances having known and stable characteristics. In most well-designed studies, a matched control population of subjects is exposed either simultaneously (preferred) or serially to pure air. One requirement of such studies is the ability to clean and condition quantities of ambient air. To meet this challenge, a suitable air purification and conditioning system must be installed, tested, and successfully operated. Additionally, equipment for generating pollutants must be carefully designed or selected, tested, and then operated successfully during exposures. Furthermore, experience has shown that the manner in which pollutants are combined and mixed with clean throughput air, and the ability to suppress unwanted, animal-generated contaminants, are important to achieving a high-quality exposure.

Two basic considerations are fundamental to success in performing a good exposure: the ability to produce the desired environmental conditions, and the ability to monitor the exposure. This chapter is concerned with the first consideration, establishing and controlling the exposure. Chapter 4 covers monitoring and characterizing the exposure. Some numerical units are in the English system to reflect manufacturer's specifications.

CLEANING AND CONDITIONING THROUGHPUT AIR

Contaminants in Supply Air

The air used in inhalation studies is usually derived from either the outdoor environment or from the air inside the laboratory. Only rarely is the purity of this air sufficient to use without cleaning. Additionally, the temperature and humidity may require adjustment. Some examples of successful supply-air cleaning systems have been described by Doyle et al. (1977), Barsocchi and Knobel (1980), Phalen (1997a,b), and DOE (2003).

The devices used for filtering and conditioning ambient air may produce substantial resistance to airflow and require an air compressor in order to provide the needed flow rate. The compressor is usually powered by either an electric motor or an internal combustion engine. Air from a compressor is usually stored in a high-pressure holding tank before being piped into the laboratory. Typically, compressed air from such systems is dirty, containing a large variety of contaminants (Table 3.1). Aside from atmospheric pollutants, compressed air is likely to contain aerosolized lubricants, unburned fuel, fuel combustion products, water, and scales from internal surfaces of tanks, pipes, and fittings. The use of oil-less compressors will eliminate many contaminants; however, additional air purification is almost always necessary.

Gas Cleaning

Removal of Water Vapor

Table 3.2 gives the amount of water vapor in saturated air as a function of temperature. The major methods for removal of excess water vapor from a gas stream are adsorption onto surfaces, absorption with chemical binding, condensation by cooling, condensation by compression, and combinations of these methods. Each method has its own characteristics with respect to cost, capacity, and water removal efficiency. (Removal efficiency is the mass of water per liter of air remaining after the removal process has reached equilibrium.) An example of an efficient drying method is extreme cooling. Cooling to −70°C with dry ice and acetone leaves about 0.01 mg of water per liter of air at equilibrium. This moisture content at room temperature represents a relative humidity of about 5×10^{-5}%. Ordinary refrigerated air cooling units can be used to reduce the relative humidity to about 20–30% (upon rewarming to room temperature).

Removal of water vapor by absorption and adsorption is often achieved by dessicants. Such dessicants may have an indicator that changes color when the material is saturated, and/or the potential for regenerating by heating or flushing with dry air. Disadvantages include relatively long equilibration times and the buildup of heat during the dessication process. Solid dessicants include calcium sulfate, silica gel, activated alumina, anyhydrous magnesium perchlorate, activated carbon, and molecular sieves (Table 3.3). Liquid dessicants include sulfuric acid, triethylene glycol, sodium hydroxide, potassium hydroxide, calcium chloride, and glycerol (Table 3.4). The book *Controlled Test Atmospheres* (Nelson, 1971) has a useful discussion of solid and liquid dessicants.

Cooling an air stream results in a reduction of the maximum amount of water vapor that is carried in the air at saturation. As an example, consider a refrigeration unit that cools an air stream to 30°F (−1°C), followed by a heater that increases the air stream temperature to 70°F (21°C). From Table 3.2, the cooling will result in a maximum water content of 4.5 g/m^3 of air. Upon warming to 70°F, this 4.5 g represents about 23% of the saturation value of 18.5 g, resulting in a relative humidity of about 23%. Higher humidity can be obtained by adding water vapor to the dried air stream or adjusting the temperature.

Table 3.1 Commonly Encountered Contaminants in Unfiltered, Compressed Ambient Air

Source	Contaminants
Ambient air	Soil and other dusts, miscellaneous gases and vapors
Compressor lubrication	Oil mist, oil decomposition products
Fuel, fuel combustion	Hydrocarbons, carbon monoxide, carbon dioxide, nitrogen dioxide, aldehydes, water vapor, acid-forming gases
Holding tank, pipes, and other surfaces	Metal, metal oxide particles, sediment particles

Table 3.2 Amounts of Water Vapor in Clean Air at Saturation (100% Relative Humidity) as a Function of Temperature

Air temperature (T) °F	°C	Grams of water per m^3 of air (at temp. T)
0	−17.8	1.23
10	−12.2	1.86
20	−6.7	2.94
30	−1.1	4.49
40	4.4	6.55
50	10.0	9.40
60	15.6	13.27
70	21.1	18.45
80	26.7	25.31
90	32.2	34.23
100	37.8	45.74
110	43.3	60.39
120	48.9	78.86
130	54.4	101.8
140	60.0	130.2
150	65.6	165.1
160	71.1	207.3
170	76.7	258.1
180	82.2	319.0
190	87.8	391.2
200	93.3	476.2

Source: Excerpted from Nelson (1971).

Removal of Unwanted Pollutant Gases

In addition to the contaminants in compressed air listed in Table 3.1, several ambient pollutant gases may be present in sufficient concentrations to require removal. Such gases include ozone, sulfur dioxide, and oxides of nitrogen. Removal can be by adsorption onto a large surface area, or by oxidation to a product that is easily removed. For this

Table 3.3 Efficiencies in Terms of Residual Water Content of Selected Solid Dessicants Used for Drying Nitrogen Gas, Delivered at 225 cc/min and 25°C Through a Bed of Dimensions 14 mm Internal Diameter and 450 mm Deep

Dessicant	Avg. efficiency (mg/L of air)	Regeneration time (h); and temp. (°C)
Calcium sulfate (Drierite®)	0.067	1–2; 200–225
Silica gel	0.070	12; 118–127
Activated alumina	0.0029	6– 8; 175–400
Anhydrous magnesium perchlorate	0.0002	48; 245
Molecular sieve 5A (Union Carbide®)	0.0039	Not applicable

Source: Adapted from Nelson (1971).

Table 3.4 Properties of Selected Liquid Dessicants

Dessicant	% RH achieved[a] at 21°C	Solution conc. (%)	Operating temp. range (°C)
Sulfuric acid	5–20	60–70	21–49
Triethylene glycol	5–10	70–95	16–43
Sodium hydroxide and postassium hydroxide	10–20	Saturated	29–49
Calcium chloride	20–25	40–50	32–49
Glycerol	30–40	70–80	21–38

[a]The relative humidity (RH) achieved is the range of values measured over the saturated aqueous solution at 21°C.
Source: Adapted from Nelson (1971).

latter purpose, a catalyst bed or an oxidizer bed can be used to promote reactions such as:

$$O_3 \rightarrow \frac{3}{2} O_2$$

$$CO + 1/2\, O_2 \rightarrow CO_2$$

$$NO + 1/2\, O_2 \rightarrow NO_2$$

$$SO_2 + 1/2\, O_2 \rightarrow SO_3$$

$$SO_3 + H_2O \rightarrow H_2SO_4 \text{ (particle)}$$

Adsorbents for trapping pollutant gases include activated carbon, molecular sieves, and activated alumina. Oxidation can be accomplished by a platinum catalyst bed operated at elevated temperature, or a potassium permanganate bed. Potassium permanganate, an excellent oxidizer for many pollutant gases, is available impregnated on activated alumina pellets (Purafil® Inc., Doraville, GA, U.S.A.). This material is not easily regenerated, but it changes color from pink to black as it becomes spent. Removal of carbon monoxide by oxidation, if necessary, may require the use of a catalyst bed containing platinum or hopcalite (a mixture of oxides of copper and manganese) or Carulite® (Carus Chemical Company, Peru, IL, U.S.A.). Such beds may require either elevated temperatures or pressures to promote efficient oxidation.

Removal of organic vapors, other than low molecular weight species, can be accomplished by adsorption onto an agent such as activated charcoal, or by high temperature oxidation. One method, attributed to H.L. Kusnetz et al. by Nelson (1971) is passage through a 5 cm diameter tube filled with copper shavings maintained at 1,250°F. This method requires cooling of the exit gas stream followed by removal of oxidation products by an acid bubbler, a filter coated with sodium hydroxide, a charcoal bed, and a final particle filter.

Removal of Unwanted Particles

Removal of particles from a gas stream can be achieved by electrostatic precipitation, thermal precipitation, or filtration. Air for laboratory use is usually filtered because of filtration's potential for efficient removal of even the smallest particles and the simplicity of operation. Membrane filters can be efficient in removing particles, but often have the undesirable feature of relatively high resistance to airflow. Fiber filters generally have much lower resistance and are available in forms that have a very high capture efficiency for particles of all sizes. One disadvantage of high-efficiency filters is their tendency to "blind," or clog, as they become filled with captured particles. This problem is usually dealt with by addition of a lower efficiency (roughing) filter upstream, by pleating the filter to increase overall surface area, and by continuous measurement of the pressure drop across the filter as an indicator of the need for replacement. A practical reference that covers many important design considerations for particle filtering systems is the *Nuclear Air Cleaning Handbook* (DOE, 2003). When feasible, the strict air-filtering requirements of nuclear facilities should be met in the inhalation toxicology laboratory. Basic information on air filtration mechanisms has been published by Dorman (1966), Pich (1966), and Hinds (1999, Chapter 9).

Final filtering of air is often performed by a *HEPA filter* (high-efficiency particulate air filter). This type of filter, formerly called an AEC, CWS, or *absolute filter*, is a disposable, deep-pleated, dry-type fiber filter in a rigid case. These filters have very fine fibers in a matrix of large fibers. HEPA filters have very high capture efficiencies, typically 99.997% or more, for removal of all sizes of particles. This efficiency is due to the fact that large particles are effectively deposited by sedimentation and interception (i.e., direct contact with filter fibers), while small particles are pushed into contact with the filter fibers by diffusion (i.e., molecular bombardment).

HEPA filters, and other filters are commercially available in the United States in a variety of sizes and shapes (Cambridge Filter Corporation, Gilbert, AZ; American Air Filters, Houston, TX; Farr Company, Riverdale, NJ; and others).

Because HEPA filters can clog rapidly when presented with large particles and because their replacement is relatively expensive, they are usually placed behind a *prefilter*. Typical prefilters include low-efficiency panel filters consisting of oil or adhesive-coated assemblies of coarse fibers (reusable and disposable types are available), and moderate- and high-efficiency pleated or bag-type dry fiber filters. Low-efficiency filters, including furnace-type units found in the home, are relatively inefficient in capturing particles and are of limited use. Medium- and high-efficiency filters usually have excellent efficiencies for particles of a few micrometers or larger, and are recommended. Selection criteria for prefilters should include their fire and corrosive chemical resistance as well as their particle capture efficiencies and replacement costs. An alternative high-efficiency filter, Filtrete™ (3M Corp., St Paul, MN, U.S.A.) uses charged fibers to enhance efficiency at a reduced pressure drop.

Air Purification and Conditioning Systems

System configurations for purifying and conditioning air for inhalation studies depend upon several factors including initial air contaminants, desired purity, required clean airflow rates and pressures, and acceptable ranges of humidity and temperature of the air. The following two designs, used at the Air Pollution Health Effects Laboratory at the University of California, Irvine, California are examples of systems specifically designed for inhalation studies.

The system shown in Figure 3.1 was designed by William Dennison in 1974 to meet the criteria listed in Table 3.5. The system cost at that time, excluding design, was approximately $30,000 installed.

Ambient air enters the system through a pipe of 10 cm (4 in) diameter from two selectable sources, either from above the laboratory roof or from indoors near the

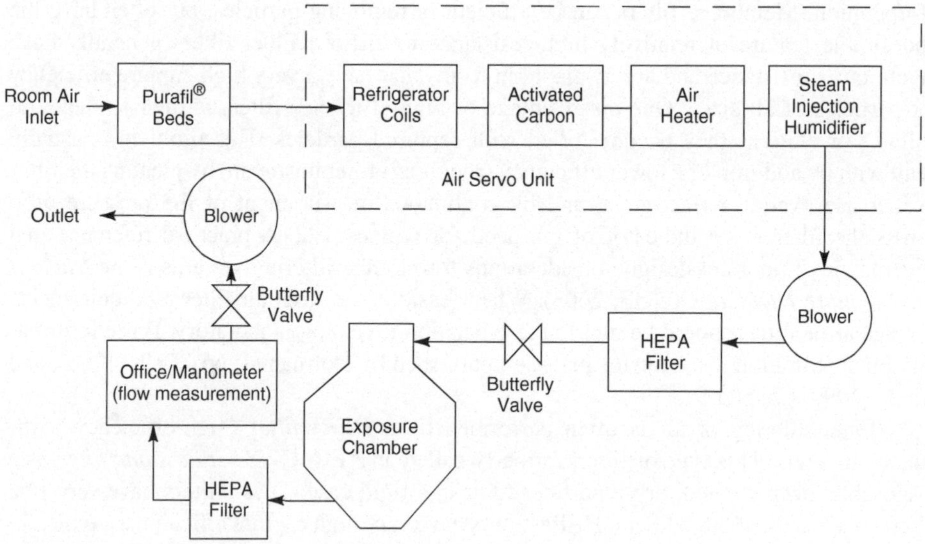

Figure 3.1. Air purification system (designed by William Dennison) used for supply of four, 1-m³ volume inhalation chambers.

laboratory floor. The ambient air source used depends primarily upon the outdoor air temperature and humidity. Ambient air is first filtered for removal of coarse particles using a disposable medium-efficiency dry-pleated filter. Filtered air is passed through two parallel columns of Purafil (Purafil Inc., Doraville, Ga, U.S.A.), 60 cm deep and 30 cm in diameter. The Purafil consists of 1/8 in diameter activated alumina pellets impregnated with potassium permanganate oxidant. The moisture in the inlet gas stream is absorbed by the alumina, dissolving the permanganate which can then oxidize sorbed contaminants. This material removes NO, NO_2, NH_3, SO_2, O_3, most hydrocarbons, and some CO from

Table 3.5 Design Parameters for an Air Purification System Used to Supply Air to Four, 1-m³ Volume Rochester-Type Exposure Chambers[a]

Maximum airflow:	100 standard ft³/min
Desired air temperature:	72 ± 3°F
Desired air humidity:	35–95% RH, controllable

Contaminant	Probable maximum inlet concentration (ppm)	Desired outlet concentration (ppm)
Nitric oxide	1	0.01
Nitrogen dioxide	1	0.01
Hydrocarbons (as C_2H_4)	100	0.5
Carbon monoxide	100	2
Sulfur dioxide	1	0.05
Ozone	0.5	0.01
Ammonia	10	0.002

[a]Particles: 99.97% removal for 0.3 μm diameter.

the air. The Purafil columns are followed by another medium-efficiency, pleated filter for removal of potassium permanganate dust. Air exits the filter holder through a stainless steel pipe of 10 cm diameter and enters an air temperature–humidity conditioning unit (Bemco PTHS Air Servo; Bemco, Inc., Simi Valley, CA, U.S.A.). This unit has a refrigeration system that cools the air to 33–35°F (0.6–1.7°C) that adjusts the water-vapor content of the air to a maximum saturation value at 35°F. An activated charcoal filter further reduces the levels of hydrocarbons. A 1,000 W electrical heater warms the air and two additional 1,500 W immersion heaters vaporize distilled water to increase the humidity to the desired level. Temperature and humidity controllers operate the heaters to maintain the set-point values in the exit air stream. Clean, conditioned air flows through a 10 cm diameter pipe, having a manually operated butterfly valve, and passes through a manifold pipe system to the top inlets of four exposure chambers. This butterfly valve is used to adjust the total output of the Bemco unit. Each chamber is also preceded by an HEPA filter. Air flows downward through the chambers and exits through a prefilter and HEPA filter followed by another butterfly valve. Blowers on the roof pull air through the chambers and exhaust it to the atmosphere. The two butterfly valves for each chamber are used to adjust the airflows and chamber pressures.

The second system (Fig. 3.2) was designed by Robert Walters in 1979 after several years of experience with the previous system. Design criteria were similar, but a system that operated at higher pressure was sought in order to achieve lower humidities, further decrease levels of ambient pollutants, and to solve other minor problems. Outdoor air is first filtered and pressurized to 100 lb/in^2 using a liquid (water) ring compressor (Nash Model CD663C, Gardner Denver Nash, LLC, Trumbull, CT, U.S.A.). Use of a reciprocating-type air compressor would have allowed higher system pressures (up to 5,000 lb/in^2) but would have meant higher maintenance, reduced reliability, oil vapor contamination of the compressed air, and the possibility of CO formation due to compressor lubricant breakdown. Further, Doyle et al. (1977) have shown the liquid ring compressor to be suitable for use in high-pressure air purification systems. Water-saturated air leaving the receiver is filtered to remove water droplets and other particles and is passed through a 31-cm diameter by 80-cm-deep bed of permangante-impregnated alumina pellets (Purafil, Inc., Doraville, GA, U.S.A.). The unit efficiently removes SO_2, NO_x, O_3, and some hydrocarbons from the pressurized air stream. Further purification and drying of the air stream are accomplished in activated alumina-filled, heatless dryer columns operated in a pressure-swing desorption regenerating mode. In this operation, high pressure air is dried by passage through one drying column while a portion of this dried air is simultaneously passed through a second idling column at atmospheric pressure to desorb moisture and contaminants. This enables the idling column to act as the drying column in the next cycle. Drying of the pressurized air stream to less than 0.1 mg/L water content improves the adsorption performance of the downstream activated charcoal bed and is essential to prevent water vapor poisoning of the following unit, a fixed hopcalite bed operating at ambient temperature.

The heatless dryer and hopcalite units are part of a system originally supplied by Del-Tech Engineering Co. (Model No. 5N12, New Castle, DE, U.S.A.) used for the purification of compressed breathing air. Similar driers are available from Twin Tower Engineering, Inc. (Broomfield, CO, U.S.A.). After passage through activated charcoal to remove heavier hydrocarbons, the air stream is passed through a filter to remove entrained hopcalite and charcoal dust, and is brought to laboratory temperature using an air-to-air heat exchanger. Delivery to inhalation chambers follows air pressure reduction, humidification, and final high efficiency (HEPA) particle filtration. The pollutant removal performance of the completed system was evaluated by monitoring ambient and processed air (Table 3.6).

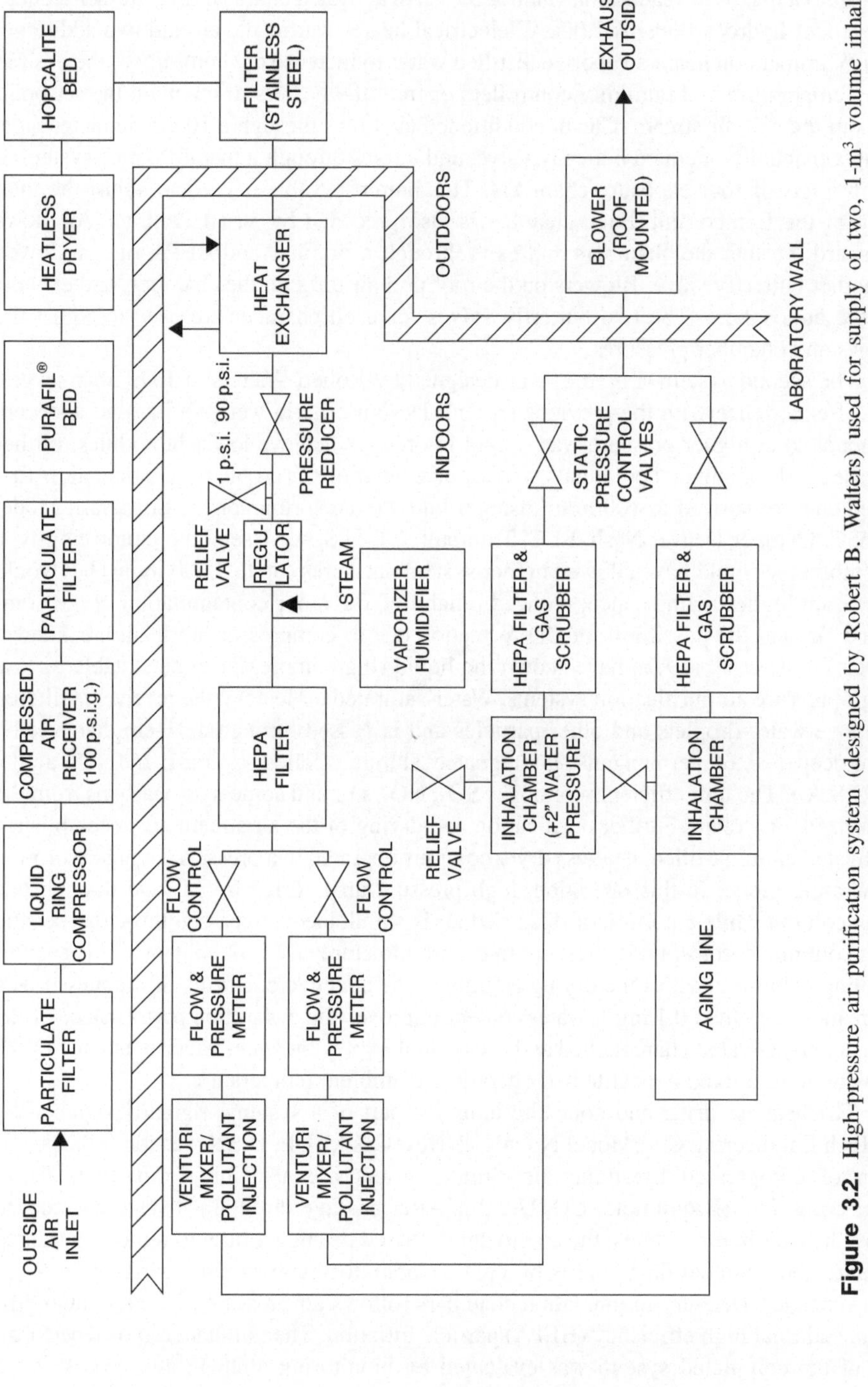

Figure 3.2. High-pressure air purification system (designed by Robert B. Walters) used for supply of two, 1-m^3 volume inhalation chambers. *Abbreviations*: p.s.i., pounds per square inch. p.s.i.g., pounds per square inch gauge.

Table 3.6 Contaminant Removal by High-Pressure Air Purification System

Compound	Ambient conc.	Purified air conc.[a]	Measurement instrument
O_3	0–0.2 ppm	<0.01 ppm	Dasibi® Model 1003-AH Ozone Monitor
SO_2	0–1.0 ppm	<0.01 ppm	Teco® Model 43 SO_2 Analyzer
CO_2	0.04%	0.02%	Perkin-Elmer® Model 1100 Medical Mass Spectrometer
N_2	78.3%	78.3%	Perkin-Elmer® Model 1100 Medical Mass Spectrometer
O_2	21.0%	21.0%	Perkin-Elmer® Model 1100 Medical Mass Spectrometer
CO	0–3.0 ppm	<1.0 ppm[b]	Ecolyzer® Model 2000 CO Analyzer
NO	0.02 ppm	<0.01 ppm[b]	Beckman® Model 952 NO_x Analyzer
NO_2	0.08	0.04 ppm	Beckman® Model 952 NO_x Analyzer
CH_4	1.4 ppb	1.4 ppm	Beckman® Model 6800 Gas Chromatograph
C_2H_6	6.3 ppb	7.1 ppb	Beckman® Model 6800 Gas Chromatograph
C_2H_4	5.1 ppb	<2 ppb[b]	Beckman® Model 6800 Gas Chromatograph
C_2H_2	4.7 ppb	<2 ppb[b]	Beckman® Model 6800 Gas Chromatograph
C_3[c]	0.46 ppm	0.20 ppm	Beckman® Model 6800 Gas Chromatograph
Particulate	10^4–10^5/cc	<100/cc	Environment-One® Rich 100 Condensation Nuclei Counter

[a]Measured at system flow rates of 600–1200 standard L/min.
[b]Lower detection limit for compound.
[c]Hydrocarbons with three or more carbon atoms.

Additional air purification systems have been described by Roehlich and Rodgers (1976), Doyle et al. (1977), and Barsocchi and Knobel (1980). Doyle's system, a forerunner to that shown in Figure 3.2, was developed for purification and humidification of air for atmospheric chemistry research at the University of California, Riverside, California. Barsocchi and Knobel's design was intended for the filling of high-pressure cylinders of dry, ultrapure *zero air*. The air purifier of Roehlich and Rodgers was developed for removing automotive exhaust from the breathing air in tunnel booths and other working areas subject to heavy vehicular traffic.

AEROSOL GENERATION

General Considerations

Selection of aerosol generators for inhalation studies is made using several criteria, including the following:

- aerosol characteristics including chemistry, shape, size, surface area, dispersity, and number of particles per unit volume of air;
- aerosol generator stability over the time of use;
- waste of input materials (some materials are expensive or rare);
- hazards to laboratory personnel or subjects;
- introduction of unwanted foreign materials;
- required maintenance;
- sensitivity of the characteristics of the aerosol to line voltage, input air pressure, barometric pressure, temperature, and relative humidity; and
- description of the aerosol generator in the open literature by scientists not associated with its sale.

Persons not experienced in aerosol generation must be very careful when selecting generators for inhalation studies. When in doubt, one is wise to contact aerosol experts.

Several excellent general references are available on the topic of aerosol generation (Mercer, 1973; Corn and Esmen, 1976; Tillery et al., 1976; Willeke, 1980; Koch, 1994; Loffert et al., 1994; Moss and Cheng, 1995; Hinds, 1999; Lange and Finlay, 2006; Hickey, 2007). Ma-Hock et al. (2007) discuss the related but special problems of generation and characterization of nanomaterials.

Monodisperse Aerosols

Monodisperse (uniform size) particles are used for calibrating laboratory instruments, performing aerosol deposition studies, examining the influence of particle size on biological effects, and for a variety of other basic investigations. With some exceptions, including microorganisms, pollen and spores, monodisperse aerosols are only made in the laboratory. The sizes of some viable particles which, within a genus, are near monodisperse are listed in Table 3.7. Three primary methods are used for production of monodisperse aerosols: growth under controlled conditions, precision reduction of bulk material, and segregation on the basis of particle size from a distribution. Selected methods of obtaining aerosols with geometric standard deviations near 1 are listed in Table 3.8.

Sinclair and LaMer (1949) described a condensation generator for monodisperse aerosols. Since then, several generator modifications have been reported, many of which

Table 3.7 Stokes Diameters[a] for Viable Particles, Which Within a Genus Are Nearly Monodisperse

Particle	Stokes diameter (μm)
Viruses	0.015–0.45
Bacteria	0.3–15
Fungi	3–100
Moss spores	6–30
Fern spores	20–60
Pollen (airborne)	10–100

[a]Definition: Diameter of perfect sphere of the particle's density with same terminal settling velocity in still air.
Source: Adapted from Jacobsen and Morris (1976), Chapter 4.

are reviewed by Mitchell (1995a). In these generators the material of interest is vaporized and the vapor slowly condensed onto seed nuclei particles. If the vapor is uniformly mixed with the nuclei and condensation is slow and diffusion well controlled, then the average diameter, D, of particles leaving the generator is given by

$$D^3 = 6M/\pi\rho N \qquad \text{(eq. 3.1)}$$

where M is the initial vapor concentration per unit volume of air, N is the initial number of nuclei per unit volume of air, and ρ is the density of condensed vapor. This equation assumes that no vapor is lost to the walls of the cooling section and that the nuclei are negligibly small with respect to D, the final particle diameter.

A successful generator of this type, described by Prodi (1972), has been used for aerosol deposition studies in humans (Tarroni et al., 1980). In this device, nitrogen gas at

Table 3.8 Selected Methods for Production of Monodisperse Aerosols

Method	Comment[a]
Dispersion of monodisperse powders	Examples include lycopodium and other spores and pollen. Generators include fluidized bed type (up to 10^7)
Sinclair-LaMer® type generators	Vapor is condensed on seed nuclei by controlled cooling (up to 10^7)
Spinning disk and top generators	Liquid is fed at a constant rate onto a spinning flat surface. Aerosol particles are formed at rim of surface (up to about 10^5)
Vibrating reed, jet, or orifice	A liquid stream is chopped into uniform pieces by high-frequency mechanical disruption (up to about 10^3)
Hydrosol formation	Chemical or physical reactions, for example, polymerization or crystal formation in an aqueous or other liquid medium (up to 10^7)
Segregation means	Use of particle spectrometers that segregate on the basis of size-related properties such as electrical mobility or terminal settling velocity (up to 10^7)

[a]*Note*: Approximate number of particles generated per cm^3 of air is given in parentheses. Beyond about 10^7 particles/cm^3, coagulation is rapid leading to production of aggregates.

a pressure of 2.5 atm is used to nebulize a dilute sodium chloride solution. The nebulized aerosol is dried by passage through a dessicator containing silica gel, forming the nuclei particles. The air stream from the nuclei generator is split, one part passing through a thermostated, heated bubbler containing melted paraffin or carnauba wax, and the other stream bypassing the bubbler and mixing with the first stream in a heating column. There, the paraffin or wax is completely vaporized and then slowly cooled to promote condensation on the nuclei. Airflow in the heating column is downward to reduce thermal air mixing. By varying flow rate through the bubbler and the bubbler temperature, particles having count median diameters ranging from about 0.2 to 2.0 μm can be produced. The geometric standard deviation of the particles is less than 1.1. Commercial versions are available (TSI, Shoreview, MN, U.S.A.; PALAS, Karlsruhe, Germany).

Walton and Prewett (1949) presented an analysis of the conditions for production of monodisperse particles that are thrown off of the edge of a rapidly spinning surface. Since that time a variety of spinning disk and spinning top liquid aerosol generators have been described (Mitchell, 1995a). These generators operate at rotational speeds ranging from a few hundred up to about 250,000 revolutions per minute to produce monodisperse droplets with diameters in the range of 6 μm to 3 mm. These generators also produce small secondary particles (satellites) that form as the main droplets separate from the edge of the spinning surface. Subsequent drying of primary particles can be used to produce monodisperse particles as small as 1 μm in diameter. Lippmann and Albert (1967) designed a spinning disk generator for producing radioactive insoluble aerosols for human and laboratory animal deposition and clearance studies. In this generator, satellite particles were separated out and collected on a filter.

A relatively simple method for obtaining monodisperse aerosols in a wide variety of sizes involves the aerosolization and drying of aqueous suspensions of commercially available microspheres (Duke Scientific Corp., Palo Alto, CA, U.S.A.). Such suspensions can be aerosolized using a variety of aerosol generators, with compressed-air nebulizers being the most frequently used. Although the commercial particles are quite uniform, several considerations must be applied in their use. First, the nominal diameters as supplied by the manufacturer may differ slightly from the sizes received. Second, when liquid suspensions are aerosolized, some droplets will contain more than one solid particle and some will contain none. Thus, after drying, in addition to individual monodisperse particles, one will also obtain some agglomerates and some small residue particles (due to the presence of stabilizer in the suspension). As the suspension is made more concentrated, the agglomerate fraction increases, and as it is made more dilute, the residue fraction is greater (Fuchs, 1973). Raabe (1968) has given dilution factors for suspensions that allow one to keep the singlet/agglomerate ratio below a given value. As an example, using a nebulizer with a volume median droplet diameter of 5 μm and a geometric standard deviation of 1.6 to aerosolize a suspension of 1.0-μm diameter latex spheres, the percentage (by volume) solids (spheres) should be below 0.3 (i.e., 1 part in 3000) in order to keep the ratio of singlets to agglomerates above 20 : 1. In this case the ratio of small, stabilizer residue particles to single latex particles will be about 50 : 1.

Kotrappa and Moss (1971) used the Stöber spiral-duct aerosol centrifuge (Stöber and Flaschbart, 1969) to size, separate, and collect insoluble polydisperse aerosols into narrow size groups of particles that were subsequently resuspended in water and nebulized as monodisperse aerosols. This technique, which can be performed with any suitable aerosol spectrometer and insoluble polydisperse aerosol, can produce aerosols with geometric standard deviations of less than 1.1 in the aerodynamic diameter range from 0.3 to about 5 μm. This technique was used by Raabe et al. (1977) to produce a variety of aerosol sizes for aerosol deposition studies in rodents.

Monodisperse droplets can be made in relatively small quantities by the vibrational breakup of a liquid stream. In a device constructed by Raabe and Newton (1970), using the work of Fulwyler (1965), a liquid stream exiting from a small orifice (10–15 μm in diameter) is broken into droplets by a vibrating piezoelectric crystal. Vibration of the orifice at 300,000 cps leads to droplets that vary in volume by less than 1%. A similar aerosol generator is described by Strom (1969). A vibrating orifice monodisperse aerosol generator is commercially available (TSI, Shoreview, MN, U.S.A.) after the design by Berglund and Liu (1973).

Further descriptions of generators for monodisperse aerosols are given by Fuchs and Sutugin (1966), Raabe (1979), Mitchell (1995a), Hinds (1999), Roth et al. (2004), and Baker et al. (2008).

Polydisperse Aerosols

A variety of techniques have been used for controlled generation of polydisperse aerosols. Even a simple listing of these techniques would be cumbersome. Certain methods are especially useful in inhalation studies because they produce aerosols with properties and in the size ranges similar to those inhaled by people in various environments. Examples include the Wright (1950) dust feeder for dispersion of dry compressed powders, the air-blast nebulizer for aerosolization of liquids, the fluidized-bed vibrating generator for dispersion of loose powders, and the exploding wire generator for vaporizing solid metals (Mitchell, 1995a).

Droplet Generators

Droplets are used in at least four ways in inhalation studies: (1) the droplet itself may be passed into an exposure system for inhalation; (2) the droplets may carry a suspended solid that is inhaled after evaporation of the droplet; (3) a solution may be aerosolized, leaving a residue particle upon drying; and (4) the droplets may be used as input to another stage of an aerosol generator (e.g., as input to the heater of a condensation aerosol generator). Table 3.9 lists two suitable polydisperse droplet generators. Each generator has its unique characteristics and either one may be preferable for a given application. For example, if one desires a large number of droplets per liter of air, and can tolerate large droplets, an ultrasonic nebulizer may be selected. Compressed air nebulizers, on the other hand, provide smaller droplets at a lower output rate.

Compressed air nebulizers are commonly selected for inhalation studies due to their relatively low cost, stable operation, and droplet size characteristics. Table 3.10 gives the operating characteristics of some commonly used compressed air nebulizers. In this type of generator, an air jet formed by the passage of air under high pressure through a small

Table 3.9 Selected Polydisperse Droplet Aerosol Generators

Type	Operating principles	References
Air-blast nebulizer	High-velocity air directed over a liquid feed tube produces breakup of entrained liquid filament	Mercer et al. (1968b), May (1973), Loffert et al. (1994), Lange and Finlay (2006)
Ultrasonic nebulizer	Vibration of a piezoelectric crystal forms a fountain of liquid that emits droplets from its tip	Mercer et al. (1968a), Mitchell (1995a), Moss and Cheng (1995), Lange and Finlay (2006)

Table 3.10 Operating Characteristics of Compressed Air Nebulizers

Nebulizer	Applied pressure (lb/in^2)	Airflow (L/min)	Aerosol out (μL/L air)	Water vapor out (μL/L air)	Volume median drop diameter (μm)	GSD
Dautrebande with open vent	5	11.2	1.0	9.7	–	–
	10	14.9	1.4	9.6	1.7	–
	20	21.2	2.3	8.6	1.4	1.6–1.7
	30	27.3	2.4	8.2	1.3	–
3-Jet Collison	15	6.1	8.7	12.6	–	–
	20	7.1	9.0	14.8	–	–
	30	9.4	9.0	19.4	About 2	About 2
	40	11.4	9.3	23.5	–	–
	50	13.6	10.4	27.9	–	–
De Vilbiss® No.40 with closed vent	15	12.4	15.5	8.6	4.2	–
	20	16.0	14.0	7.0	3.2	1.8–1.9
	30	20.9	12.1	7.2	2.8	–
Lovelace	20	1.34	(34)[a]	12	6.9	1.7
	30	1.81	(22)[a]	11	4.7	1.9
	40	2.28	(15)[a]	9	3.1	2.2
	50	2.64	(19)[a]	11	2.6	2.3
(Lovelace Nebulizer chilled to 0°C)	20	1.34	55	1	–	–

[a]Calculated.

Abbreviation: GSD, geometric standard deviation.

Source: Adapted from Mercer et al. (1968b), and Raabe (1979).

orifice shatters a liquid stream. The liquid stream is drawn from a reservoir into the path of the air jet by the reduced pressure within the jet, or as a result of positive pressure in the reservoir. The shattered liquid forms large and small droplets. The larger droplets impact on a nearby surface (an adjacent wall, a baffle, etc.) and coalesced fluid drains into the liquid reservoir. Smaller droplets follow air streams around the impaction surface and exit from the nebulizer. The air that entrains the particles will be essentially saturated with the vapor of the fluid in the reservoir. If an aqueous solution or suspension is nebulized, the concentration of the dissolved or suspended material will steadily increase with time. Three methods are used to control this effect: the reservoir may be cooled to reduce evaporation; the reservoir may be enlarged or continuously renewed; or the feed air can be saturated with liquid vapor.

Although the majority of the droplets impact within the nebulizer, usually about 10^6 to 10^7 particles per cm^3 of air exit as useful aerosol. These droplets typically have a distribution of sizes that is describable by a lognormal distribution function with a geometric standard deviation between about 1.5 and 2.2. Freshly nebulized droplets are usually highly electrically charged and are often passed through a bipolar ion field for discharging (Liu and Pui, 1974; Ji et al., 2004).

When a solution is nebulized and the droplets dried to a residue aerosol, the resultant dry particles will be smaller than the original droplets. The theoretical relationship between the mass median droplet diameter (D_d) and the mass median residue particle diameter (D_p) for spherical residue particles is

$$(D_p)^3 = (D_d)^3(C)\ (\rho_d/\rho_p) \qquad \text{(eq. 3.2)}$$

where ρ denotes density and C the weight fraction of solute. The geometric standard deviation of the residue aerosol will be that of the droplet aerosol.

Dry Dust Generators

Aerosolization of dry dusts is achieved by several means including: (1) scraping of material from a plug of packed powder; (2) use of a fluidized-bed elutriator; (3) entrainment of powder by means of a venturi jet; and (4) dispersion of powder or erosion of a plug by an air jet.

Table 3.11 lists some dry dust aerosol generators. The dust loaded into the generator can have a variety of forms including fibers, spheres, plates, etc. The dispersed phase will have a size distribution that may differ from that of the nondispersed material due to

Table 3.11 Selected Generators of Dry Dust Aerosols

Generator	Operation principle
Wright dust feeder	Powder which has been compressed into a plug is scraped free by a blade and entrained in a dry airflow
Timbrell type fibrous dust generator	Similar to Wright dust feeder but blades rotate at high speed over compacted asbestos or fibrous glass plug
Fluidized bed	Upward flowing air fluidizes (suspends) and carries particles upward into a vertical elutriator. New powder is continuously fed into the fluidizing section
Turntable dust feed	Powder flows from a hopper (agitated) onto a turntable which rotates toward an aerosolizing air jet

breakup of particles, or agglomeration of primary particles. Reviews of dry dust generators have been published (Mitchell, 1995a; Hinds, 1999).

The fluidized-bed elutriator is often used in inhalation studies because of its stability over long periods of generation and its favorable elimination of larger agglomerates and primary particles by elutriation. Modern fluidized-bed elutriators usually incorporate two important features: continuous feeding of material to the generator and discharging of the aerosol within the elutriation column. Continuous feeding of material into the generator is often necessary because material in the generator initially loses the fine particle fraction, leaving the larger particles in the generator. Thus, unless material is continuously supplied to the generator, the emitted size distribution will change with time. Fluidized bed generators impart electrical charges to the aerosols they emit and thus passage through an ion field is recommended. Figure 3.3 shows a generator described by Marple et al. (1978) incorporating such features. Mitchell (1995a) reviews recent designs.

The Wright dust feeder (Figure 3.4), described originally by B.M. Wright (1950), has been used to generate aerosols from a wide variety of dry materials. The material is first formed into a plug which is then scraped by a blade and aerosolized by entrainment in a flowing air stream.

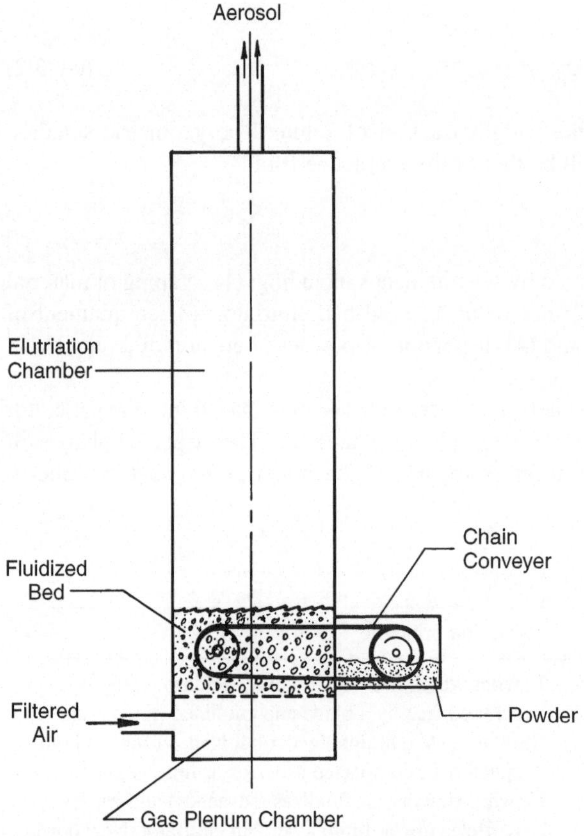

Figure 3.3. A fluidized bed elutriator aerosol generator with an ion source for discharging of aerosols. *Source*: With permission from Marple, V.A., Liu, B.Y.H., and Rubow, K.L., *Am. Ind. Hygiene Assoc. J.*, 39(26), 1978, http://www.aiha.org

Figure 3.4. The Wright dust feeder in which a packed dry plug is scraped and an aerosol formed by entrainment of removed material in a flowing air stream. *Source*: Photograph supplied by L. Leach of the University of Rochester.

A final example of a polydisperse aerosol generator is the exploding-wire device (Figure 5.13, Chapter 5). This generator produces large amounts of a chain-agglomerate metal fume aerosol similar to that formed during welding or metal smelting. In this device a large electrical current, 10,000 A or more, is rapidly passed through a few milligrams of wire. The wire is rapidly and explosively heated forming an ionized metal vapor (plasma) that condenses to form branched chains composed of very tiny (submicrometer) primary particles. This type of aerosol generator does not produce continuous aerosol but it can be used to produce significant deposition of material in the lung during a brief inhalation period (Phalen, 1972), and it can be used for producing nanopowders (Kotov, 2003).

GAS GENERATION

General Considerations

The generation of gases for inhalation studies typically involves controlled injection of a concentrated gas into the throughput air of an exposure system. Methods of generation of gases include the following:

- metering the output from compressed gas tanks,
- injection of volatile liquids,
- vaporization of a liquid, or sublimation of a solid by control of the temperature,
- diffusion of the gas through a permeation tube, and
- creation of the gas by controlled chemical or physical reactions.

In a sense, gases are less difficult to generate than particles because stabilizing particle size is not a consideration. Also, it is usually relatively easy to obtain uniform mixing of gases in an air stream. However, the purity of the gas must be maintained, including the suppression or removal of unwanted particles. Additionally, the gas generator must have a stable output over the course of the study. Many gases are rapidly absorbed by moisture or tubing, and in some cases can penetrate through tubing or container walls. Reactive gases can also be lost on surfaces with which they come into contact. Ozone, for example, can be rapidly depleted by fresh surfaces, and several hours of surface conditioning by contact with the gas may be necessary.

Useful references on the controlled generation of gases include: the book *Controlled Test Atmospheres* by Nelson (1971); a NIOSH research report *Development and Validation of Methods of Sampling and Analysis of Workplace Toxic Substances* by Gunderson and Anderson (1980); and reviews of Tillery et al. (1976), Wong (1995), and Kleinman and Phalen (1997).

Common Techniques

Compressed Gas Cylinders

Tanks containing known concentrations of a given gas at an elevated pressure are obtained commercially or prepared in the laboratory. Preparation of such tanks is usually done by first evacuating the tank and then either introducing the concentrated gas or injecting a known amount of the liquid with a syringe. This is usually followed by the addition of clean compressed air. Mixing can be achieved by rotating the tank or alternatively heating and cooling an end of the tank. Some gases are not stable for long periods when stored; special tank materials such as aluminum can extend shelf-life.

Metered removal of the tank gas is often achieved by a pressure regulator and a flow-rate controller. Typical flow-rate controllers include needle-valve equipped rotameters, critical orifices, capillary tubes, or porous plugs. If a rotameter is not used, passage of the gas mixture through a wet test meter, dry gas meter, heated wire anemometer, or soap bubble meter will serve to measure the flow rate. Such devices may alter the measured gas by addition of water vapor or other contaminants or by absorption of the gas within the measuring device. Table 3.12 gives the range of applicability of various flow-rate measuring methods. The *Compressed Air and Gas Handbook* (Rollins, 1988) is a useful resource for those planning to prepare or use compressed gases.

Table 3.12 Ranges of Selected Flow-Rate Measuring Devices

Device	Common useful range
Rotameters	1 mL/min to 10,000 L/min
Wet test meters	1–80 L/min
Dry gas meters	5–5,000 L/min
Orifice meters	5 mL/min to 5,000 L/min
Heated-wire anemometers	0.1 mL/min to 10,000 L/min
Soap-bubble meters	0.5 mL/min to 6 L/min
Mass flowmeters	0.005–3,600 L/min[a]

[a]Approximate range.
Source: Adapted from Nelson (1971).

Syringe Injectors

Motor-driven syringes are convenient for injecting volatile liquids either directly into an air stream or into a heated vessel to promote volatilization. These devices can be used to produce constant liquid flow rates down to a fraction of a microliter per minute. These systems can be accurately calibrated by weighing the syringe contents before and after timed periods of operation. Glass, plastic, and relatively inert Teflon® (duPont de Nemours) syringes are available in a variety of sizes from 0.05 to 10,000 mL. Nelson (1971) discusses several drive mechanisms for motor-driven syringes.

Vaporization and Sublimation Systems

These methods of controlled gas generation require precise control of the generator temperature, since the rate of gas evolution is strongly temperature dependent. Submersion of the source reservoir in a temperature-controlled water (or oil) bath usually suffices, and the throughput air can be passed through a metal tube immersed in the same bath. Because some vapors are reactive, explosive, or flammable in concentrated form, the carrier gas is frequently nitrogen.

Permeation Tubes

These devices are usually polymeric tubes that are porous to the material (usually liquid) inside. Hundreds of organic and inorganic materials are available in permeation tubes. The gas output rate is determined by successive weighings of the tube.

The permeation rate Q is a function of the vapor diffusion coefficient (D) through the polymer, the solubility coefficient (C) of the vapor in the polymer, the thickness (t) of the tube, the available surface area (A) for permeation, and the pressure difference between the inside (P_i) and outside (P_o) of the tube.

$$Q = D \bullet C \bullet A \left(\frac{P_i - P_o}{t} \right) \quad \text{(eq. 3.3)}$$

When not in use, the life of the permeation tube can be extended by storage at low temperature. However, in some cases cold storage alters the future permeation rate, so manufacturer's specifications should be consulted regarding storage. Nearly any gas that can be liquefied can be used in a permeation tube. Permeation tube calibrators can also be custom made. Permeation tubes are described in more detail by Mitchell (2000).

Chemical and Physical Reactions

A variety of processes, both chemical and physical, lead to the production of gases. The use of such processes to generate atmospheres may be necessary when the gas of interest is unstable, which eliminates generation by the previously mentioned techniques. When producing study gases by chemical reactions, the control of several variables may be crucial, including the following:

- reactant purity,
- reactant concentrations,
- reactant supply rates,
- reactant mixing kinetics,
- temperature,

- pressure,
- relative humidity, and
- reactor geometry.

Several physical processes lead to the generation of gases, including:

- electrolysis from solutions or molten salts,
- thermal transformations,
- photochemical reactions,
- ionization and other electrical phenomena, and
- radioactive decay.

For example, ozone, triatomic allotropic oxygen, can be formed from diatomic oxygen by electrical discharge or by ultraviolet irradiation. The use of pure oxygen, rather than air, as a reactant will prevent possible formation of oxides of nitrogen. A laboratory ozone generator could then consist of compressed pure oxygen followed by a pressure regulator, needle valve, and rotameter to provide a known and controllable flow into an ultraviolet irradiator. The output from the irradiator, carried through inert tubing, is then fed into the mixing section of an exposure chamber where it is thoroughly mixed with purfied throughput air. Precise control of the voltage supplied to the irradiator and of the oxygen flow rate is necessary for obtaining steady ozone levels. Because rotameters change calibration as the gas density changes, corrections are needed (Caplan, 1985). As previously mentioned, conditioning of the tubing and chamber by flowing ozone will be necessary before stable concentrations can be achieved.

MIXED AEROSOLS AND GASES

Single-component atmospheres are seldom found in nature. Recognition of this fact, plus the steady improvements in exposure techniques, have led to increased popularity of multicomponent laboratory inhalation studies. New problems occur when such atmospheres are generated. The introduction of two or more pollutants simultaneously into a mixer can lead to unwanted reactions, especially when the pollutants are combined at high concentrations prior to dilution with the throughput clean airflow. For example, the mixing of concentrated ozone and sulfur dioxide can rapidly lead to the formation of tens of thousands of fine particles per cubic meter of air. The particles appear to be sulfuric acid droplets, which can convert to ammonium sulfate aerosols if ammonia, a common contaminant, is also present. The problem is diminished by introducing the two gases at points in the air stream that are separated from one another sufficiently to permit dilution of one gas before the second is introduced.

Similarly, when both particles and gases are introduced into an air stream, separation of the concentrated outputs of the generators may be necessary to reduce unwanted interactions. Such interactions might include saturation of aerosol surfaces with gas molecules, or chemical reactions that may produce secondary compounds. Also, excessive losses of particles on tubing walls can occur due to turbulence, thermal effects, or differences in density between the introduced gas and the air stream.

Some general principles should be considered when pollutants are mixed:

- dilute each component with air separately,
- expect and monitor for unwanted reaction products,
- ensure that the levels of agents are appropriate at the subject's breathing zone, and

- avoid large differences in temperature and humidity between the gas or aerosol generator output streams and the throughput air stream.

PRINCIPLES FOR STABILIZING THE EXPOSURE ATMOSPHERE

Generator Stability

Aerosol and gas generators of various types have differing stability characteristics with respect to their output. For particles, one can expect some variation over time in both the particle-size distribution and the number of particles generated. Such variations are caused by a variety of factors including: fluctuations in the electrical input; changes in pressure, humidity, or flow rate of supplied air; changes in generator temperature; obstruction of orifices; and changes in concentrations of solutions or suspensions to be aerosolized. Air-jet nebulizers, for example, may cool rapidly during operation due to evaporation of the reservoir liquid. Also, such evaporation can cause the reservoir liquid to become steadily more concentrated during operation.

After an aerosol leaves the generator, additional factors can act to produce variability in the efficiency with which particles reach the exposure zone. Such factors include: aerosol coagulation; deposition of charged particles onto surfaces; and losses in tubing due to sedimentation, impaction at bends, diffusion, or air turbulence. Several techniques can be used to minimize fluctuations in aerosol delivery to the subject:

- electrical discharging of freshly generated aerosols,
- use of large bore, straight, electrically conductive tubes for aerosol transport,
- stabilizing the temperature of the aerosol generator and tubing,
- reducing the evaporation rate of generator liquids by cooling, the use of presaturated supply air, or the use of a continuous supply of fresh liquid for aerosolization,
- use of high-purity, filtered liquids for aerosolization,
- stabilization of the electrical supply to equipment,
- stabilization of air pressures and flow rates,
- use of filtered air,
- use of multiple aerosol generators in parallel,
- frequent cleaning of critical aerosol generator parts,
- avoidance of "temperamental" generator components or aerosol generators,
- use of monitoring systems (for aerosol output and parameters that affect output),
- adequate dilution of freshly generated particles to reduce coagulation,
- use of tight connections, and
- vibration isolation of aerosol generators.

Exposure System Stability

Whether the exposure system involves a chamber or a smaller device, stability of the exposure atmosphere in a dynamic system will involve controlling several factors:

- pollutant generator outputs;
- airflow rate;
- pressure;
- temperature;
- humidity; and
- losses on walls, animals, animal cages, and other surfaces.

The second and third factors, airflow rate and pressure within the exposure system, are interrelated. Throughput air is usually pushed through the system by a compressor or pressure in high-pressure air tanks. In addition, a downstream air pump is commonly used to draw air through the system. In general, good air pumps have a relatively constant output. Flow of air through a typical exposure system will therefore be stable provided that no significant change in the resistance to flow occurs. For example, resistance will increase if a filter becomes loaded with trapped particles. The use of filters of large surface area and a monitor to indicate the filter pressure drop can help to maintain flow stability.

The control of temperature and humidity, discussed earlier, requires a throughput air-conditioning system. Relative humidity is a function of water vapor content and air temperature, but temperature is more difficult to control. Metal exposure systems conduct heat rapidly and thus make it difficult to operate chambers at temperatures that differ greatly from the surrounding room temperature. Even within a modern air-conditioned and heated laboratory, room temperature gradients are likely to exist. This can be due to locations of heat-generating equipment, windows, and forced-air registers. Such gradients can be diminished by the use of air-circulating fans that keep the air well mixed within the room. Another source of heat that must be considered is that produced metabolically by experimental subjects. As discussed in Chapter 5, chamber loading with animals should typically be 5% of the chamber volume or less in order to control temperature in the exposure zone.

Another factor in exposure system stability, losses on walls and surfaces, presents a stability problem primarily when dealing either with very reactive gas atmospheres or with aerosols. Reactive atmospheres (e.g., those containing ozone) that undergo chemical changes on surfaces can also produce alterations in the surface reactivity. Such a time-dependent reactivity can lead to problems when trying to stabilize the level inside of an exposure chamber. Problems with changes in surface reactivity are minimized by use of relatively chemically inert surfaces, preconditioning surfaces with the exposure atmosphere, and avoiding buildup of debris on internal surfaces.

Aerosol losses within an exposure system can be severe. Common conditions that lead to the unwanted deposition of aerosol particles on walls are electric fields, thermal gradients, turbulent flows, and fluctuations in relative humidity. Electric field fluctuations can essentially be eliminated by the use of grounded, conductive aerosol chambers and piping. In addition, electrical charges on the subject's fur, skin, or clothing can lead to large and unpredictable losses of exposure aerosol. Such losses are reduced by discharging the aerosols and avoiding excessively low relative humidity.

With due care, variations in atmospheres that are mixtures of particles and gases can be controlled to low concentrations. A paper by Walters et al. (1982) presents results from a 4-h exposure of rats to 5.0 ppm SO_2, plus 0.8 ppm O_3, plus 1 mg/m^3 sulfate aerosol (ammonium and ferric) droplets. The atmospheric characteristics are summarized in Table 3.13.

Real-Time Adjustments

Despite all efforts to build a stable exposure system, one may find that the measured pollutant concentrations at the breathing zone still appear to vary more than desired. In this case, what must be addressed is whether or not these variations are really occurring in the exposure zone, or whether they are mainly due to variations in the sensitivity of the monitors. Thus, before one can investigate the stability of an exposure atmosphere, one must have knowledge about the stability of the monitoring system. The methods for assessing the uncertainty associated with an instrument can be complex and demanding.

Table 3.13 Stability of Gas Concentrations in Mixed Atmospheres Consisting of 5.0 ppm SO_2, 0.8 ppm O_3, and 1 mg/m^3 Sulfate Aerosol (Ammonium and Ferric) Droplets[a]

	Ozone			Sulfur dioxide		
Length of exposure (h)	$\bar{x}$ (ppm)	sd (ppm)	sd/$\bar{x}$	$\bar{x}$ (ppm)	sd (ppm)	sd/$\bar{x}$
4	0.81	0.04	0.05	5.0	0.2	0.04
4	0.80	0.04	0.05	5.1	0.1	0.02
4	0.79	0.04	0.05	5.0	0.1	0.02
4	0.82	0.03	0.04	5.0	0.1	0.02
4	0.80	0.02	0.03	5.0	0.1	0.02
4	0.82	0.02	0.02	5.0	0.1	0.02

[a]*Note*: 24 samples of each gas concentration were acquired.
Source: Adapted from Walters et al. (1982).

Two approaches can be used: establishing very constant test environments followed by multiple measurements and statistical analyses; or testing each component (e.g., pumps, filters, sensors) of the measuring instrument for stability and then calculating the effect of each component's reasonable drift or fluctuation on the output measurement.

Assuming that the monitoring method is sufficiently refined as to introduce negligible uncertainty, and that the monitor provides accurate real-time measurements, one can make periodic adjustments of either the pollutant generators or the exposure system airflows and thereby improve the constancy of the exposure concentrations. Such adjustments can be made by hand, or automatically if adequate technology is available (Carpenter et al., 1979). Whether hand or automatic adjustments are made, some problems must be solved. The first problem arises from the fact that an adjustment in the system does not instantly change the concentration in the exposure breathing zone. In fact, several minutes may pass before the adjustment results in a change at the subject's breathing zone. Further, in a well-mixed exposure zone, the drift toward the new selected concentration will be exponential in form, requiring several air exchanges before the new concentration stabilizes (Phalen et al., 1994b). One method for dealing with this problem is to place a second monitoring system near the generation system and then make adjustments on the basis of those concentration readings. This method does not eliminate the need for measurement of the exposure atmosphere characteristics at the exposure breathing zone.

Another problem that must be overcome when making generator or flow adjustments involves making the right magnitude of adjustment to achieve the desired change at the exposure zone. The ability to do this implies several things: (1) that the system has been calibrated and the relationship between adjustment magnitude and response is quantitatively known; (2) that the desired range of adjustments is realizable (i.e., that the airflows or generator outputs are not already at maximum or minimum); and (3) that adjustments will not lead to large "overshoot" and throw the exposure concentration too far in the opposite direction.

Another problem in making automatic, real-time adjustments relates to losses in sampling line integrity. For example, if a sampling line leading from a chamber to a monitoring instrument becomes contaminated, clogged, or develops a leak, the monitor can give an erroneously low or high reading. In this case the exposure level can be proper, but a corrective adjustment in the concentration will be made anyway. It is wise to check the sampling and monitoring system integrity before adjusting airflows or pollutant generators.

4

Characterizing Exposures

INTRODUCTION

Adequate characterization of an exposure is essential in inhalation toxicology, and thus deserves both insightful planning and meticulous attention to detail. The fundamental objective is to establish relationships between observed biological responses and measurable properties of an exposure.

First, one must know which physical and chemical properties of the exposure are relevant to the expected effects. For particles, the composition, number concentration, size, and size distribution come immediately to mind. Less obvious, but often relevant, particle properties include the surface area, state of electrical charge, surface character, hygroscopicity, aspect ratios, fractal dimensions, dissolution rates in lung fluids, and amounts and types of absorbed or adsorbed materials. Selection of the parameters for measurement is not trivial. Consider the size parameter: Which sizes are important, geometric, aerodynamic? Should size measurements be based on count, surface, volume, or mass considerations? For a size distribution, is the mode, mean, or median most relevant?

Although gases are usually easier to characterize than particles, other issues arise: How short should the sampling intervals and instrument response times be? Should concentration data be summarized by the mean and standard deviation, the median and geometric standard deviation, or the peak value recorded? The answers will depend upon the study objectives.

Assuming that one has identified the relevant physical and chemical properties for measurement and has selected the appropriate equipment, one is then faced with additional questions: Where should one sample? How many samples must one acquire? How will sampling artifacts and interferences be controlled?

Is it any wonder that much of the progress in inhalation toxicology has been in showing the inadequacy of atmospheric characterizations in past studies? One is tempted to believe that a good inhalation study may involve expending as much effort in exposure characterization as is spent in measuring the biological responses.

In the following sections, some counsel will be offered on atmospheric characterization. One should bear in mind that exposure characterization is still evolving and no simple or certain universal rules can be given.

THE BREATHING ZONE

Usually, one wishes to characterize the material in an inhalation study as it is inhaled by a subject. Because it is often impractical to place a sampler inlet at the subject's nose or mouth, the breathing zone must be considered. The *breathing zone* is that volume within which the subject breathes. It extends from the lowest to the highest elevations commonly reached by the nose or mouth during the exposure, and laterally includes the full range of travel of the head. For a caged rat this may be the entire volume of the cage. For a masked animal it may be the small volume in front of the nares. The major point is that the breathing zone is not above or below the cage, it is not in a pipe leading to a mask, nor is it at the outlets of the aerosol/gas generators.

Implicit in this discussion of the breathing zone is the assumption that it has a uniform composition. If it does not, one should consider additional confinement (or repositioning) of the subject, mixing the atmosphere, or sampling at sufficient sites so that an average composition can be described. When the breathing zone is variable in composition, one must suspect that animal subjects may preferentially breathe from areas that contain less exposure material. After all, one of the functions of the sensory apparatus appears to be to limit toxic exposure, either by altering breathing patterns or by producing other avoidance behaviors.

WHAT SHOULD BE MEASURED?

Particle Parameters

All of the particle properties that significantly contribute to biological effect in the experiment at hand should ideally be measured. Table 4.1 lists candidate properties that should be considered, but the list is certainly not exhaustive. The first four are nearly always needed when a particle exposure is to be related to biological effects. Electrical charge is often controlled by use of aerosol dischargers.

In specific cases, the other listed properties may be important for interpreting an inhalation experiment. As an example, hygroscopicity can alter the amount and distribution of inhaled material that is deposited (Martonen, 1982; Ferron et al., 1988; Ferron and Busch, 1996; Asgharian, 2004). A well-known example of the importance of particle shape on toxicity is that of asbestos. Such particles, by virtue of their persistence and extreme length, appear to frustrate phagocytosis and can be retained in the deep lung for long periods. For ultrafine and/or nanoparticles, particle size, count, and surface area should be considered (Kreyling et al., 2006, 2007). Practical considerations will limit the number of particle properties that can be measured in any given investigation. Thus, good judgment on the part of the investigators will be essential to the success of the sampling strategy.

Gas Parameters

The problem of characterizing experimental gases (including vapors) is relatively straightforward in comparison to particles. In general, one must know one of the following for each relevant gas that is present: (1) the concentration (e.g., ppm) and the total atmospheric pressure; (2) the amount per unit air volume (mg/m^3); or (3) the partial pressure (torr, mmHg, etc.). From any of these data one has a measure of both the number concentration of gaseous molecules present and the maximum concentration gradient at the air–tissue interface. As for aerosol particles, these measurements must be made in the breathing zone. Although gases tend to distribute more uniformly than particles in an air volume, local inhomogeneities in concentration may exist due to the presence of sinks, poor mixing, and leaks in the containment.

An important part of characterizing a gaseous atmosphere is a determination of the number and sizes of particles that are present. In general, regardless of the precautions taken, particles will be present that may influence the biological effects of the gases. The ability of particles to influence the toxicity of a gas has been implicated in studies of both humans and laboratory animals (Amdur, 1957; McJilton et al., 1976; Bell et al., 1977).

Table 4.1 Aerosol Particle Properties That May Relate to Their Biological Effects

Property[a]	Comment
Chemical composition	Will influence biological response
Mass concentration (particles per unit mass air)	Will influence total dose deposited and biological response
Aerodynamic size	Will influence inhalation and deposition pattern
Size distribution	Will influence all of the above
Geometrical size	Will influence deposition, clearance, dissolution, and sensory irritancy
Electrical charge	Will influence coagulation rates and deposition on chamber walls, animal fur, and in animals
Hygroscopicity	Will determine growth rates in the respiratory tract and hence deposition pattern
Surface area	Will determine adsorbed gases and biological responses
Particle shape	Will influence deposition and responses
Dissolution rate	Will determine persistence in lung fluids and tissues
Water solubility	Will influence particle stability in the respiratory tract
Deliquesence	Will influence particle size in the respiratory tract
Irritancy	May influence breathing patterns and modify dose
Specific gravity	Will influence deposition pattern and clearance rates
Number concentration (particles per unit vol. air)	May influence deposition pattern via "cloud effect" and will influence effects
Antigenicity	May influence irritancy, induce tissue responses, and clearance rates
Odor	May induce avoidance behavior
Taste	May induce avoidance behavior
Radioactivity	Can influence electrical charge and dissolution rate characteristics as well as toxicity to lung cells
Surface character	May influence the interaction with macrophages and other lung cells
Temperature	Particle temperature will be similar to surrounding gas which will influence deposition pattern

[a] *Note*: Many of these properties are interdependent. For example, for spherical particles, the aerodynamic diameter is a function of geometric diameter and density.

In these studies, in which sulfur dioxide or sulfur dioxide plus ozone were the gases of interest, the presence of particles, intentionally added or spontaneously formed, appeared to produce effects greater than that attributable to either the gases or the particles alone. A probable mechanism for the observed effects is the enhanced delivery of adsorbed or absorbed gases to the deep lung. These classic studies make excellent supplemental reading on the topic of particle and gas interactions.

The biological importance of transient concentration spikes in gas toxicity, especially for ozone, is of interest. In view of this unsettled matter, it is wise to record concentration maxima in inhalation studies. This type of monitoring implies the use of instruments that continuously sample, have a short cycle times, and provide frequent concentration data. Additionally, a sufficient number of samples should be taken so that the standard deviation of the gas concentration during the exposure can be calculated and reported.

Environmental Parameters

The toxicity of essentially all inhaled materials is influenced by the environmental conditions during exposure. Any environmental parameter that produces a change in the ventilation rate will influence the dose received by the subject. Furthermore, when environmental conditions produce a costress, the susceptibility to injury can be altered. For further reading on the effects of such environmental stresses as heat, cold, and exercise, one should see a dedicated issue of *Environmental Research* (vol. 92 (1), 2003) which contains nine papers from a symposium, "Combined Impact of Exercise and Temperature Stress on the Physiological Response to Toxic Agents." Heat, cold, and exercise are all significant modifiers of toxic responses. Broadly considered, environmental stresses can be quite subtle. For example, animals may react to being moved from a housing area into a laboratory, even though the conditions are hospitable in both places. Also, animals tend to adjust to common environmental pollutants so that animals raised in a clean environment may be more, or sometimes less, responsive to experimental pollutants. Nutritional deficiencies can also alter the response of subjects to pollutants. Such deficiencies can be related to factors other than food composition. Rats, for example, can be stressed to the point of weight loss by being placed in a situation where their feces are not accessible for reingestion (Thomas and Roe, 1974). Gordon (2003) presents the argument that environmental stress should be included in toxicology studies that are intended to simulate real-world exposures.

Table 4.2 lists some of the most important environmental parameters that should be controlled and/or measured. The importance of proper control of these parameters cannot be overemphasized. Unless the investigator personally experiences the exposure environment under conditions simulating an actual exposure, he or she may be totally unaware of serious environmental deficiencies. The investigator should enter the chamber in order to uncover any serious deficiencies in the environmental quality.

A special mention of nuisance air contaminants is required because some studies have been compromised by excessive unintentional environmental contaminants. Table 4.3 lists several contaminants that can commonly be found in exposure chambers. Means for controlling such contaminants were discussed in Chapter 3.

INSTRUMENTATION FOR AEROSOL CHARACTERIZATION

Comment

The selection and use of an appropriate array of aerosol instruments are not trivial tasks. Untrained or inexperienced persons are at the mercy of the often biased claims made in

Table 4.2 Environmental Parameters of Importance in Inhalation Studies

Parameter	Comments
Temperature	Influences physical activity, ventilation rates, and several other biological characteristics
Relative humidity	Influences thermal regulation and hence ventilation, and abnormalities in respiratory tract mucus
Atmospheric pressure	Influences ventilation as well as cardiac and hematapoietic functions and the uptake of gases by tissue
Illumination	Influences state of activity and can cause stress due to altered diurnal patterns.
Air contaminants	Can act as copollutants, lead to adapted or sensitized states, and influence ventilation and other physiological functions
Noise	A known costress for mammals. Often present due to human activity, equipment, or flow of air through small orifices or over sharp edges
Vibration	A known costress that may be produced by motors or other equipment
Overcrowding	A known stressor, especially serious if more than one species is present in a chamber

Table 4.3 Common Environmental Contaminants Found in Exposure Chambers

Contaminant	Comments
Ammonia	Produced by bacterial action on urine. Levels can rapidly increase over the duration of an exposure
Carbon dioxide	Crowding of animals or low airflows can cause a buildup of this exhaled metabolic byproduct
Water vapor	Crowding of animals or low airflows can lead to rapid buildup
Oil mist	Present in output air from oil-lubricated air pumps
Animal dander, fur, and excreta	May be present in excessive amounts due to poor housekeeping, overcrowding, or unhealthy animals
Organic vapors	May be residual from cleaning agents
Viable aerosols	May be present due to inadequate housekeeping

advertisements and the manufacturer's literature. Aerosol characterizations represent some of the most difficult of physical measurements, and even persons with advanced degrees in physics or engineering may not have had sufficient training to undertake them with confidence. It can be safely stated that no instrument, in the hands of a novice, will consistently provide reliable data.

Sampling

Before discussing aerosol sizing instruments, it is necessary to consider devices that collect samples for subsequent analysis. Several sample collection devices are listed in Table 4.4. Aerosol sampling methods include filtration, impingement, and precipitation. One can also consider aerosol centrifuges and cascade impactors, to be presented later, as size-selective samplers. These, and other samplers can be found in several general references (Willeke and Baron, 1993; Cox and Wathes, 1995; Mitchell, 1995b; Vincent, 1995; Hinds, 1999; Cohen and McCammon, 2001; Ruzer and Harley, 2005).

A particle air filter allows the passage of air but traps particles, with some aerosol size dependency. Typically, air is drawn through a filter using a pump or some other method for producing a pressure gradient. Filters are usually of two types: pads of compacted fibers; or membranes having many holes and/or channels penetrating through (Crook, 1995). Electrically charged fiber filters, are also available. Fiber filters designed for particle collection usually have fiber diameters in the micrometer range. The fibers are composed of a variety of materials including glass, viscose, cellulose acetate, asbestos, cotton, or other mineral and organic substances. Membrane filters are usually made of cellulose esters having pores of controlled diameter between about 0.01 and 10 μm. Membrane filters typically offer considerably more resistance to airflow than do fiber filters, but microscopic examination is easier for membrane filters.

Filter samples taken for gravimetric analysis of airborne particle mass concentrations must be obtained with great care to eliminate artifacts. Such artifacts include changes in filter efficiency and in throughput airflow rate due to clogging by the sampled particles, inefficient collection due to air leaks around the filter or through tears in the filter, changes in the filter's weight due to loss or uptake of water to or from the sampled particles or the air, gain or loss of deposited mass due to chemical reactions, loss of volatile materials, errors due to loss of pieces of filter due to handling, errors due to loss of sample in handling, and nonrepresentative sampling due to nonisokinetic conditions. From this list it is apparent that obtaining a proper filter sample requires a knowledge of the physics

Table 4.4 Aerosol Sample Collection Devices for Subsequent Analysis[a]

Sampler	Principle	Artifacts and limitations
Filters	Air passes through holes, channels, or around fibers, but particles are trapped by interception, impaction, diffusion, sedimentation, or electrostatic interaction	Flow rate may change as filter loads, air flow may be nonisokinetic, filter media may be damaged, filter holders may leak, and reactions may occur on filter
Electrostatic precipitators	Particles are charged and attracted to an oppositely charged collection surface	Very tiny particles may lose charge prior to precipitation. Particles may deposit in nonrandom attitudes, or overlap
Thermal precipitators	Particles are driven onto a relatively cool collection surface by a thermal gradient	Aerosol particles may be altered by high temperatures
Elutriators	Particles settle under the action of gravity onto a collection surface	Particles smaller than about 1 μm in diameter do not readily settle because of Browian motion
Impingers	Aerosol is passed through a tube and collected in a liquid	Collection efficiency depends on diameter, agglomerated or fragile particles may break up
Cyclones	Particle-laden air is given a rotational motion leading to inertial collection of particles	Collection efficiency diminished for particles smaller than about 0.5 μm in diameter

[a] *Note*: Several good, general references cover aerosol sample collection (Willeke and Baron, 1993; Cox and Wathes, 1995; Vincent, 1995; Hinds, 1999; Cohen and McCammon, 2001; Budyka and Ogorodnikov, 2005; Hopke, 2005; Ruzer and Harley, 2005).

of airflow, the possible chemical reactions that can occur on the filter, and several other factors that may be specific to the application at hand.

Impingement sampling involves collection into a liquid, which is often water. One cannot assume 100% collection efficiency, especially for small particles. For a review of impingers see Crook (1995) and Hering (2001).

Sampling for transmission electron microscopy usually begins with depositing a representative particle sample on a carbon-coated, small copper screen called an electron microscope grid (Fig. 4.1). Also shown is a portion of a diffraction-grating replica that is placed in the electron microscope and photographed to establish a magnification factor during particle sizing. Alternatively, for scanning microscopes, samples may be deposited on membrane or fiber filters or metal surfaces. A variety of criteria for electron microscopy must be met.

- The collection surface should usually be flat.
- The collection surface must be a good conductor of heat and electricity.
- Sufficient numbers of particles must be collected.
- Particle overlap on the collection surface should be negligible.
- A representative sample must be obtained.
- The sampling process must not significantly change the particles.
- The sampling surface must be clean and relatively free of unwanted particles.

Electrostatic precipitators designed for sample collection are frequently used for collecting aerosol samples for electron microscopy. In an electrostatic precipitator, aerosol particles are given a charge by a unipolar ion field, usually produced by a corona discharge.

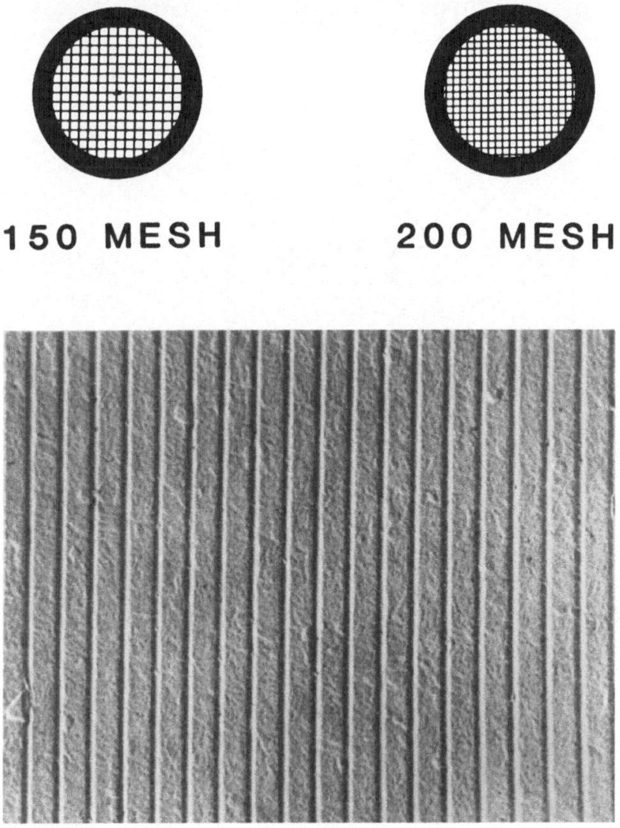

Figure 4.1 Electron microscope grids and diffraction-grating replica. *Source*: Photo supplied by Ted Pella.

The particles are then collected on an electron microscope grid that is given an opposite charge or is at ground potential. In one successful design by Morrow and Mercer (1964), a sharp metal corona-discharge point held at a negative 7 kV is placed 1 cm above the grid which is held at ground potential (Fig. 4.2). Air is drawn between the grid and ion source at 70 cm^3/min and particles are efficiently deposited on the grid. Care must be taken to keep the discharge needle quite sharp in order to provide an adequate ion current. Adequate samples are obtained in about 1 min when particle concentrations are about 100,000/cm^3 of air. Sampling times will be proportionately longer or shorter for different particle concentrations. This device provides a relatively unbiased sample for aerosols containing particles down to about 0.05 μm in diameter. Design, performance, and use data for this and other electrostatic precipitators are described by Mercer (1973), Cheng et al. (1981), Yeh (1993), and Hinds (1999).

Similar considerations apply to thermal precipitators (Tsai and Lu, 1995; Wen and Wexler, 2007). Thermal precipitation is useful for sampling ultrafine particles, which are difficult to charge. However, significant heating may alter some samples.

Size Analyzers

Aerosol-measuring instruments are based on exploitation of size-dependent properties of aerosols. Examples of size-dependent aerosol properties include terminal settling velocities,

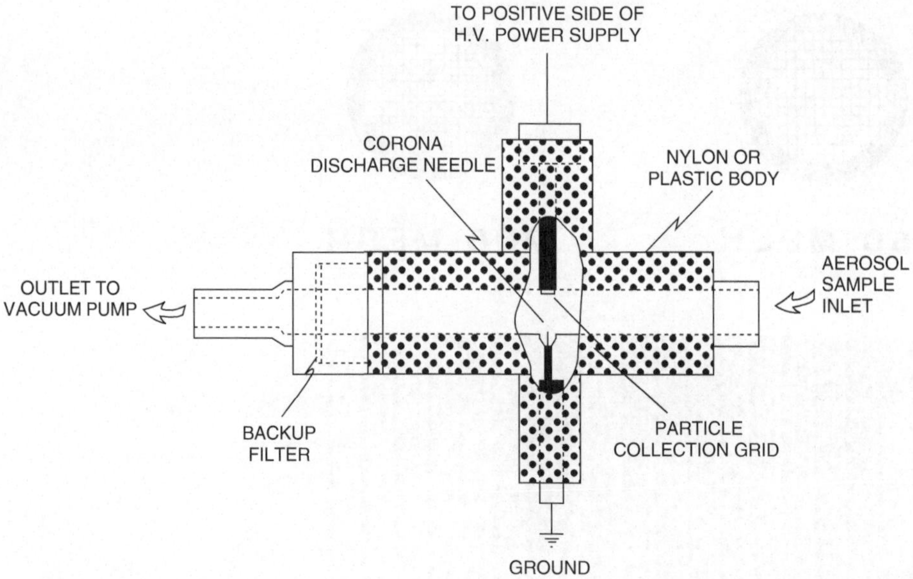

Figure 4.2 A point-to-plane electrostatic precipitator. *Abbreviation*: H.V., high voltage. *Source*: Redrawn from Morrow and Mercer, (1964).

light-scattering patterns, mobilities, and abilities to serve as centers for condensation of supersaturated vapors. It is a well-known fact that detection or measurement is always associated with some perturbation of the system under measurement. This perturbation may be negligible for a macroscopic body but may be appreciable for a very tiny one such as an aerosol particle. An intense light beam, for example, can produce wild gyrations in the motion of particles or may rapidly evaporate liquid ones down to minute residues. Now that the reader is sufficiently warned, some methods that have been developed to characterize aerosols will be examined.

Several references are useful for a practical understanding of the great variety of aerosol-sizing instruments (Willeke and Baron, 1993; Mitchell, 1995a,b; Hinds, 1999; Cohen and McCammon, 2001; Burtscher, 2002; Abdel-Salam, 2006). These sources provide insight into the variety and uniqueness of various aerosol-sizing instruments. Maynard (2000) reviewed methods for analyzing ultrafine particles. Before any instrument is used, it is necessary to read additional literature on the device itself, or on studies in which that device has been used.

Ideally, in any inhalation study, a breathing zone air sample should be examined using a microscope. Microscopy is a sensitive way of detecting unwanted contaminants that could influence an experiment in chaotic ways. Fine particles, and perhaps larger ones, will inevitably be present, and unless the experimenter has some knowledge of their sizes, shapes, and numbers, he or she is out of touch with the experiment. Some years ago, in our laboratory, we were examining an air sample taken from an exposure chamber that should have only had clean air with a trace of an air-pollutant gas. In the electron microscope, to our great surprise, we saw a number of fibers that looked like asbestos. These particles were, of course, an unwanted contaminant, and had we not taken an air sample for electron microscopy we would not have known they were present. We never discovered their origin; perhaps it was a contaminated gas feed-line or a deteriorating fiber filter. The problem was eliminated by changing generator feed and sampling lines, replacing all of the air filters and cleaning the exposure system. A useful review of biological particle microscopy methods by Morris (1995) pertains to aerosol samples in general.

A very important aerosol parameter, mass median aerodynamic diameter (MMAD), should be measured in any aerosol experiment involving particles larger than about 0.5 μm in diameter. A reliable, calibrated cascade impactor with a back-up filter is an ideal instrument for obtaining this parameter. Assuming one has properly taken an impactor sample and avoided the many possible artifacts listed in Table 4.5, one must determine the collected mass on each of the impactor collection stages as well as the back-up filter. In some cases this can be done gravimetrically, but usually a chemical analysis is required. In all cases, one must be sure that the entire mass collected on each stage is recovered and quantitatively assayed. In some cases, where the aerosol is liquid and the impaction surface is hard, for example, one may have few problems. On the other hand, if the impactor collection surface is a filter or has been coated with oil or grease to prevent particle bounce (Chang et al., 1999; Maricq et al., 2006), elaborate means may be necessary to remove all of the collected deposit. Although the median aerodynamic size can be obtained using a cascade impactor, effects such as particle bounce and electrostatic repulsion will tend to artificially broaden the estimated size distribution. Sizing of several hundred particles from electron microscope photographs can be used as an adjunct to provide a more reliable geometric standard deviation. Several excellent references on theoretical and practical aspects

Table 4.5 Aerosol-Measuring Instrumentation for Inhalation Studies[a]

Instrument	Principles	Artifacts and limitations
Optical microscope	Image formation using visible light, and glass optics	Lower limit of resolution, about 0.5 μm, can lead to grossly distorted size distributions
Transmission electron microscope	Projected image formation using electron beam, and magnetic optics	Evaporation of sample due to vacuum and local heating. Shallow depth of focus and small field of view can bias size distributions. Coating of small particles with column contaminants can lead to overestimate of size
Scanning electron microscope	Image reconstruction using scanning electron beam scattering and signal processing	Evaporation of sample due to vacuum and local heating. Poorer resolution than with transmission electron microscopy
Cascade impactor	Impaction stages in series collect successively smaller fractions on basis of aerodynamic size	Cutoffs for stages can be broad. Particle bounce and re-entrainment produce bias. Wall losses can be appreciable and low pressure can cause evaporation of liquids. Lower cutoff at about 0.3 μm aerodynamic diameter except in low pressure models. Amount of sample collected is usually small before overloading occurs. Particle charges can produce artifacts
Condensation nuclei counter	Supersaturated vapor causes particle growth, particle number determined by intensity reduction of a light beam	Lower limit of detection near 0.002 μm. Coincidence errors occur at high particle concentrations. Calibration is difficult

[a] See Refs in Table 4.4.

Continued

Table 4.5 Aerosol-Measuring Instrumentation for Inhalation Studies—cont'd

Instrument	Principles	Artifacts and limitations
Centrifuge spectrometer	Rotation produces increased sedimentation velocities across a clean air field where particles deposit segregated with respect to aerodynamic size	Inlet losses can be large for particles above a few micrometers. Lower cutoff at about 0.3 μm. Gas phase of aerosol must match clean air sheath in density or distortion of deposition occurs
Optical spectrometer	Size of electrical pulse generated by scattered light is measured and classified for individual particles	Calibration is specific for a given index of refraction. Particles interfere at greater than about 10^2 to 10^3/cc of air. Size range about 0.3 μm to about 10 μm. Deviations from particle sphericity produce loss of resolution. Liquids can evaporate
Diffusion battery	Aerosol is drawn through small channels in parallel at various airflow rates. Measurements of penetrating fractions lead to size distribution	Inappropriate for particles greater than a few tenths of a micrometer in diameter. Analysis is increasingly difficult as geometric standard deviation increases
Charge spectrometer	Aerosol is drawn between oppositely charged plates and deposits are analyzed to yield number of charges on particles	Operational difficulties include maintaining stable laminar airflow and prevention of arcing. Analysis of polydisperse aerosols is difficult
Mobility analyzer	Particles are given known charge and selectively deposited with respect	Instruments may have internal loses that are particle-size dependent. Calibration is sensitive to many environmental factors. Lower size limit is undefined. Upper size limit is about 1 μm
Surface area measurement device	Particles are degassed and then allowed to adsorb gas on their surfaces, amount of absorbed gas is determined	Requires large quantities of collected particles. Method is sensitive to surface characteristics and presence of cracks and voids in particle. Low pressures and high temperatures used may alter particle surface area during measurement
Elutriator	Particle travels horizontally and settles through clean air onto a surface, position on the surface is related to aerodynamic diameter	Impractical for submicrometer particles due to their large diffusion coefficients. Charged particles may not deposit at proper locations

of impactors are available (Lodge and Chan, 1986; Young, 1995; Cohen and McCammon, 2001; Hering, 2001; Leung et al., 2005).

The previous sizing techniques, electron microscopy, and cascade impaction have a major shortcoming; they do not usually give a final answer rapidly. Typically, many hours or even days will pass before one has a reliable result. Thus, these techniques are not generally useful for real-time adjustments of aerosol generators. To solve this problem, instruments that give real-time responses may also be required. In most cases, it is necessary to check their accuracy periodically under the specific conditions of an experiment, since their calibration (typically using ideal particles) may not apply well to the studied aerosols. Examples of real-time monitors include optical particle spectrometers (*optical particle counters*), *condensation nucleus counters, low-pressure piezoelectric impactors,* and *mobility analyzers.* For a review of particle-size analyzers see Mitchell (1995b) and Cohen and McCammon (2001). When using these instruments, great care must be exercised, and one must have a thorough understanding of possible artifacts in each specific application. Real-time monitors can be very useful for controlling and characterizing atmospheres in inhalation studies.

In special cases, additional aerosol instrumentation will be required (Table 4.5). For example, one may need information on the surface areas or aerosol electrical charge states of particles. Thus, specialized instrumentation will be necessary. In such cases, experienced personnel and a knowledge of the aerosol literature are essentials for success.

Isokinetic Sampling

The airflow patterns entering sampling devices can be complex and lead to collection of a nonrepresentative sample. A review of experimental work with some theoretical treatments has been published by Fuchs (1975). Mercer (1973), and Hinds (1999) described conditions under which particles of various sizes may be sampled isokinetically. It is very difficult to sample particles isokinetically with diameters of 20 μm or greater, especially when sampler inlets are tubular and sampling flow rates are less than about 1,000 L/s. Using Davies' (1968) data, Figure 4.3 depicts the effects of particle size and sampling flow rate on selection of sampler pipe diameters. In each pair of curves the top curve gives the maximum permissible sampler diameter and the lower curve the minimum one.

INSTRUMENTATION FOR GAS CHARACTERIZATION

Methods used for obtaining samples for analysis of gaseous materials include grab sampling, absorption sampling, and continuous sampling. In each method, the objectives are to acquire a representative sample that is not altered by the sampling process, and that is suitable for subsequent analysis. Some introductory references are recommended for information on sampling and analysis of pollutant gases and vapors. *Air Sampling Instruments* edited by Cohen and McCammon (2001) has several chapters covering gases and vapors, along with instrument suppliers. Methods for analysis of vapor-phase organic pollutants are described in an introductory fashion in Appendix B of *Vapor-Phase Organic Pollutants* published by the Committee on Medical and Biologic Effects of Environmental Pollutants (1976). Much of what follows in this section is taken from these sources.

The collection efficiency of a gas sampler or sampling procedure must be known in order to quantitatively determine an airborne concentration. The collection efficiency, C, is:

$$C = \frac{\text{Amount collected or sampled}}{\text{Amount present in sampled volume of air}} \quad \text{(eq. 4.1)}$$

The collection efficiency can be less than 1 due to factors such as absorption or adsorption in sampling lines, losses or conversion due to chemical reactions, and collection of gases by airborne particles. Estimation of the collection efficiency is typically accomplished by measuring the concentrations in standardized calibration mixtures. Great care must be taken in selecting appropriate sampling-line materials and, as necessary, keeping the lines clean.

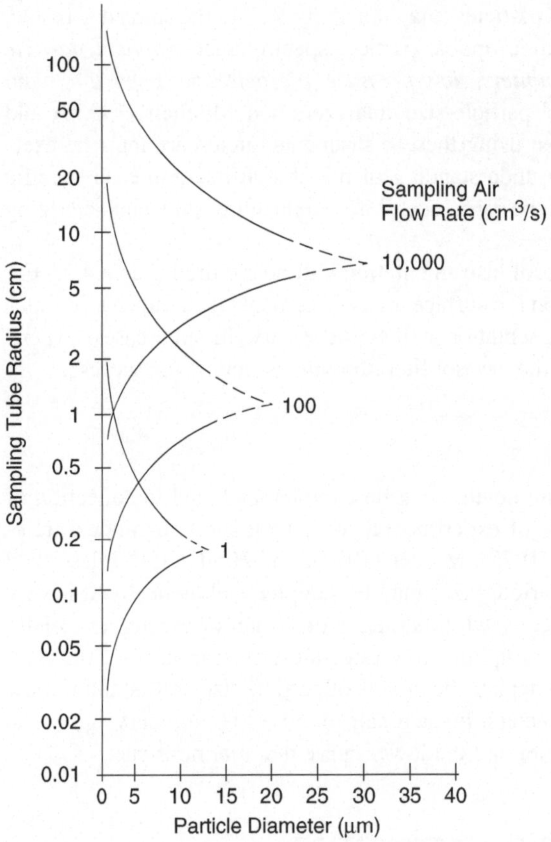

Figure 4.3 Permissible maximum and minimum radii of tubes for sampling aerosols in calm air. Three sampling flow rates are included. Dashed segments are extrapolated *Source*: Drawn from data given in Mercer, 1973.

Grab sampling refers to the rapid acquisition of an air sample that is then transported to analytical equipment. Evacuated containers, calibrated syringes, plastic bags, and a variety of other containers are used to hold grab samples. Problems encountered include porosity of the container, sample contamination by container materials or contaminant air residing in the container, and degradation of the sample within the container.

Adsorption sampling and *absorption sampling* refer to the capture of a sample on a high specific surface material such as activated charcoal, or capture of the sample in a liquid solvent such as water. The collected sample is then either driven from the capture medium for analysis or analyzed *in situ* on or in the medium. A related type of sampler, the *cold trap*, collects a gas by low-temperature liquefaction. Several types of adsorption and absorption media are in use, including glass beads (wet, dry, or chilled), activated charcoal, silica gel, ion-exchange resins, impregnated filters, liquids in bubblers, and a variety of other specialized materials.

The analysis of a sample of gas may be performed by traditional quantitative wet chemistry followed by gravimetric, photometric, or other analytical method. Other phenomena or analytical techniques that have been exploited for analysis of gases include: chemiluminescence (chemically stimulated light emission); electrical conductivity of a solution; heat of combustion; thermal conductivity; coulometry (measurement of charge generated by chemical reactions); flame ionization; gas chromatography; mass spectrometry; spectrophotometry; polarography (electrolysis); radioactivity; and mass spectrometry. Table 4.6 lists some gases and representative analytical techniques applied to them. Cohen and McCammon's (2001) book has several chapters on these and other methods.

The trend in developing real time instrumentation for gases includes improvements in specificity, portability, and sensitivity to low concentrations. Reading the current scientific literature is essential to the proper selection of gas-monitoring instrumentation.

ELIMINATING MEASUREMENT INTERFERENCES

General Principles

Interference occurs when the output data of an analyzer for one material is significantly altered by the presence of another material in the sampled air. In an air sample taken during inhalation exposure, several materials, intentional or unwanted, are often present.

Three basic techniques can be applied to the problem of interferences: selection of detection methods that do not suffer from significant interferences; removal of interfering species from the air stream that enters the sampler; and correction of analyzer readings to remove that portion of the reading due to interfering species.

Selection of detectors that do not suffer from significant interferences has become easier in recent decades. In general, various absorption and emission spectra are unique to a given material, and monitoring instruments based on such spectra tend to be relatively free of interferences. One problem that arises when airborne gases are analyzed spectroscopically is that the presence of particles can lead to unwanted scattering and absorption of light and other electromagnetic radiation. Such interfering particles may be present in the breathing zone, even when they are not intentionally generated. Also, particles may be formed, or depleted, within a sampling line or an analytical instrument due to changes in pressure, electrical charges, temperature, humidity, or other factors.

Removal of interfering species from a sample air stream may be necessary in order to obtain valid instrumental readings. Particle filters can be placed in gas sampling lines, but problems associated with in-line particle filters include: reaction or capture of the sampled gas on the filter material, or on particles trapped by the filter; reduction of the sampler airflow rate due to the resistance of the filter; dilution of the sample by air leaks that occur

Table 4.6 Representative Methods for Collection and Analysis of Gases and Vapors

Gas or vapor	Sorption medium	Analysis	Interferences
Ammonia	25 mL 0.1 N sulfuric acid	Nessler reagent	None reported
Benezene	Activated charcoal	Chromatography	Other similar hydrocarbons
Formaldehyde	1% 10 mL sodium bisulfate	Color of product	Phenols, some other aldehydes
Nitrogen dioxide	20–30 ml Saltzman reagent	Reacts with absorbing solution	Ozone in fivefold excess; peroxyacyl nitrate
Ozone	1% potassium iodide in 1 N potassium hydroxide	Color of iodine liberated	Other oxidizing agents
Styrene	15 mL spectrograde isooctane	Ultraviolet radiation analysis	Other aromatic hydrocarbons

Continued

Table 4.6 Representative Methods for Collection and Analysis of Gases and Vapors—cont'd

Direct Reading Instruments			
Operating principle	Applications and remarks	Range	Sensitivity
Chemiluminescence	Measurement of NO in ambient air selectivity and NO_2 after conversion to NO by hot catalyst. Specific measurement of O_3. No atmospheric interferences	0–10,000 ppm	Varies: 0.1 ppb to 0.1 ppm
Colorimetry	Measurement and separate recording of NO_2-NO_x, SO_2, total oxidants, H_2S, HF, NH_3, Cl_2, and aldehydes in ambient air	Parts per billion and parts per million	0.01 ppm (NO_2, SO_2)
Coulometry	Continuous monitoring of NO, NO_2, O_3, and SO_2 in ambient air. Provided with strip chart recorders, some require attention only once a month	Selective: 0–1.0 ppm overall, or to 100 ppm (optional)	Varies: 4–100 ppb dependent on instrument range setting
Infrared analyzer (photometry)	Continuous determination of a given component in a gaseous or liquid stream by measuring amount of infrared energy absorbed by component of interest using pressure sensor technique. Wide variety of applications including CO, CO_2, Freons, hydrocarbons, nitrous oxide, NH_3, SO_2, and water vapor	From parts per million to 100% depending on application	0.5% of full scale

Source: Modified from Phalen (1984), p. 116.

in the filter holder; and introduction of new materials into the sample due to chemical reactions or physical desorption of materials on the filter. Intentional removal of gases from the sampled air can be achieved by coating the walls of a sampling line with specific absorbers, or passage of the sample through a bed or filter containing an absorber. One must be careful in designing such *scrubbers* (*denuders*) to insure that they do not retain the sampled species. Examples of successful designs are described later in this chapter.

In many cases, it is necessary to correct instrumental readings for interfering materials. Two general methods can be considered. One may monitor the concentration of the interfering material separately and correct the output of the primary instrument using data on the relationship of concentration of the interfering substance and the size of the interfering signal. Alternately, one may experimentally recalibrate the primary instrument under actual exposure conditions with the interfering species present. The first technique is usually superior when the interfering material may vary significantly in concentration; the second technique is better when the interfering material has a relatively constant concentration. In order to have confidence in one's ability to correct for, or eliminate, interferences, it is wise to intentionally generate known amounts of interfering species and thus test the correction technique before using it in an actual inhalation study.

Gas/Vapor Denuders

Gas and/or vapor denuders may be used to collect or to remove gases (or vapors) from an airstream. Here, one is mainly interested in removing unwanted gases from a sampling line in order to prevent interferences with subsequent analyses. Such denuders are typically tubes coated with, or otherwise containing, a medium that absorbs, adsorbs, or destroys the unwanted gas. Since Cheng (2001) reviewed the design, use, and some suppliers of denuders, we will only consider a few examples.

Stevens et al. (1978) and Kleinman et al. (1981) used ammonia gas denuders to prevent neutralization of acid particles on a sampling filter. The device described by Stevens consisted of a set of 16 parallel glass tubes (30 cm length, 0.5 cm i.d.) coated with phosphorous acid. The acid adsorbs and neutralizes ammonia and other basic gases that could neutralize the collected acids.

A gas diffusion denuder for removal of sulfur dioxide and hydrogen sulfide was described by Coburn et al. (1978). The denuder consisted of a stainless steel tube (1.5 m length, 3 mm i.d.) coated on the inside with PbO_2. Under the flow rate conditions used, the denuder had a gas capture efficiency of nearly 100% and a submicron particle diffusional capture efficiency of less than 2%.

Dasgupta et al. (1980) described a sulfur dioxide diffusion denuder that also permitted penetration of submicrometer aerosols. The denuder was a 30-cm long, stainless steel tube (1.25 cm i.d.) inside of which was a filter paper strip impregnated with sodium carbonate. At a flow rate of 500 mL/min, sulfur dioxide removal was essentially 100% without any significant loss of aerosol particles.

Another sulfur dioxide denuder, intended for use with oxidant samplers, was developed by Saltzman et al. (1965). Several methods were attempted, the best being a loosely packed bed of glass fiber filter paper impregnated with chromium trioxide and sulfuric acid.

SAMPLING PROTOCOLS

Ideally, all of the relevant atmospheric characteristics would be continuously monitored throughout an inhalation exposure. In practice, this ideal is usually an impossibility for several reasons. Many samplers do not provide continuous data. The total airflow rate through a large battery of samplers can exceed the total exposure system throughput airflow rate. And, the instrument costs and manpower requirements for an ideal, complete monitoring scheme are nearly always prohibitive. In reality, practicality, ingenuity, and scientific insight are all elements in designing valid sampling protocols that are also cost effective.

At a minimum, the primary experimental materials being investigated, as well as any other materials or parameters that might reasonably be expected to modify the measured responses, must be measured. The mean concentration is usually the most important datum relating to the potential biological effect of a material, but the standard deviation, range, and peak value may also be important. In some studies, the rate of increase or decrease in concentration of a material may be measured, as such phenomena can also influence biological responses. When a continuous data record is made, each of the foregoing exposure parameters can, in principle, be obtained from the record. When discrete samples are obtained in order to determine a mean concentration value, the minimum number of samples (N) that need to be taken can be calculated given two additional input parameters; the required precision with which one must know the mean (i.e., the standard error of the mean, SEM), and the expected standard deviation (SD) of the measured concentrations.

$$N = (SD/SEM)^2 \qquad \text{(eq. 4.2)}$$

This equation assumes that the concentration measurement data are randomly distributed about the mean and that each measurement is independent of the others.

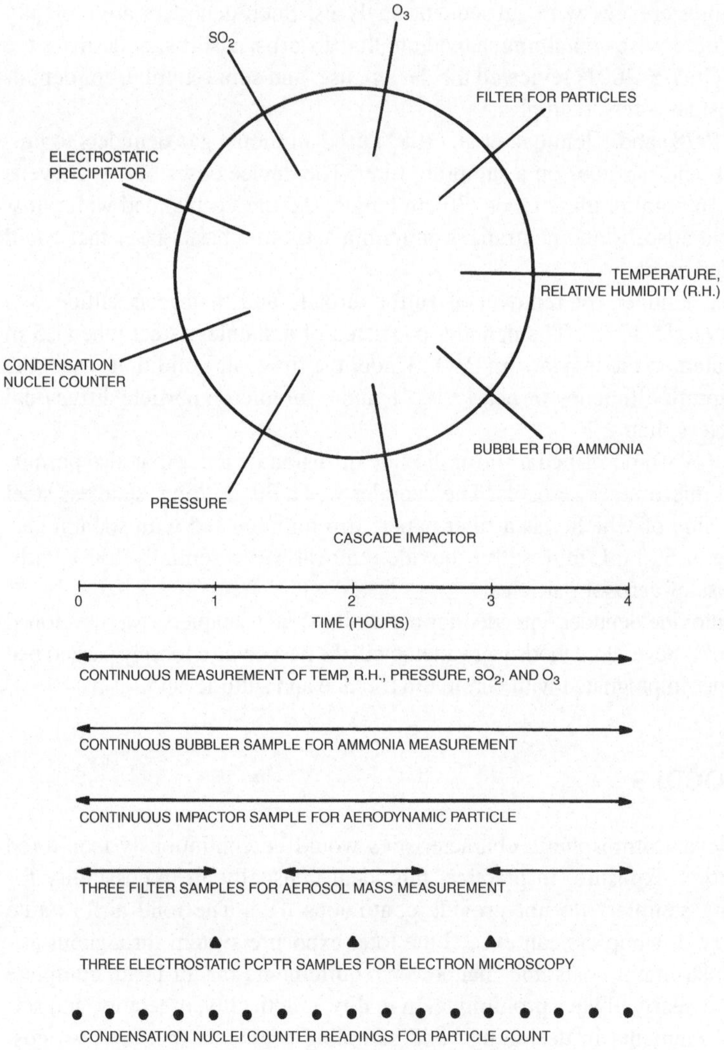

Figure 4.4 Sampling instruments and sampling protocol used in a 4-h chamber exposure of rats to a mixture of ozone, sulfur dioxide, and ammonium sulfate particles. Gas and particle samples were taken from the breathing zone. *Abbreviation*: PCPTR-precipitator.

When discrete samples are acquired, one cannot determine the maximum and minimum concentrations, as they may have occurred between sample periods or may have been masked by the finite duration of sampling. When discrete samples are acquired, one should report the number of samples taken, and the duration of each sampling cycle.

When the time required to acquire a sample for analysis is comparable to the total duration of the inhalation exposure, one should consider beginning the sample at the start of the exposure and continuing the sampling throughout the exposure period. In this way the sample will represent an average which is representative of the conditions that existed throughout the exposure.

Figure 4.4 depicts an actual sampling protocol used at the author's laboratory in order to characterize a chamber inhalation exposure of rats to a mixture of ozone gas, sulfur dioxide gas, and ammonium sulfate aerosol. This protocol allowed for both characterization and continuous, real-time control of the exposure atmosphere. Nine separate monitors were used during the actual toxicological exposure.

5
Methods for Exposing Subjects

INTRODUCTION

Perhaps the most consistently under-designed component in an inhalation experiment is the exposure system. This is unfortunate, considering the expense and permanency of most laboratory exposure systems. Design, fabrication, installation, and checkout of an inhalation chamber, with an air purification system, can easily exceed $100,000, and may require more than a year to complete.

In general, exposure systems may be classified into five types, based on the exposure of the subject (Table 5.1):

- whole body (or immersion),
- head only, including helmet types,
- nose or mouth only,
- lung only, in which entubation or tracheostomy is performed, and
- partial lung, which usually involves entubation or cannulation.

The available resources and objectives of a particular investigation will determine the exposure method chosen by the investigators (Lippmann, 1980; Phalen et al., 1994b).

The primary purpose of an inhalation exposure system is to provide a controlled delivery of airborne material to the respiratory system. Durations of exposure range from acute exposures lasting minutes, to repeated or continuous exposures lasting years. Exposure may be to airborne material in physical states ranging from subatomic (e.g., ions) to complex mixtures of gases and particles. In each instance, unwanted interactions of the exposure atmosphere with the exposure system must be considered. In most inhalation studies, it is desirable to eliminate or limit confounding exposures through skin, eyes, food, and other nonrespiratory pathways (Pauhlun, 2003). Also, the metabolic heat produced by animals, which can produce stress, must be considered (Roberts and Wong, 1998).

A humane exposure is essential. Exposure systems tend to become more complex when used for longer term exposures, especially when they must also serve as housing for animals. Criteria for animal housing are set forth in *The Guide for the Care and Use of Laboratory Animals* (ILAR, 1996). Although these criteria apply strictly to animal housing, chronic exposure systems should follow their principles when possible. The air supply to any exposure system should be clean, except for intentional materials. The air should be maintained at a comfortable temperature, relative humidity, and velocity, and should not contain excessive amounts of animal waste products, including dander, ammonia, and carbon dioxide.

Table 5.1 Comparison of Various Inhalation Exposure Methods

Mode of exposure	Advantages	Disadvantages	Design considerations
Chambers (whole body)	Variety and number of subjects Chronic studies Minimum restraint Large database Controllable environment	Messy Multiple routes of exposure: skin, eyes, oral (food, water) Variability of dose Cannot pulse exposure easily Inefficient Poor contact between subjects and investigators Expensive Animal by-product contaminants	Clean air Inert materials Losses Even distribution in space and time Sampling Animal care Observation Noise, vibration, humidity Air temperature Safe exhaust Loading limits
Head only	Good for repeated exposure Limited routes of entry into subjects More efficient dose delivery	Stress to subjects Losses can be large Seal around neck Labor in loading/unloading subjects	Even distribution Pressure fluctuations Sampling and losses Air temperature, humidity Subjects comfort Subjects restraint
Nose/mouth only	Exposure limited to respiratory tract Uses less material (efficient) Containment of material Can pulse the exposure	Stress to subjects Seal about face Effort to expose large number of subjects	Pressure fluctuations Body temperature Sampling Seals Subjects comfort Losses in plumbing/masks
Lung only	Precision of dose One route of exposure Uses less material (efficient) Can pulse the exposure	Technically difficult Anesthesia or tracheostomy Limited to small numbers By-passes nose Artifacts in deposition and response	Air temperature and humidity Stress to animal Physiologic support
Partial lung	Precision of total dose Localization of dose Can achieve very high local doses Unexposed control tissue from same animal	Anesthesia Placement of dose Difficulty in interpretation of results Technically difficult Possible redistribution of material within lung	Stress to animal Physiologic support

For conscious animals, the exposure environment should be humane with respect to noise, vibration, lighting, and freedom of movement. If possible, scientific personnel should enter an exposure chamber during operation, and note any undesirable environmental stresses before animals are exposed inside the chamber. Environmental stressors tend to modify the responses of the subjects and increase the difficulty of drawing valid conclusions from a study. Anesthetized animals may require special physiological support with respect to lung ventilation, maintenance of body temperature, and humidification of air entering the respiratory tract.

Control and characterization of the exposure atmosphere implies accurate monitoring and sampling in the subject's *breathing zone*, either continuously or frequently enough to define the exposure history. Determination of the actual dose delivered to the animal by inhalation usually necessitates a direct assay of exposed tissue: The simple product of airborne concentration and duration of exposure is seldom adequate, especially when airborne particles are present, or when the atmosphere produces abnormal breathing patterns.

The sections that follow cover the basic types of exposure systems, methods for providing aging of exposure atmospheres, the problem of ammonia as contaminant, methods for determining the dose, and the ethical responsibilities of the investigator.

BASIC TYPES OF EXPOSURE SYSTEMS

Chamber Systems

Design Considerations. Modern exposure chambers are usually of the dynamic type in which a continuous flow of throughput air is maintained. Static chambers in which the air is not flowing are usually unsuitable for all but brief exposures. In most cases, the quality of the throughput air must be rigorously maintained by an air purification and conditioning system.

Functions performed in sequence by a typical throughput air purification and conditioning system include: coarse particulate filtration of ambient air, pollutant gas removal by an adsorbent bed, second-stage particulate filtration, catalytic oxidation of low molecular weight hydrocarbons, dehumidification, warming to desired temperature, humidification to the desired relative humidity, fine particle filtration, and metering of air into the chamber at the desired flow rate. More detailed information on the design of such a system was presented in Chapter 3. A chamber exposure system with air purification is shown in Figure 5.1

If the chamber is operated at a temperature other than that maintained in the room housing the chamber, thermal insulation must be considered. This is especially true for metal chambers because they rapidly conduct heat through their walls, making it difficult to maintain a significant temperature difference between the chamber and ambient air.

The most widely used material for construction of dynamic chambers is stainless steel. It is sufficiently inert for most experimental atmospheres, durable, does not build up localized electrical surface charge, is sterilizable, and can be used to form a variety of shapes and sizes. Metal chambers may have glass or plastic windows for observing the subjects. Disadvantages of stainless steel and other metals are their expense, poor thermal-insulating properties, and the fact that iron, chromium, nickel, and other components of the alloy can catalyze some chemical reactions.

Glass and plastic are also used for construction of chambers—glass because it is transparent and very inert, and plastic because it is transparent, inexpensive, and easy to work with as a construction material. Most plastics are less chemically inert than glass and

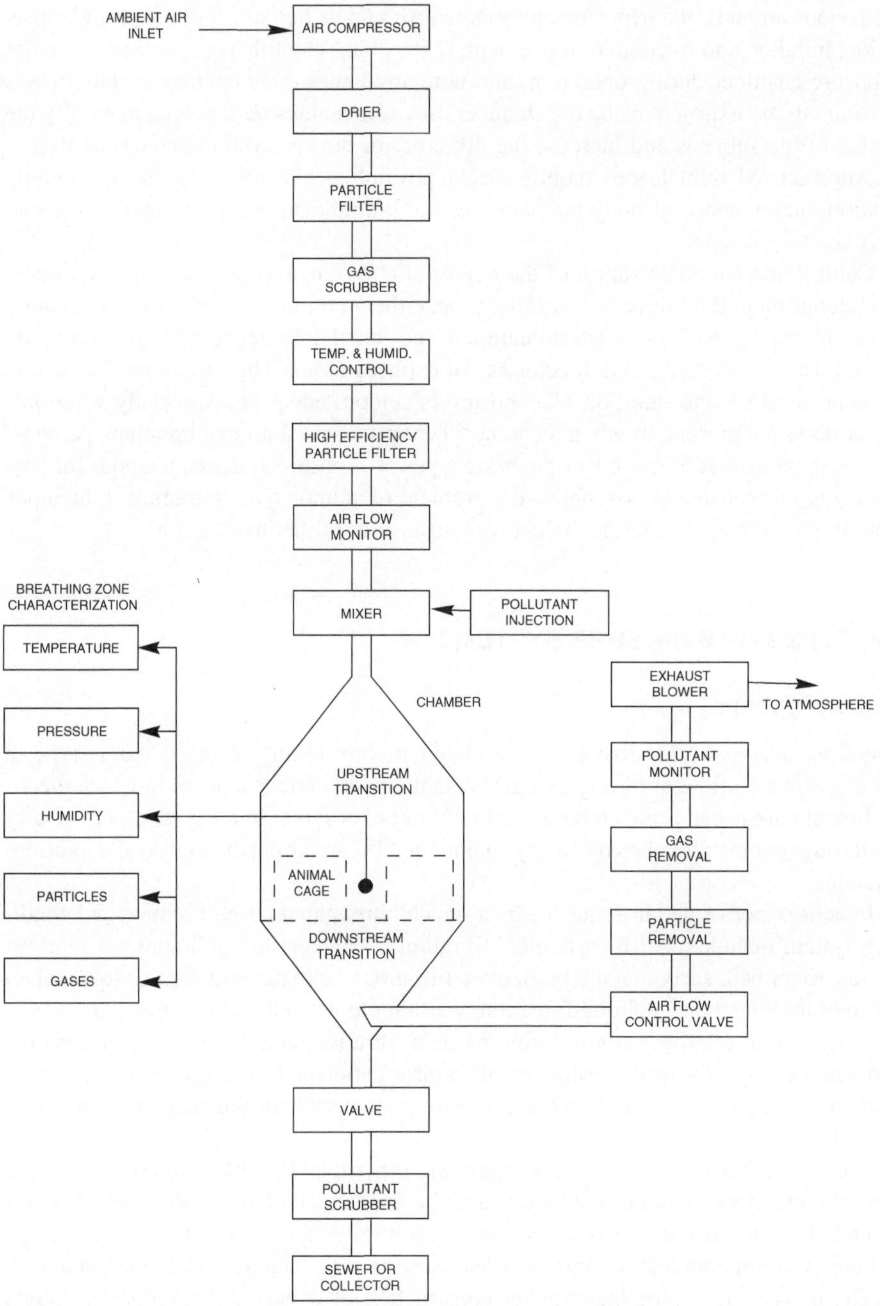

Figure 5.1 Components of a simplified chamber exposure system.

therefore tend to age more rapidly with use. A major disadvantage of glass or plastic derives from its poor electrical conductivity. Such surfaces can have a conductivity that is of the order of 10^{-20} that of steel. This means that islands of high static charge can build up on surfaces and remain for long periods. Such charge islands arise through contact charging when dissimilar materials are brought into close contact and then separated. Charge islands may be positive or negative, have potentials of several kilovolts, and avidly attract particles. This electrical effect can be seen from the studies of Cooper et al. (1979) in which aerosol concentrations were monitored over periods of time in aluminized plastic bags having cracked surfaces that prevented the neutralization of positive and negative charge islands (Fig. 5.2). Lai (2006) performed similar investigations and found that antistatic spraying reduced losses. Losses of particles in glass or plastic chambers can also be controlled by other coating or washing techniques (van Dingenen et al., 1989). Another solution to the problem of loss of charged particles involves discharging the aerosol particles with a bipolar ion source, as described in Chapter 1.

Advantages of the chamber exposure system include adaptability to a large variety and large numbers of animals exposed, capability for housing animals for long periods under exposure conditions, lack of restraint or anesthesia during exposure, and the existence of a relatively large database for this type of exposure. Chambers can be sealed and are thus appropriate to use with very toxic materials and for exposures under conditions of nonambient pressures, temperatures, and humidities. Inhalation chambers are the exposure method of choice in many studies. Large chambers also photograph well, demonstrate a high level of commitment of the laboratory, and appear to impress visitors.

Several disadvantages are associated with inhalation exposure chambers. The exposure is messy; airborne material contacts the totally immersed subjects through the skin and eyes, it covers the fur, food, caging, and anything else present. The dose received in chambers is often highly variable. Animals tend to avoid exposure by huddling together, covering their noses with their own fur, or burying their noses in corners of the cage. The desired level of exposure does not stabilize quickly, nor can the concentration be varied rapidly in large chambers operating at typical flow rates. Stabilization of proper exposure levels may require an hour or more (Hinners et al., 1966; Phalen et al., 1994b; Wong, 1999) and the effects of adjustments made on generating equipment are not seen immediately. Automatic feedback control of generators, air conditioners, and motors may produce

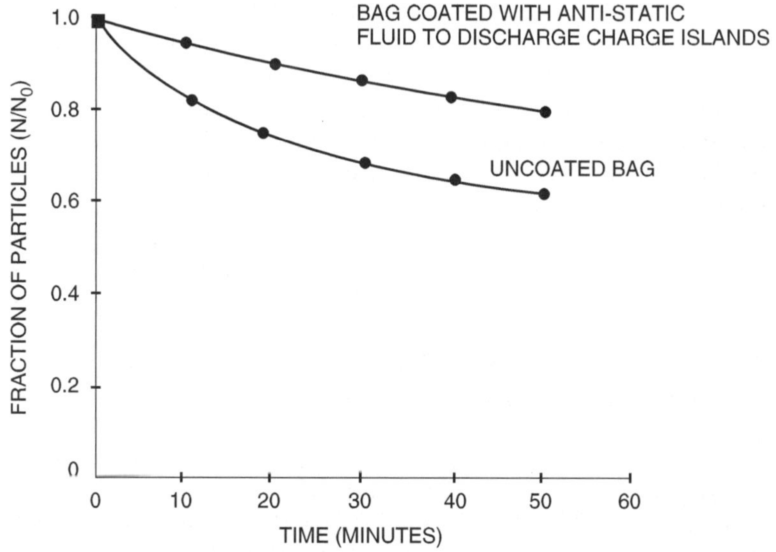

Figure 5.2 Fraction of particles, 0.1 μm diameter, having a unit electrical charge that remain airborne in an aluminized mylar bag. *Source*: Redrawn from Cooper et al., 1979.

cycling of the level of exposure. Also, large volumes of air may be required in dynamic chambers, resulting in a waste of exposure material and often great expense in conditioning the main throughput air. Animals in chambers are somewhat remote and it is difficult to be aware of their condition, or to monitor physiological parameters. The cost of chambers is relatively high, especially when they are constructed of expensive materials.

Inert materials which do not significantly change, deplete, or augment the desired atmospheres must be chosen. A discussion of this topic is found in Fraser et al. (1959). It may be convenient to cover the walls of a chamber with stainless steel foil or another protective material. Losses to chamber surfaces, especially of particulate materials, can be a severe problem. Losses can be diminished by use of large volume chambers, nonturbulent airflows, and electrically conductive interior surfaces. An unlucky designer may inadvertently end up with an unwanted, oversized aerosol collector.

Spatial uniformity of airborne materials is usually achieved by fitting the chamber with cone- or pyramid-shaped entry and exit sections (Fraser et al., 1959; Leach et al., 1959; Hinners et al., 1966; Lippmann, 1970; Yeh et al., 1986). More recently, computational fluid dynamics modeling has been applied to designing aerosol injectors that provide exposure uniformity in small chambers (Oldham et al., 2004). Uniformity is also achieved by proper introduction of materials into the chamber air. Venturi mixers and premixing chambers can be useful. Systematic rotation of cages during exposure may also be desirable. Temporal uniformity of exposure requires not only stable generators and airflow systems, but also chamber construction materials that do not change their reactivity during exposure. Plastic, for example, has variable reactivity to ozone over a period of hours at moderate airborne levels.

Samples for characterization of the exposure atmosphere should be taken from the *breathing zone* of subjects through lines that do not compromise the sample. Use of movable, large-bore, metallic sampling tubes is advisable. Samples must be taken during the actual exposures, as the presence of occupants may reduce the exposure concentrations. Sampling is a serious problem and reports stating "no effect" can make one question the sampling methods.

Provision for animal care and observation should be considered in chamber design. Internal flushing, watering, and feeding systems must be carefully designed to prevent accumulation of exposure material and animal wastes on their surfaces. Brown and Moss (1981) described an exposure chamber suitable for permanently housing up to 360 mice. Large chambers may be provided with air-lock type entries so that continuous exposure may be maintained (Fraser et al., 1959; Thomas, 1965; Lippmann, 1970). Observation windows should be limited in size to prevent the undesirable effects of static charge buildup. And the level of illumination in the chamber should be considered, with the subjects' diurnal cycles in mind.

Attention should be directed to control of environmental noise and vibration as well as temperature and humidity. Valves and motors are sources of noise that can be reduced by placing soft, in-line air filters between them and the chamber. Motors attached to chamber surfaces by hard mounts can cause excessive vibration.

Animal loading limits are based upon consideration of reactivity of animal surfaces with the exposure atmosphere, and generation of heat and water vapor by the animals. Experience indicates that no more than 5% of the chamber volume should be occupied by animals (Fraser et al., 1959; Lippmann, 1970). Bernstein and Drew (1980) have shown that the heat generated by animals in a stainless steel exposure chamber is primarily dissipated through the chamber walls rather than by the throughput airflow. In these studies, the chamber volume was 380 L, the chamber flow rates were either 100 or 170 L/min, and the loading with rats varied from 8 to 40 animals. They estimated that about 150 L/min airflow per 450-g rat would be required to remove the heat generated by rats with an increase of only 1°C in temperature. The required flow rate could be unreasonably high if several rats were in a chamber at once. Problems noted for such high airflow rates included drying of animals, the need for excessive exposure material, and the burden of postchamber air cleanup. In most cases, the room air surrounding the chambers should be maintained at a temperature low enough to conduct animal-generated heat away from the chamber.

System reliability involves consideration of failure of critical components, ease of routine maintenance, and replacement of components with finite lives such as filters and seals. Filters can be continuously monitored for pressure drop during chamber operation and replaced when they become excessively occluded. A safe exhaust system is important, especially when toxic, explosive, or otherwise offensive atmospheres are investigated. This may involve addition of filters, electrostatic precipitators, wet scrubbers, or even collection of all of the throughput.

Many other considerations enter into the design of chambers, including animal placement, compatibility of species, caging design, placement of generators, etc. Original designs should be drawn with great care, and with the help of persons having experience in exposure chamber operation.

Several reviews on exposure systems address general principles of design and operation. The Public Health Service monograph by Fraser et al. (1959) is perhaps still the most complete general discussion of inhalation chambers. Topics treated in this work include chamber shape and air movement, cleaning, construction materials, caging of animals, illumination, aerosol applications, sampling, warning devices, and chamber loading with animals. Nelson (1971) reviewed air purification, flow rate measurements, and static and dynamic exposure system designs. In describing the Thomas dome, Thomas (1965) provided information on some considerations for chronic exposure chambers that operate below atmospheric pressure. Emphasizing aerosol exposure, Lippmann (1970) briefly discussed exposure systems and methods of generating aerosol atmospheres. Hoffman and Wynder (1970) discussed several exposure systems for handling cigarette smoke. Karg et al. (1992) described a large (19 m^3) chamber capable of exposing five beagle dogs at once. Selected design considerations for chambers have been discussed by Carpenter and Beethe (1980) from the point of view of the engineer: factors such as airflow system design, characteristics of pyramidal chamber end cones, and costs are discussed. A review of environmental inhalation chambers, including some early designs was published by Drew and Laskin (1973). An early paper by Silver (1946) covered several practical issues in inhalation chamber operation, including subject loading factors, and buildup and decay of atmospheric concentrations. *Generation of Aerosols and Facilities for Exposure Experiments,* edited by Willeke (1980), contains eight chapters on exposure facilities and systems: chapters written by Lippmann, Carpenter and Beethe, Bell et al., Crider et al., Ferin and Leach, and Hinners et al., describe several aspects of inhalation chamber design and operation. More recent reviews of exposure chambers are available (Gardner and Kennedy, 1993; Cheng and Moss, 1995; Wong, 1999).

It is useful to consider some of the major problems that plague chambers which are designed by the inexperienced. In most cases the remedy is obvious (Table 5.2). Before designing or selecting an exposure chamber, one must become familiar with the published literature, especially some of the reviews cited above.

Successful Systems for Laboratory Animals. An impressive variety of exposure chamber designs have been used in inhalation studies. Each one reviewed here has been successful enough to generate publishable toxicology data, or to be published as a scientific instrument. Barrels and large glass jars, often used for exposing animals, are adequate for some purposes but are not included in this review.

Table 5.2 Common Problems Experienced in Inhalation Chambers

1. Excessive buildup of animal-generated ammonia due to poor housekeeping, too small a chamber volume, or an airflow inadequate for the number of animals
2. Excessive losses of aerosol particles or gases to surface of an exposure system due to impaction, electrostatic attraction, eddy flows, or chemical reactivity of surfaces
3. Difficulty in cleaning due to unsealed crevices, hard-to-reach recesses, or lack of a bottom drain
4. Uneven spatial distribution of particles within the chamber due to excessive losses near surfaces (especially true for plastics, glass, and other electrical nonconductors), or poor mixing within the chamber volume
5. Inability to adequately control chamber throughput air with respect to temperature, humidity, or purity. Under-designed purification and conditioning systems or poor temperature control in the chamber room are the main causes of this problem

One widely used chamber is the University of Rochester, hexagonal cross-section chronic exposure chamber. This chamber has a tangential inlet, a cone to allow even distribution of the exposure materials, an exposure section designed to accommodate several animals of various species, and a lower cone for removal of the chamber air. Designed over 50 years ago, the chamber has aerodynamic characteristics which are superior to many modern chambers. The Rochester chamber design (Figs. 5.3 and 5.4), described in detail by Leach et al. (1959), has been used by several laboratories and is still acceptable for modern studies. The 1.3-m^3 volume version was originally used for the simultaneous exposure of 4 monkeys, 8 dogs, and 40 rats using two levels of cage. Current accepted practice discourages putting mixed species in such close contact with one another.

A similar chamber having a tapered inlet and outlet, but with a rectangular cross-section, has been described by Hinners et al. (1966) and by Laskin et al. (1970). The Hinners or Laskin design is easier to fabricate and appears to have performance characteristics similar to the Rochester design.

A large, rather distinctive chamber suitable for lifetime exposure of nearly any species has been described by Thomas (1965) (Fig. 5.5). The Thomas dome is hemispherical, 3.7 m (12 ft) in diameter, and is capable of containing atmospheres that are explosive, or at nonambient pressures. These domes, used at The Toxic Hazards Research Unit, Wright-Patterson Air Force Base, Ohio, U.S. have been used to simulate submarine and spacecraft environments and for the study of a great variety of airborne toxicants.

A 1.8-m diameter, all-plastic, spherical chamber used for the exposure of great numbers of small animals (e.g., 100 hamsters) is described by Stuart et al. (1971). In this chamber, as well as some described previously, special open-mesh wire caging is used to permit unobstructed contact between animals and the atmosphere. Such caging should have dividers to separate individual animals.

A chamber in which airflow is horizontal rather than from top to bottom was used by Ferin and Leach (1980) (Fig. 5.6). The chief advantage of this mode is that several levels of animal cages can be used and the catch pans below each level do not interfere with exposure of animals below. A design by Karg et al. (1992) uses horizontal airflow for exposing five dogs. Four such chambers are used for simultaneous exposures of control and exposure groups.

Another solution to the problem of exposing several layers of animals involves staggering the cages in a chamber that has airflow from top to bottom. The most sophisticated design, that of Brown and Moss (1981), is shown in Figure 5.7.

Figure 5.3 Photograph of University of Rochester chronic inhalation exposure chamber. *Source*: Courtesy of L.J. Leach.

Other large chamber designs for the chronic exposure of laboratory animals have been described by Leong (1976). The 14.5-m chambers, built for the Dow Chemical Company and for the International Research and Development Corporation, are cubical with flat bottoms and pyramidal tops. Air enters from the top and exits either through exhaust openings at each corner of the floor or through holes in two horizontal pipes at floor level. The chambers are designed to hold racks of animals, accommodating 400–500 rats in each exposure unit.

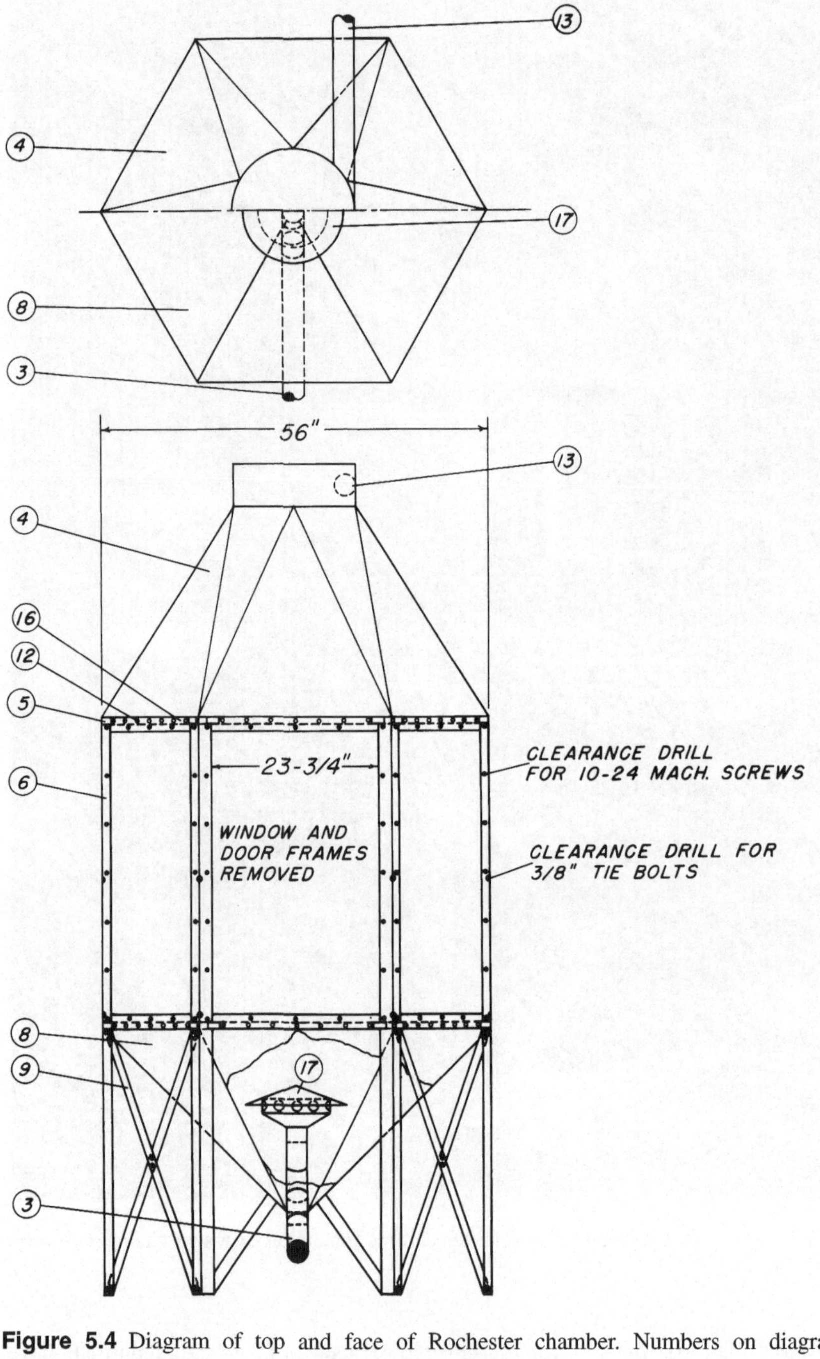

Figure 5.4 Diagram of top and face of Rochester chamber. Numbers on diagram refer to construction details. *Source*: Courtesy of L.J. Leach.

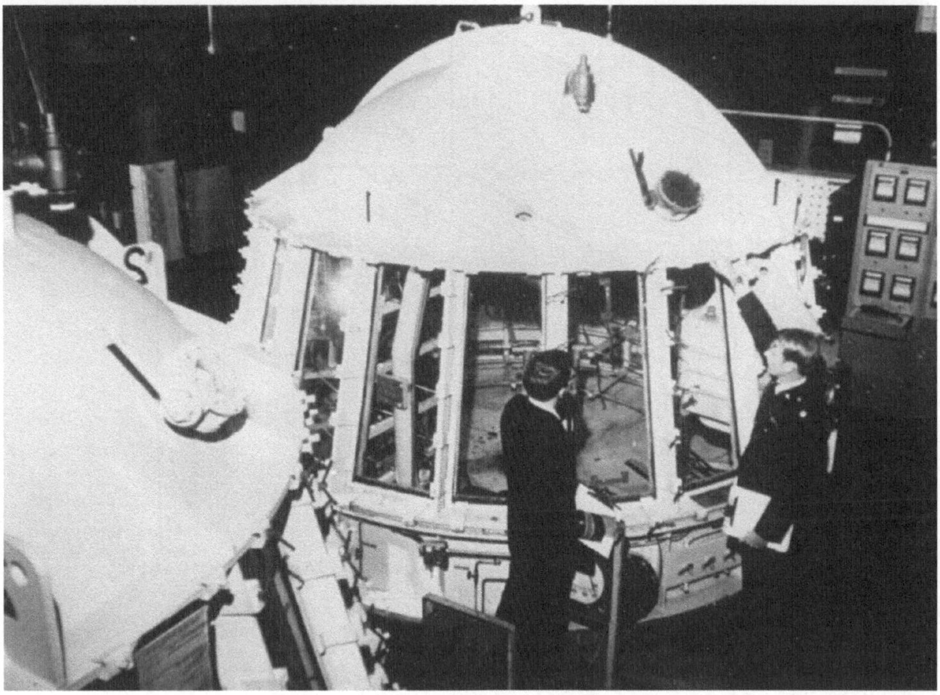

Figure 5.5 The Thomas dome chambers. *Source*: Courtesy of A.A. Thomas.

Austin et al. (1978) described a relatively small (0.25 m^3) chamber for exposing rats to toxic gases (Figure 5.8). The system, designed to hold four polypropylene rat cages, establishes horizontal flow into a side window of each cage, with air exiting from the wire cage tops. Food and water are supplied to five rats in each cage. Trays under the wire-bottom cages collect excreta and allow for continuous exposure with only 5-min shutdown per day for maintenance. When the system was tested at 35 L/min throughput air, 15 min of washout was required to reduce the concentration of a test gas to trace levels. Thus, some recirculation of air existed. Tests with rats indicated normal weight gain and food and water consumption in comparison with a conventionally housed group. A more modern portable small cage for exposing mice to concentrated ambient aerosols is described by Oldham et al. (2004).

Automatic control of an exposure chamber and its sampling system can be used to increase reliability and decrease operating costs. One such system has been described by Carpenter et al. (1979). The system uses a microprocessor interfaced to solenoid valves that control airflow to the exposure chamber, the atmosphere generators, and sampling devices. A paper tape provides the control input to the microprocessor. Other automatic system designs were reported by Wong and Moss (1996) and by McKinney and Frazer (2008). The Wong and Moss system—which controls exposure parameters, data logging, and password access—was used in a 90-day, whole body exposure of rodents to chloroform. The McKinney and Frazer system maintains a tight range of ozone exposures for up to 8 h.

Frostling (1973) published a design for a slowly-rotating drum aerosol chamber in which particles could be kept in suspension for lengthy periods. Losses due to settling of a polydisperse, 0.8 μm mass median diameter aerosol were reduced to 10% of the value

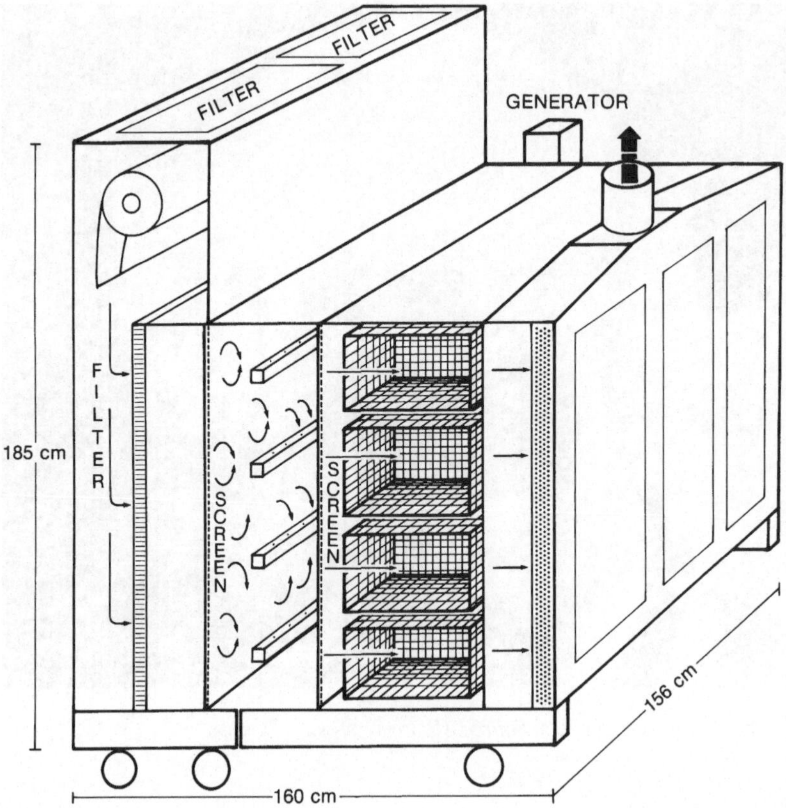

Figure 5.6 Horizontal flow exposure system of Ferin and Leach. *Source*: From Ferin, J., and Leach L.J., *Generation of Aerosols*, Willeke, K., Ed., Ann Arbor Science, Ann Arbor, MI, 1980, p. 517. With permission.

obtained using a nonrotating chamber. Asgharian and Moss (1992) developed a theoretical approach for designing a rotating drum capable of holding aerosols for several days: The design can be used to supply exposure systems, or to remove large particles from a distribution. The model predicts that 50% of 1–2 μm diameter particles could be stored for 6 months.

An all-plastic chamber designed for long-term exposure of rodents to gases was described by Montgomery et al. (1976). This system, with a capacity of six small animals, fits into a glove box so that relatively toxic gases can be safely handled. The plastic construction should probably only be used with gases or charge-neutralized aerosols.

An exposure chamber, designed for use with diesel exhaust, metallic oxides, and asbestos dust, was described by Wehner et al. (1972). A spherical version was later described (Wehner et al., 1978) (Fig. 5.9). Open-mesh wire caging with individual sections for rodents was used in this chamber.

A system designed to automatically record the activity of mice during exposure to cannabis smoke was described by Weinberg et al. (1977). Activity was monitored by counting the number of times animals broke a light beam passing through the exposure zone.

Mautz et al. (1985) developed a 10-track enclosed treadmill for exposing rats to aerosol/gas mixtures (Fig. 5.10). In this system, exercise level can be controlled and

Figure 5.7 Airflow patterns in a chamber designed by O.R. Moss. Source: Adapted from Brown and Moss (1981).

metabolic gases measured. The system was used to demonstrate the effect of exercise on intensifying responses (Mautz et al., 1988).

Interest in environmental particulate air pollution has stimulated the design and use of particle concentrators to supply exposure systems (Ghio and Huang, 2004; Lippmann et al., 2005). In these systems, the gasses are not concentrated. Such systems are also adaptable to human studies (Utell and Frampton, 2000).

Successful Systems for Human Subjects. Chambers for exposing human volunteers may contain response-measuring instrumentation that is operated by the subjects, or by the experimenters, inside the chamber. Such chambers, or more accurately exposure rooms, can have unique problems. First, the larger volume means increased requirements for quantities of pure air and intentional pollutants. Second, the presence of equipment can lead to the generation of heat and new pollutants, and to losses of pollutants on surfaces of the apparatus. Also, the health and safety of the subjects must be considered (Folinsbee et al., 1997).

The design and operation of one system for exposing human volunteers to controlled levels of common air pollutants was described by Hackney et al. (1975) and Bell et al. (1980). The chamber has a complex air-purification and conditioning system, provision for pulmonary function testing at rest and during exercise, and high-capacity nebulizers for establishing an aerosol in the chamber (Fig. 5.11). The chamber, a room with a floor area of 25.2 m^2 and a volume of 62.3 m^3, acts nearly like an ideal, constant flow–stirredtank reactor in which incoming air mixes completely with air in the chamber. Chamber walls are stainless steel and exposure materials are introduced downward through a perforated

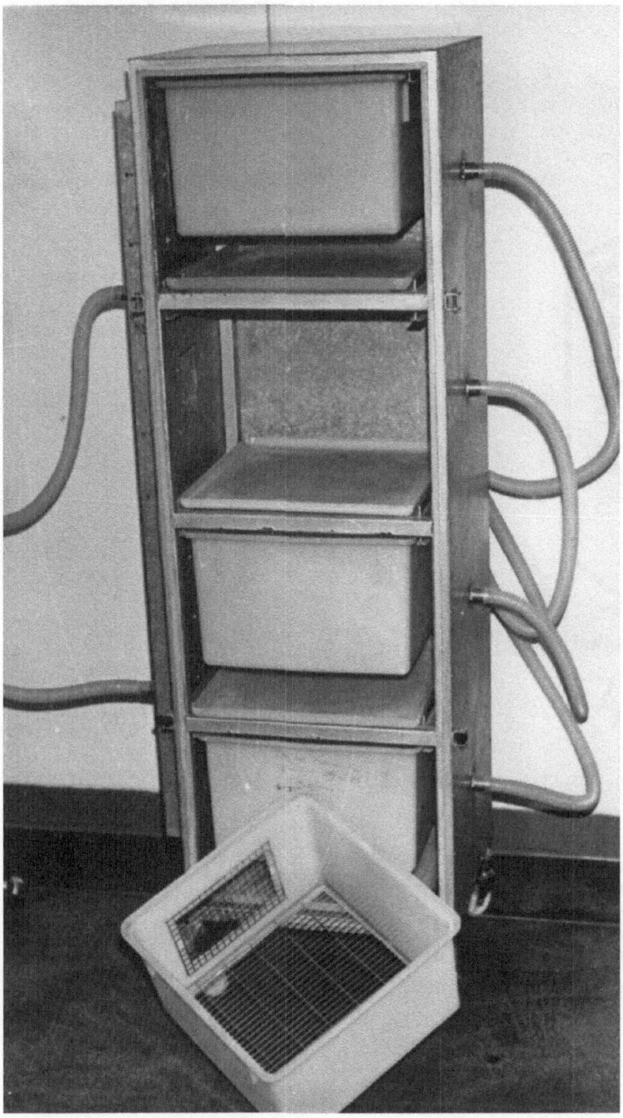

Figure 5.8 Small chamber for exposure of rats to gases. Source: Courtesy of J.C. Austin.

ceiling, and exit through a grating at the bottom of one wall. The operating airflow of 14 m^3/min is totally exhausted with no recycling. Gases and fine particles have a mean residence time of about 4 min in the exposure room.

Other human exposure chambers with capabilities for measuring responses during exposure have been described by Utell et al. (1984) and by Aris et al. (1990).

A movable laboratory for controlled clinical studies to ambient air pollutants in humans was described by Avol et al. (1979). The system, a modified trailer, can be used to expose human subjects to purified or ambient air. The laboratory contains pulmonary function testing equipment, exercise ergometers, atmospheric monitoring equipment, a restroom, medical examination room, and air-conditioning and air-cleaning equipment. Measurements of air inside the exposure chamber in the purified air operation mode

Figure 5.9 Spherical chamber for inhalation exposure of rodents. *Source*: Reprinted from Wehner, et al., *Environ. Res.* 16: 393-407, 1978. With permission from Elsevier.

indicate that ozone and particulate levels are reduced to less than 15% of outside levels, and NO_2 to about 30% of outside levels. Carbon monoxide and hydrocarbon levels inside approximate those outside.

Head-Only Exposure Systems

Design Considerations. The head-only type of exposure has two main advantages: its applicability to repeated brief exposures, and limitation of the pathways of entry of material into the subject. It is difficult, if not impossible, for a subject to avoid inhalation exposure in head-only systems. However, the head or neck of the subject is usually firmly restrained, so stress may be significant. Other disadvantages include losses of material to the head (especially if fur becomes electrically charged), difficulty in achieving a good neck seal without interfering with blood flow or ventilation, and the additional difficulty and time required in handling subjects.

Two basic types of systems are used: individual helmets; and group exposure, with subjects' heads exposed in a chamber or other airflow system. Design considerations for head-only exposure systems include the following. When the head-only exposure is from a chamber, distribution of material in the chamber must be uniform. Helmet exposures may also require large airflows to prevent condensation of expired water vapor, buildup of expired products, or depletion of the exposure atmosphere. As subjects inhale and exhale, pressure fluctuations may occur in the system. Raabe and Yeh (1976) described placement of a spirometer in parallel with the subjects, which reduces pressure fluctuations and also provides a record of the breathing pattern during exposure. As in chamber exposures,

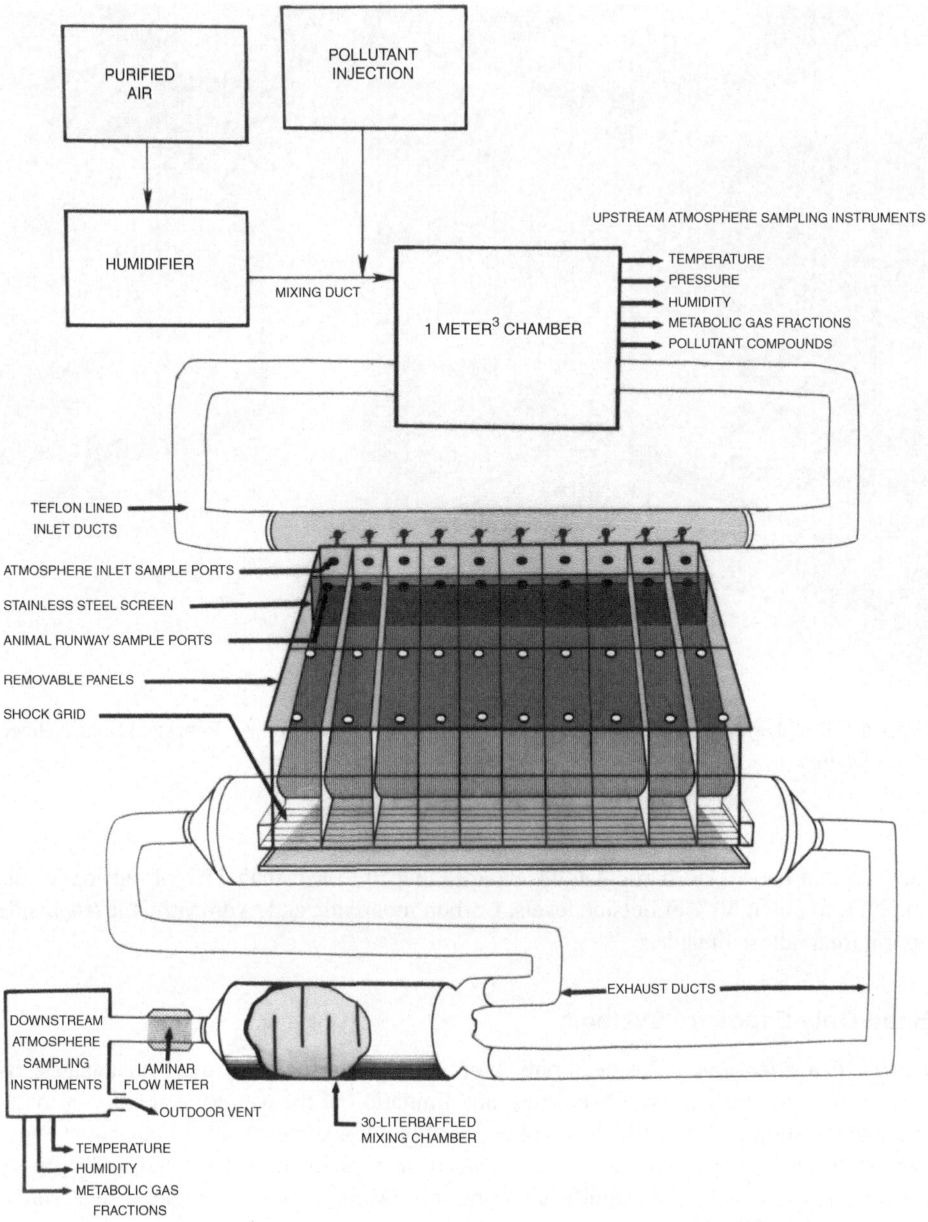

Figure 5.10 Ten channel treadmill for exposing small animals to particles and gases, with measurement of metabolic gases. Source: From Mautz et al., *J. Appl. Physiol.*, vol. 58(2) 1985, p. 674). With permission from the American Physiological Society.

losses and sampling of the exposure atmosphere must be carefully considered to insure adequate definition of the exposure. Environmental considerations include proper air temperature and humidity, a proper level of carbon dioxide, and control of noise and vibration. Physical comfort may require a tolerable seal around the neck. Two seal types are inflatable collars (Thomas and Lie, 1963; Kirk et al., 1975) and thin rubber membranes having a small hole that stretches to accommodate the neck (Esparza et al., 1979). The seal must

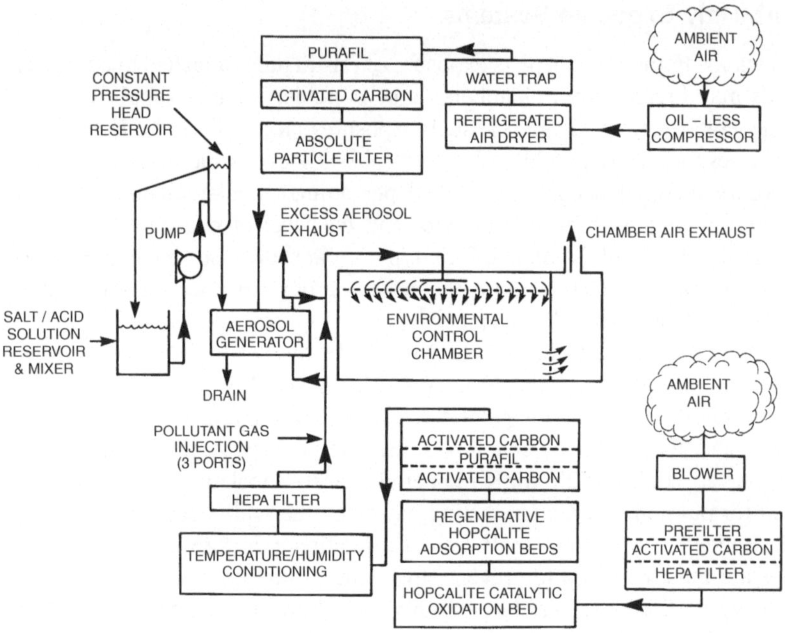

Figure 5.11 Human exposure chamber system at Rancho Los Amigos Hospital. *Source*: Courtesy of E.L. Avol.

be soft and conform to the noncircular cross-section of the neck. A wide, soft support beneath the neck may be necessary to prevent choking the animal.

During exposure of unanesthetized animals, restraint additional to that about the neck may be necessary. Comfortable slings and padded stocks may be useful in preventing undue stress to the animal subjects. A struggling, restrained animal may rapidly develop very high body temperature, and anesthesia or sedation should be considered.

Successful Systems. Head-only exposure systems are typified by the system described by Kirk et al. (1975) for exposing guinea pigs to radioactive gases. The animals bodies are in plethysmographs, which can be used for recording breathing patterns. A rubber seal about the neck isolates the head inside the exposure zone. In the operation of this system, the animals could also be given a body-immersion in the radioactive gas while breathing fresh air. A similar system, for exposing rodent heads only to aerosols, is described in some detail in a report by Thomas and Lie (1963). Stuart et al. (1971) used a head-only exposure system for exposing dogs to radioactive particles and diesel exhaust. In this system, the animals were in a sitting position with their heads protruding inside a chamber. An interesting system involving individual exposure helmets was used by Scheimberg et al. (1973) to expose nonhuman primates to aerosolized medicinals. Vick et al. (2007) used head-only helmets to expose dogs to insulin powder for 6 months. In all of the above systems, the animal subjects were either physically restrained or anesthetized.

A head dome that permitted natural breathing for exercising human subjects during aerosol exposure was described by Bowes et al. (1990). A pneumotachograph and thermocouple monitored breathing and the onset of oronasal breathing.

Nose- or Mouth-Only Exposure Systems

Design Considerations. Figure 5.12 depicts exposure of rats in tubes attached to a dynamic chamber. Animals may breathe from chambers, large conduits, smaller supply tubes, or directly from a generator (e.g., a cigarette). Such exposures largely limit the initial entry of materials to the respiratory tract and oral cavity (possible eye irritation is thus eliminated), less exposure material need be generated per animal, containment of material within the exposure apparatus is straightforward, and the level of concentration of the exposure material may be rapidly changed. This type of exposure does involve potential stress due to close confinement. Getting comfortable seals around the face and the need for extra handling of subjects are other disadvantages.

General design considerations are similar to those of the head-only systems. Mask design presents some additional special problems. Masks should be comfortable, seal tightly, and allow for collection or drainage of saliva. Successful masks or nasal tubes are usually handmade and the literature contains some useful designs (Muggenberg and Mauderly, 1974; Stavert et al., 1982b; McKirnan et al., 1986). Inhalation exposure masks may also be used for pulmonary function testing, provided the seal is good and the dead space is small. Rodents and other small animals require exposure tubes for nose-only exposures. Care must be taken to prevent stress due to improper tube sizes or heat buildup (Narcisco, et al., 2003). If possible, rat tails should protrude from the rear of the tube, for heat dissipation. Well-designed tubes permit low-stress exposures, especially if the subjects are acclimated by brief training sessions.

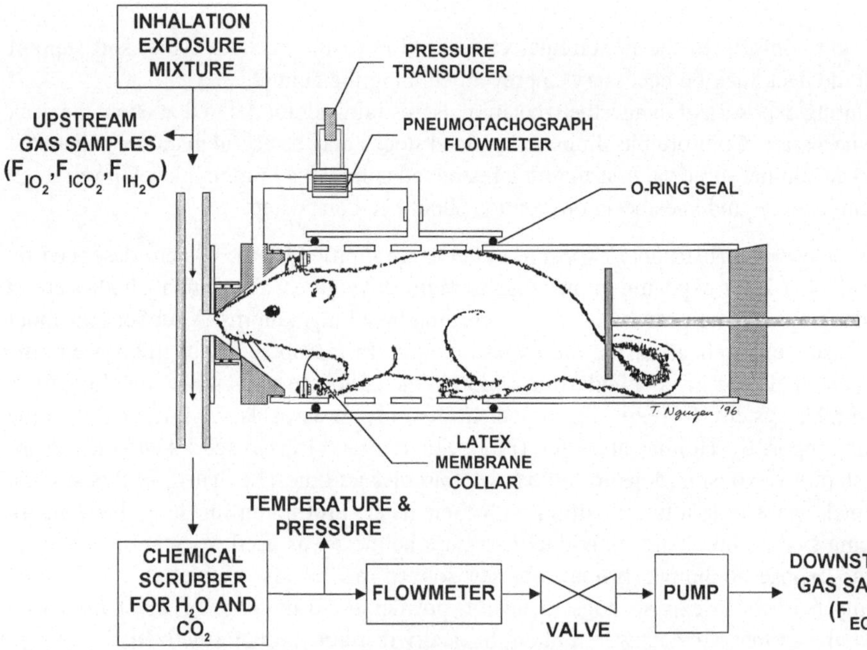

Figure 5.12 Schematic rodent nose-only exposure system with capability for assessing pulmonary function. *Source*: Adapted from Mautz (1997).

Successful Systems. Inhalation exposures via either the nose or mouth are usually accomplished using masks, catheters in the nose, or individual animal tubular containers with one end open to the exposure atmosphere. Mask-type exposure, usually limited to relatively large animals, is described for dogs (Bair et al., 1969; Dubin and Morrison, 1969; Poynter and Spurling, 1971; Cuddihy and Boecker, 1973; Stavert et al., 1982b). Masks suitable for inhalation exposures are described for ponies by Mauderly (1974a) and for pigs by McKirnan et al. (1986). Nasal tubes for inhalation exposure of donkeys are described by Albert et al. (1974). Restraint of masked, unanesthetized large animals is typically by sling (Boecker et al., 1964; Mauderly et al., 1971). For exercise exposures, masked dogs can run on a treadmill (Stavert et al., 1982a).

Tubular holders and delivery systems for nose-only exposure of rodents have been described (Raabe et al., 1973; Smith and Spurling, 1974; and Phalen, 1997a). Perforated metal holders for small rodents can reduce the stress due to buildup of body temperature during exposure. Mauderly and Tesarek (1973) and Mauderly (1986) described alterations in pulmonary function induced by the restraining devices used in inhalation exposures. Nose-only exposure systems have been designed for automation (Pauluhn, 1994a), and high-containment of aerosols and gases (Raabe et al., 1973; Cannon et al., 1983; Hoskins et al., 1997; Roy et al., 2003). Large chambers have also been modified to accept nose-only exposure tubes (Prasad, et al., 1988; Warheit et al., 1991).

Lung and Partial Lung Exposure Systems

Design Considerations. Three basic techniques (endotracheal entubation, tracheostomies, and the use of airway catheters) are used to bypass the upper airway and expose the lung or its subunits to aerosols and gases. Endotracheal tubes are usually made of flexible rubber and are passed through the oral cavity, or a tracheostomy, into the trachea and sealed by inflation of a balloon that surrounds the tip. Specialized endotracheal tubes, having a double lumen, permit isolation of the right and left halves of the lung. Endotracheal tubes are usually inserted under general anesthesia, and even then a desensitizing anesthetic spray, such as lidocane, may be used to suppress the gag reflex.

The advantages of endotracheal tubes are the elimination of exposure of the upper airway, a high degree of control over the dose delivered, and the efficient containment of exposure material. This latter advantage can be important when one is studying an extremely hazardous or expensive material. An indwelling tube can also be used to monitor the breathing pattern with a flow sensor such as a pneumotachograph; or to collect the expired air.

Disadvantages of entubation include: loss or depression of natural respiratory tract defenses due to bypass of upper airways or the effects of anesthesia; mechanical trauma of the larynx and trachea; interference with normal airflow by the tubing; loss of normal humidification and thermal regulation of the inspired air; and the additional time and manpower required for anesthesia, placement of the tube, and monitoring of the subjects. Special problems that should be considered include precise control of the temperature and humidity of inspired air, and the possible need for additional physiological support for maintaining adequate ventilation and body temperature.

Tracheostomy involves surgical penetration of the trachea followed by insertion of a connector or a tube. This procedure is highly invasive, always requiring surgical skill and adequate anesthesia. Proper humidification of inspired air is necessary to prevent drying of respiratory tissues. If properly performed, the operation may be either reversible or permanent. Infection can be a problem and good postsurgical nursing care by trained persons is essential. Otherwise the advantages and disadvantages of this method of exposure are similar to those described for endotracheal entubation. Airway catheters are usually substantially smaller in diameter than the trachea and are passed through the oral cavity or through a tracheostomy. These tubes may extend deeply into the lung to expose a single lung lobe or even a specific portion of a lobe. Radiography or fluoroscopy can be used as a placement aid. Precise localization of the delivered dose is possible. The exposure material may be delivered in a steady airflow, or in puffs synchronized with the inspiratory efforts.

Successful Systems. An exposure system that involves use of a spirometer for holding an aerosol, endotracheal tubes for two dogs, and balloons for collecting expired air was used by Phalen and Morrow (1973) for exposures to radioactive, metal fume aerosols (Fig. 5.13). This system assures safe containment of the radioisotope, permits measurement of the volume of air expired, and allows for the determination of aerosol deposition efficiencies in the animals.

An endotracheal tube system used for exposing dogs to fresh cigarette smoke was described by Auerbach et al. (1970). An endotracheal tube was used by Bianco et al. (1974) to expose dogs to radioactive tantalum dust for the purposes of performing contrast radiography and measuring the rate of deep lung clearance of this highly insoluble material. In these studies, the injection of dust into the tube was synchronized with inspiration in order to enhance the amount of material deposited deep in the respiratory tract. The method was successful in depositing gram quantities of material, a feat that would be impossible in a chamber-type exposure.

A technique for performing chronic tracheostomies in laboratory animals was described by Thilenius and Vial (1963). After recovery from surgery and complete healing, the tracheal opening was protected by a collar so that it may be used repeatedly to gain

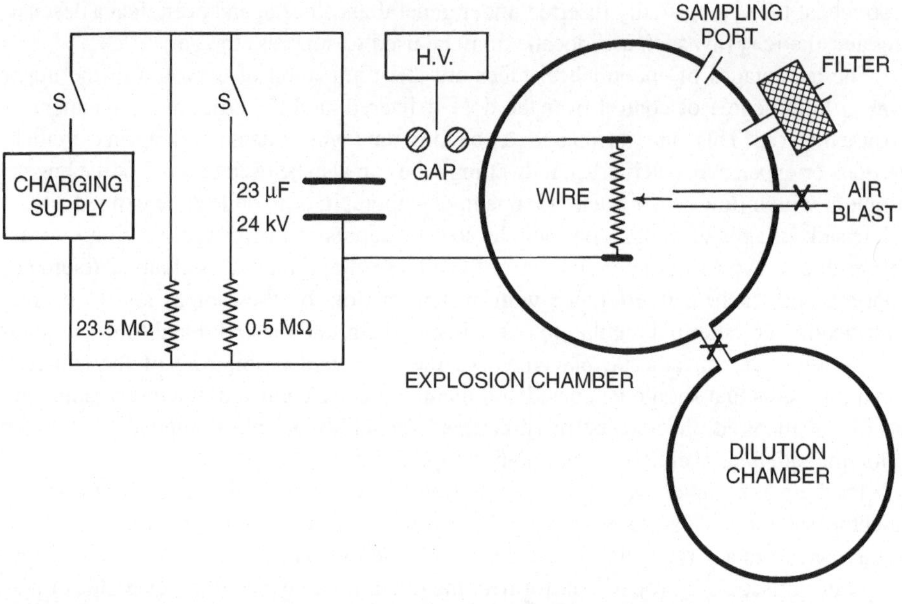

Figure 5.13 System for exposing dogs in pairs via endotracheal tubes to exploded-wire aerosols. Dilution chamber is a large, 40-L spirometer. *Abbreviation*: H.V., high voltage.

access to the lung. One undesirable effect, the alteration of brain temperature in dogs with tracheostomies, was subsequently reported by Baker et al. (1974).

An interesting use of a dual-lumen tube that is connected to two one-way valves was described by Battista et al. (1973). By separating the inspired and expired air with such a system, the effective dead space of the exposure apparatus was reduced. The tube was passed through a chronic tracheostomy.

Stavert et al. (1991) described a tracheal tube passed through the mouths of rodents in order to bypass nasal airways, and Flavin et al. (1986) adapted tracheostomized rabbits to aerosol exposures using an infant ventilator.

Intratracheal Instillation

Intratracheal instillation of materials remains a popular alternative to inhalation for exposing animals for several reasons. It is a means of exposing the respiratory system that requires a small amount of the study material; there is essentially no waste. It is therefore an inexpensive mode of exposure. Money is also saved because chambers and/or masks and elaborate generation equipment are not required. Instillation does not require the often-complex technical support that is involved in dealing with aerosols. Extremely high concentrations of materials can be placed into contact with lung tissue, and the dose delivered can be precisely controlled. It is one of the safer methods, with respect to handling and containment, for exposing animals to very hazardous materials.

The technique, for all its advantages, suffers from one serious flaw: the distribution of dose to respiratory tissues is artificial. As demonstrated by Brain et al. (1976) intratracheal instillation tends to lead to less uniform deposition than inhalation and tends to favor the dependent (lower) portions of the lung due to gravitational settling of the instilled material. The Brain group exposed rats and hamsters to radiolabeled particles, both aerosolized, and in aqueous suspension for their instillation. After the animals were killed the lungs were each cut into 54 pieces that were individually assayed for radioactivity. Statistical comparisons of the activities of the inhalation-exposed vs. the instillation-exposed animals were unequivocal, showing that the distribution of dose was strikingly different. The investigators also noted that upon microscopic examination of lung slices, instillation of the material produced heavy deposits in medium-sized bronchi, seldom reaching the alveoli. In contrast, inhalation of the aerosol led to considerable deposition in the small airways of the lung. Similar differences between inhalation and instillation were described by Leong et al. (1998).

Another problem with the instillation technique is that the high local concentrations of toxicants, or their carrier liquids, can cause local tissue damage. This may lead to local hemorrhage by mechanisms not directly associated with the material under study. Although instillation is an acceptable form of exposure in many instances (e.g., screening and mechanistic studies), it can not ultimately substitute for inhalation exposure. On the other hand, in some instances instillation may be the only practical means of exposure.

AGING THE ATMOSPHERE

Although all inhalation exposure systems deliver airborne material to the subject some finite time after generation, the age of an atmosphere at the time of inhalation is usually short. An animal drawing smoke from a cigarette attached to a mask may inhale smoke approximately 1 s after it is formed, and large chambers may deliver an atmosphere to animals minutes after generation. In some instances it is desirable to purposefully age the material under study for a longer time. Examples include mixtures of particles and gases, and materials that undergo radioactive transformations. In such cases, an aging or delay line may be used.

A vessel enclosing compounds undergoing chemical reaction, such as an aging atmosphere, is called a *reactor*. For the purpose of analysis and design of reactors, a number of ideal reactor models have been developed. These models include the batch, plug flow, and back-mix reactor types.

For the ideal *batch reactor*, all reactants are initially placed into the reactor at $t = T_0$, and the reaction process concluded before products are withdrawn at $t = T_f$. Complete mixing within the reactor is assumed for theoretical treatment; this being accomplished usually by a stirrer. Assuming the time for filling and emptying the reactor is negligible, the mean residence time, T′, for the contents of the reactor is:

$$T' = T_f - T_0 \quad \text{(eq. 5.1)}$$

Batch reactors may be constant-volume (rigid container) in which the pressure can change, or constant pressure (nonrigid container) which allows the volume to change. Bag-reactors are commonly used in laboratories studying air chemistry.

A second ideal reactor is the *plug-flow* type, which allows complete mixing of materials that enter the reactor at the same time, but no mixing of materials that enter at different times. In engineering terms there is complete radial mixing but no axial back mixing. As a result, composition of the atmosphere is variable along the length but constant across any cross-section perpendicular to the flow direction. Real systems that approximate plug-flow may be of the multiple-tube design or may have periodic barriers such as screens or metal plates to induce plug flow (Walters et al., 1982).

The third ideal reactor is the *back-mix* or stirred tank type. In this reactor, there is steady continuous flow and perfect mixing so that material entering rapidly assumes the composition of material throughout the vessel. In this case the exiting products have the same composition as the mixture with the reactor.

Engineering aspects of these three types of reactors are discussed by Perry and Chilton (1973). Figure 5.14 illustrates some reactor configurations suitable for use in inhalation studies.

The aging of combustion products, specifically those produced in cigarette smoking, was discussed by Hoffmann and Wynder (1970). During aging, one sees a general decrease in unstable and metastable organic radicals, which are considered by some to be potential carcinogens. However, also during cigarette-smoke aging, a conversion of NO to NO_2 and an increase in *N*-nitrosamines are seen, which could lead to an increase in their carcinogenicity.

An aging line coupled to a chamber exposure system has been described by Walters et al. (1982). This system can age aerosols up to 90 minutes without significant loss of aerosol particles.

AMMONIA AS A CONTAMINANT

The problem of ammonia buildup in exposure chambers has been noted repeatedly by investigators. The major source of ammonia appears to be the action of urease in bacteria from animal feces on the urea found in urine. Factors that control the amount of ammonia in chambers are the animal loading factor (volume of animals/chamber volume), the chamber airflow rate, and the length of time that animals are in the chamber. Temperature, humidity, and the presence of other reactive chemicals, such as acid vapors or droplets, will also influence the amount of ammonia present.

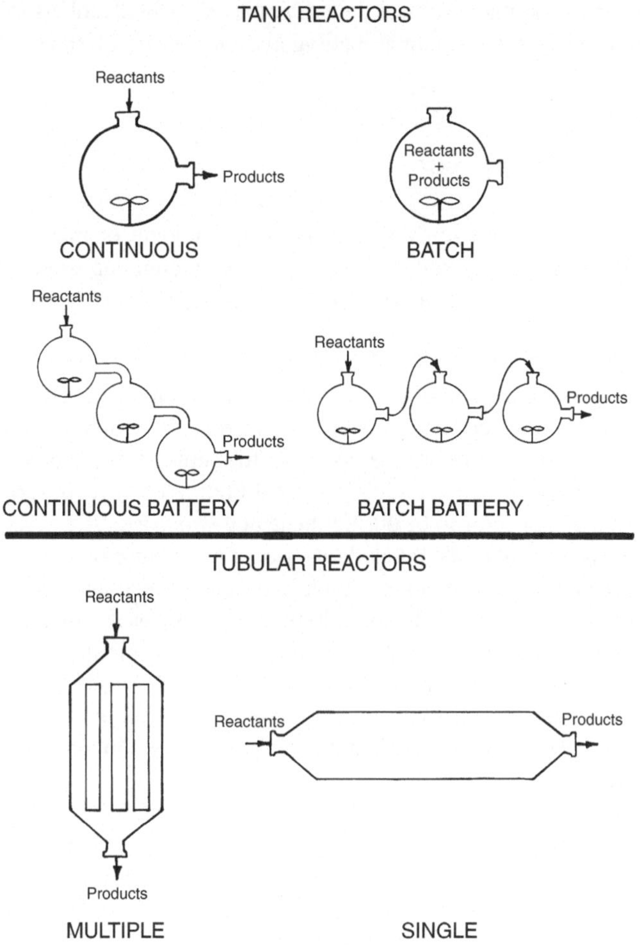

Figure 5.14 Basic types of chemical reactors that can be applied to inhalation exposure systems.

In general it is desirable to reduce the ammonia levels in chambers to low levels, since ammonia may have effects on animals as well as inducing chemical interactions with the exposure atmosphere. This is accomplished by lightly loading chambers with animals, maintaining high airflow rates, and thoroughly cleaning chambers and cages before each use. Avoiding inaccessible or hard-to-clean areas that collect urine and feces and designing for smooth airflow to prevent air stagnation are useful. Ideally, plug airflow, without back mixing of air, should be provided when ammonia must be suppressed to very low levels, as in studies involving sulfuric acid mists. This may be accomplished by interposing fine-mesh wire screens above and below the animal exposure zone, which flattens the transverse profile of air velocity and reduces vertical mixing of air in the breathing zone.

Barrow and Dodd (1979) published data on the ammonia levels in a 99-L glass chamber operated at various airflow rates and various loadings with rats. Using initially very clean chambers and open-mesh wire cages, they found the ammonia levels shown in Table 5.3. The percentage loading factor (% LF) is 100 times the ratio of animal volume to chamber volume. Since their chamber was rectangular with an inlet pipe and an outlet pipe at opposite ends, the air change of one chamber volume only renewed 65% of the total air in the chamber. From their data, one may estimate the ammonia levels in chambers of similar design using the relationships between animal loading and number of air changes per hour.

DETERMINATION OF THE INHALED DOSE

One of the major problems in an inhalation study is that the actual amount of material deposited in the subjects is seldom accurately known. Worse yet, the amount deposited is often variable from one subject to another, or from one time to another, even in the same subject.

The uncertainties in estimating the dose in inhalation studies are often so severe as to cast doubts on studies in which no biological effects are observed. When no effects are seen, one must ask whether or not the intended dose has been delivered. Three approaches are used in estimating the actual amount of material deposited in the subject: direct assay; use of airborne concentrations along with uptake models; and calibration of the exposure apparatus with a tracer substance, either prior to or during the actual exposure.

Direct assay refers to a measurement of the inhaled material or an induced metabolic product in tissue or in excreta. Such a measurement is relatively simple when a radioactive or other easily measured material is under study. An alternative direct assay method involves measuring the airborne concentrations of the material in both the inspired and the expired air of the subject: The difference, together with the volume breathed, is used to calculate the delivered dose. When a mask or other apparatus intervenes between the atmosphere and the subject, losses in this device must be taken into account. Losses in masks or endotracheal tubes can be appreciable.

The use of models for predicting inhaled dose is often the best that can be done in a given study. In the case of airborne particles, data exist for predicting uptake as a function of particle aerodynamic size for a variety of species (see Chapter 9). Referring to the literature will provide more detailed information on regional particle deposition, for example, separate nasopharyngeal, tracheobronchial, and parenchymal deposition probabilities (ICRP, 1994; NCRP, 1997; Brown et al., 2005).

As an example of the use of these deposition models, consider a calculation of the deposited amount of particles inhaled by rats. If the exposure involved unsedated rats

Table 5.3 Ammonia Concentration in an Inhalation Chamber

Animal loading (%)	Chamber airflow (L/min)	No. air changes per hour	Hour of sample (ppm NH_3 ± SE)		
			2	4	6
1	13	8	0.38 ± 0.08	0.48 ± 0.07	0.46 ± 0.13
1	26	16	0.20 ± 0.01	0.24 ± 0.02	0.45 ± 0.06
1	40	24	0.19 ± 0.04	0.24 ± 0.05	0.22 ± 0.03
3.1	13	8	0.84 ± 0.14	1.13 ± 0.14	1.11 ± 0.27
3.1	26	16	0.60 ± 0.09	1.04 ± 0.23	1.60 ± 0.22
3.1	40	24	0.19 ± 0.02	0.33 ± 0.05	0.39 ± 0.05
5.1	13	8	1.23 ± 0.18	1.51 ± 0.16	2.42 ± 0.38
5.2	26	16	0.66 ± 0.06	1.23 ± 0.20	2.05 ± 0.41
5.2	40	24	0.46 ± 0.08	1.02 ± 0.11	1.30 ± 0.27

Source: Adapted from Barrow and Dodd (1979).

weighing about 250 g which were exposed nose-only to a radiolabeled, monodisperse aerosol having an aerodynamic diameter of about 1.7 μm, an exposure of 20 min, and an airborne concentration of radioactivity of 1 μCi/L of air, then the radioactivity deposited per rat (A) can be calculated using the relationship: A = P·C·V·T, where P is the deposition probability in the rat for particles of 1.7 μm aerodynamic diameter, C is the airborne concentration in microcuries per liter of air, V the minute ventilation (volume of air breathed in each minute) in liters, and T is the number of minutes of exposure. From Raabe et al. (1977) P is found to be 0.22 (i.e., 22% of the inhaled particles deposit). The minute ventilation of the unanesthetized rat is assumed to be 0.16 L/min according to Crosfill and Widdicombe (1961). Therefore:

$$A = (0.22)\cdot(1)\cdot(0.16)\cdot(20) = 0.70\ \mu\text{Ci} \qquad \text{(eq. 5.2)}$$

A third means of estimating inhaled dose involves calibrating the exposure system with an easily assayed material in order to determine the rate of deposition per unit time of exposure. Before this value can be reliably used to compute the deposited dose for another material, one must consider any differences between that material and the calibration material. Differences in water solubility of gases, or particle size for aerosols, will require corrections to be made. The ideal tracer would, of course, be identical to the material to be studied later. Hartings and Roy (2004), described a method for providing known doses to monkeys exposed under a high containment condition. And Nadithe et al. (2003) compared predicted and actual doses of particles delivered to mice in nose-only exposures. Their actual deposition of 8 ± 4% was close to the predicted value of 4.4–5.9%. With care, an inhalation exposure system can be calibrated for exposure dose.

ETHICAL RESPONSIBILITIES OF THE INVESTIGATORS

Ethics is the discipline concerned with the standards of right and wrong, and with values that modify or govern behavior. As a scientist, one has a strong ethical duty to add meaningful new information to the growing body of knowledge. As users of research subjects, one also clearly has ethical obligations to minimize their pain, suffering, and distress.

In an inhalation experiment, the subject is often critically dependent on the exposure system for delivery of a breathable atmosphere in a comfortable environment. Animals can be placed in situations of high concentrations of CO_2, extremes in humidity and ambient temperature, or cruel confinement, without the knowledge of the investigator. For example, improperly fitted collars, slings, and other restraint devices can cause considerable discomfort. Padding should be used liberally and sedation or anesthesia considered in many cases. Horses, donkeys, ponies, pigs, and other animals seem to require some freedom of movement of the head when restrained in an unanesthetized state for more than brief periods. Large and small animals can overheat within minutes in close-fitting body enclosures unless provision is made for cooling. Alternatively, animals that are anesthetized, sedated, or prevented from exercising by other means, can suffer from hypothermia (a drop in body temperature), even in a room that feels comfortable to the investigator.

Ethical experimentation and proper scientific procedure require that the investigator have intimate knowledge of stresses placed on the animal. Similarly, reliability of procedures and critical mechanical components can have ethical ramifications. The author recalls one animal, fortunately anesthetized at the time, being given about 15 successive, forced inhalations by a respirator while an exhalation valve was inadvertently left in a closed position. The life of this valuable animal could have been spared by use of a pressure relief valve or even a pressure gauge in the exhalation line.

Training of animals, or at least putting them through a few short trial sham exposures, can help them to relax and thus improve their performance during actual exposures, as well as improve the quality of resultant data. Relaxed, well-trained animals can provide more reliable and less variable data.

There is no excuse for handling an animal roughly or cruelly. Such practices can do a great disservice to the field by raising the ire of colleagues and of groups that oppose the use of animals for research purposes. Also, roughly treated animals are difficult to handle and apt to fight back in the laboratory. Training of those handling animals is a must. Before an investigator works with a species for the first time it is important to become familiar with the nutritional, environmental, and psychological needs of that species, as well as the proper techniques for handling and administering anesthetics.

6
Testing for Toxicity

INTRODUCTION

An organism can respond to pollutant exposure in as many ways as its total number of properties and functions. Thus, if one can define and measure a property or function, it can be used as a biological endpoint in an inhalation study. Furthermore, no matter what is selected for measurement, there will be number of airborne materials that, in sufficient concentration and exposure time, will produce a statistically significant change in the endpoint. But until the role of the measured property is understood in terms of its relationship to harm or disease, one does not know how to treat the result. In other words, it may be a health effect or just basic scientific information.

The respiratory tract has defenses, including changes in breathing pattern, excess mucus secretion, and efficient healing that act to limit and resolve damage. Thus, many of the changes seen in laboratory investigations after short-term inhalation exposure are transient and cannot be always classified as adverse. Transient changes can, however, harm compromised individuals, or indicate that the material may produce progressive damage and disease if the exposure is prolonged or repeated.

Inhalation toxicologists have developed a number of meaningful measures of injury. For simplicity these measures can be categorized into broad groups, for example, anatomy, physiology, behavior, and biochemistry. These categories represent specific scientific disciplines that have their own tools, and their own states of sophistication.

In order to select meaningful endpoints, it is useful to consider the vast number of possibilities from which to choose. When all of these are considered, one must conclude that it is impossible to measure more than a small fraction. This point is clear when one considers a list of levels of complexity for biological systems.

- Atomic
- Molecular: biochemical relationships
- Macromolecular: including secondary and tertiary structures of proteins (about 100,000 in the body), nucleic acids, carbohydrates, and lipids
- Cellular: about 200 types of cells are found in mammals, and their cellular components include membranes, mitochondria, endoplasmic reticula, lysosomes, vesicles, Golgi apparatus, chromosomes, nuclear pores, centrioles, etc.
- Tissue: functional units made up of cells
- Organ: brain, heart, lung, eye, ear, gonad, kidney, liver, skin, spleen, pancreas, stomach, etc.

- Organ systems: cardiovascular, neuromuscular, cardiopulmonary, reproductive, etc.
- Individual performance: reproduction, growth, survival, mobility, strength, learning, cognition, etc.
- Population: survival, population numbers, social behavior, etc.

In this chapter, only a small fraction of these levels of organization will be discussed. The emphasis is on biochemical, cellular, tissue, and organ levels of biological organization, primarily related to the respiratory tract. As a further limitation, responses that are detrimental in that they are likely to diminish either the quantity or quality of life are selected. Thus, only a small sampling of potentially useful endpoints is considered.

QUANTITATION

Qualitative observations of responses to inhaled materials can provide important insights. However, *quantitative data* are essential to scientific studies for a number of reasons. Dose–response curves can be constructed from quantitative data, quantitation allows for estimation of uncertainties, statistical testing, and ranking of treatments. Additionally, without quantitative data, mechanistic modeling and extrapolation beyond a specific study are virtually impossible.

Many observations that have traditionally been qualitative can be made quantitative. For example, the description "several focal lesions containing thickened alveolar walls and infiltrating cells" can be expressed quantitatively as number of foci per unit area, size of foci, mean number of cells per lesion, mean alveolar wall thickness, etc. Similarly, the observation of "ataxia or lethargy" can be expressed as number of times the subject stumbles, ratio of time spent lying down or standing, number of responses per unit time to a stimulus, or mean time for righting when a subject is placed lying face up. When observations are quantitative, one builds a database with which to compare past and future studies and also generates data that can be verified and extended by other investigators.

ANATOMICAL CONSIDERATIONS

Respiratory Tract Regions and Common Diseases

It is useful to simplify the respiratory tract by identifying major compartments, that is, regions that exhibit internal anatomic similarity (Table 2.2). Three airway regions, the *extrathoracic airways*, the *tracheobronchial airways*, and the *pulmonary* or *parenchymal airways*, are characterized by their exposures to air pollutants, their tissues, injury patterns, and disease states.

Extrathoracic (Head) Airways

The nose and nasopharynx begins at the anterior nares (nostrils) and includes the respiratory airway down to the level of the larynx. The nasal cavity, which is roughly triangular in cross-section, contains a central septum and turbinates that form narrow channels through which air flows. The forward portion of the nose is lined by cells not unlike those of the body skin, the remainder being lined with ciliated or olfactory epithelium. The most common disease state of the nasopharynx, rhinitis, is characterized by inflammation; nasal membranes are swollen and excessive secretion or excessive dryness are both possible.

Rhinitis, which includes the common cold, is associated with infection by viruses, bacteria, or fungi. Rhinitis can also be produced by allergens and by irritating gases such as chlorine or formalin vapor or by prolonged exposure to a variety of dusts. Rhinitis often precedes the formation of benign growths or polyps. Other responses of the nasopharynx include ulceration and cancer, which can be produced by some gases or particles (Barrow, 1986).

The oral and laryngeal regions contain the oral cavity, oral pharynx, and larynx. Irritation and infection can occur in the nasal cavity, oral cavity, pharynxes, and larynx. Doty et al. (2004) published a review of a large number of assessments for upper respiratory tract irritancy including anatomical, physiological, and psychophysiological (relating to olfaction and other sensation). Diseases of the nasal cavity and methods of evaluation are also covered in *Toxicology of the Nasal Passages* (Barrow, 1986).

Tracheobronchial Airways

The human tracheobronchial region begins below the larynx and includes the trachea and 16 to 20 generations of cilia-covered bronchial airways down to and including the terminal bronchioles. Interspersed in the ciliated cells are goblet cells and mucous glands that secrete protective mucus. Muscle bundles that control airway caliber are also present throughout the bronchial tree.

The responses of the tracheobronchial region to inhaled materials include constriction of muscles, hypersecretion of mucus, swelling or edema of bronchial walls, infection, and cancer. Strong bronchial constriction can be induced in asthmatics and in normal persons by a variety of inhaled substances. The most potent are usually organic materials that first produce a state of immunologic sensitization and later induce an allergic response. Representative causative agents include dusts, pollens, mold spores, animal danders, certain chemicals, and grain dust. Direct stimulation of bronchoconstriction in nonsensitized subjects occurs with sulfur dioxide, ammonia, cigarette smoke, sulfuric acid mists, a number of sulfate salts, and so-called "inert dusts" (Bates, 1989; Rom, 1992; Brooks, et al., 1995; Gehr and Heyder, 2000; Salem and Katz, 2006; Donaldson and Borm, 2007).

Bronchitis with inflammation and excessive secretion of thick tenacious mucus that may or may not be effectively removed by cough can be produced by many agents. Most commonly, viruses or mycoplasma, and less commonly bacteria, are causative factors, with common air pollutants probably playing a potentiating role. In sufficient concentrations, sulfur dioxide, chlorine, and nitrogen dioxide can lead to this condition. In the previously healthy person, bronchitis is not commonly fatal; cough and treatment with antibiotics usually produce complete recovery. A serious, often fatal form, bronchiolitis obliterans (silo-fillers disease), progresses to a condition where nodules of scar tissue completely occlude the bronchial tubes. Most commonly, inhalation of nitrogen dioxide from nitric acid spills, exposure to burning nitrocellulose, or working in silos are precipitating factors. This disease is characterized by shortness of breath, unremitting cough, and cyanosis (poor oxygenation of tissues).

The most common tumor of the respiratory tract of man, bronchiogenic (usually squamous) carcinoma, arises from the cells lining the tracheobronchial tree. Such growths usually occur in the central bronchi involving a mainstem (lobar) bronchus or its largest branches, and upon enlargement completely obstruct bronchi and lead to collapse and infection of tissues supplied by the diseased airway. Although 20 years may lapse between a single exposure or onset of prolonged exposure to a causative agent and the development of disease, once established, bronchial carcinoma invades extensively and can produce metastases, that is, growths throughout the body. A prime causative agent is tobacco smoke, but asbestos, nickel, chromium, arsenic, beryllium, coal tar, radioactive particles, vinyl chloride, and polyurethane are among the airborne agents that appear to produce the disease in humans and experimental animals (Crofton and Douglas, 1975; Laskin et al., 1976; Witschi and Last, 2003). Irritants may act as cofactors in the production of squamous carcinoma.

A less common class of bronchial tumors of man, the bronchial adenomas and adenocarcinomas, appear to derive from glands of the bronchi rather than epithelia of central bronchi per se. The causative agents are less certain than those of bronchial carcinoma. There are also a number of rare forms of bronchial tumors, including lipomas, papillomas, cystadenomas, and melanomas, which are probably not directly caused by common air pollutants. Chronic inhalation bioassays for carcinogens are described by Hahn (1999).

Pulmonary (Gas Exchange) Airways

The pulmonary or parenchymal region represents the functional gas-exchange sites of the lung. It includes respiratory bronchioles, alveolar ducts (ducts whose walls are composed totally of alveoli), and alveolar sacs, that is, terminating alveolar ducts. The basic unit of this region, the alveolus, is a thin-walled, polyhedral sac with one face open to the air. The interior walls are covered by pulmonary epithelial cells. A dense network of small capillary blood vessels intimately surrounds the alveoli where exchange of oxygen and carbon dioxide between the air and blood occurs. Alveoli in humans are about 150–300 μm in diameter, being much larger than the inhaled particles that expose the deep lung. In addition to the alveolar epithelial (type I) cells, alveolar macrophages (mobile cells that engulf and digest foreign bodies) and alveolar septal (type II) cells (which secrete surfactant) are present. Alveoli are lined by a thin layer of surfactant that prevents their collapse and aids in fluid balance.

Responses of the parenchymal region which are produced by inhaled substances include inflammation, edema (fluid accumulation), fibrosis, and cancer. The term *pneumoconioses* refers to parenchymal diseases caused by dusts. It is perhaps best characterized by a failure of the self-cleansing mechanisms of the lung that results in the development of specific pathologic responses, depending on the type of material which is deposited and retained. A large variety and number of materials produce parenchymal diseases (Bates, 1989; Lippmann, 1992; Witschi and Last, 2003).

Gases that are relatively insoluble in water are generally not well-absorbed in the upper airways, and thus expose the deep lung. Ozone (triatomic oxygen), a component of photochemical air pollution, is a good example of gas which is sufficiently insoluble to pass beyond the tracheobronchial region to produce damage in the lung parenchyma. In concentrations near 1 ppm, ozone has many effects including killing of pulmonary alveolar epithelial cells and damaging alveolar capillary endothelium, which produces hemorrhage and edema, that is, filling of airspaces and interstitial tissues with blood and/or plasma ultrafiltrate. Even in concentrations below 1 ppm, ozone destroys cilia, interferes with the ability of phagocytic cells to inactivate microorganisms, produces edema with thickening of alveolar walls, and causes various biochemical and breathing-pattern changes (Costa, 2003). A major feature of this injury pattern is that it indicates the vulnerability of the thin alveolar epithelial cell and the closely neighboring capillary endothelial cells.

Tissue destruction followed by inflammatory changes in the pulmonary region is produced by highly irritant gases including ammonia, chlorine, and phosgene. Inhalation of such gases can lead to acute pulmonary edema, a condition that may resolve, or after severe exposures progress to fibrosis of the deep lung. Similar responses have been caused by the inhalation of a variety of dusts including cadmium, osmium tetroxide, bauxite fumes, beryllium, and a variety of organic materials including molds from hay, sugar cane, sawdust, bark, and cheese. In many of the organic dust diseases, a hypersensitivity reaction apparently plays a role in the lung response, which is then called extrinsic allergic alveolitis. In such reactions, an allergic response, involving an antibody that appears after repeated exposure, participates in the formation of biochemical products that injure cells of the lung (Crofton and Douglas, 1975; Bates, 1989).

Morphologic Evaluations

Because the respiratory system is composed of a large and varied set of tissues, one must devise careful sampling techniques for morphologic (structural) studies. The problem of performing an adequate postmortem examination and sampling of respiratory tract tissues was discussed in reviews by Dungworth et al. (1976, 1995) and Dungworth (1994). Such an examination begins with a critical gross observation of tissues before fixation for subsequent microscopic evaluation. Typically, after deep anesthetization, the exposed animal is killed by exsanguination (bleeding), and the thorax carefully opened to prevent puncture of the lungs. The lungs with the trachea attached are removed and a search made for abnormalities such as edema, adhesions to the visceral (chest wall) or pleural surface, hemorrhage, scarring, and possible tumor nodules. The upper trachea should be opened and its interior similarly examined, but enough undisturbed trachea should remain attached to permit subsequent hook-up to a fixative perfusion system. The next step according to Dungworth et al., involves tying off major blood vessels, dissecting them and the heart away, and weighing the lungs. This weight can be used to quantify possible edema or hemorrhage.

The nasal region, pharynx, and larynx may be opened, carefully examined, and pieces taken for fixation and further preparation for subsequent microscopic examination. In this phase of the examination, the locations and character of sites of inflammation, of tumors or nodules, and other abnormalities, are recorded.

Fixation (preservation) of tissues, including the complete lungs is performed prior to embedding, sectioning, mounting, staining, and microscopic examination. The latter preparative steps are part of the histologist's armamentarium and are too specialized to be thoroughly covered here. Rasmussen (1997) can be consulted in order to gain a better appreciation of such techniques. Fixation of the lungs must be done properly in order to provide maximum detectability of abnormalities and to permit quantitation of any abnormalities seen using a microscope. Artifacts of fixation include underfixation, uniform and nonuniform shrinkage, and formation of precipitates. Various lung fixation methods that can be considered include: air drying; freeze drying; freeze substitution; formalin vapor inflation; intratracheal infusion of formaldehyde, gluteraldehyde, or another solution; and vascular infusion of fixative solutions. Although each technique suffers from artifacts, some have uniform and reproducible artifacts, which allow for accurate correction of subsequent morphometric data. Such corrections are not usually necessary for qualitative pathologic examinations.

Intratracheal infusion of a fixative, such as 10% neutral buffered formaldehyde, is commonly used to fix lungs for morphologic analysis. This method will serve as an example suitable for quantitative studies. More rapid fixatives, such as gluteraldehyde can lead to fixation that is too rapid, resulting in a greatly underinflated organ. Hayatdavoudi et al. (1980) compared final lung volumes after intratracheal fixation with either gluteraldehyde or formaldehyde solutions using various instillation schemes. Several of their findings are noteworthy. Formaldehyde (10% formalin, pH 7.4 using potassium phosphate buffer) instilled at a hydrostatic pressure of 20 cm of fixative gave adequate fixation of rat lungs as defined by loss of tissue recoil after 72 h. Increasing the hydrostatic pressure head of fixative did not lead to significantly increased volumes in fixed lungs. The fixed lung volumes after 72 h fixation at 20 cm pressure was 83% of the total lung capacity. Thus, good inflation, perhaps even greater than that occurring at end inspiration in the living animal, can be achieved.

For morphometric study purposes, it is essential that the degree of inflation of fixed lung, be reproducibly controlled. In addition, the fixation pressure should not be so low as to permit closure of airways, nor so high as to produce excessive distension. A device described by McClure et al. (1982) has been used for constant pressure fixation of lungs by tracheal infusion of fixative. As shown in Figure 6.1, the apparatus has three fixative reservoirs. The top reservoir serves as the source of fixative solution for the middle reservoir which, via a float-regulated inlet, maintains a constant pressure head above the lungs.

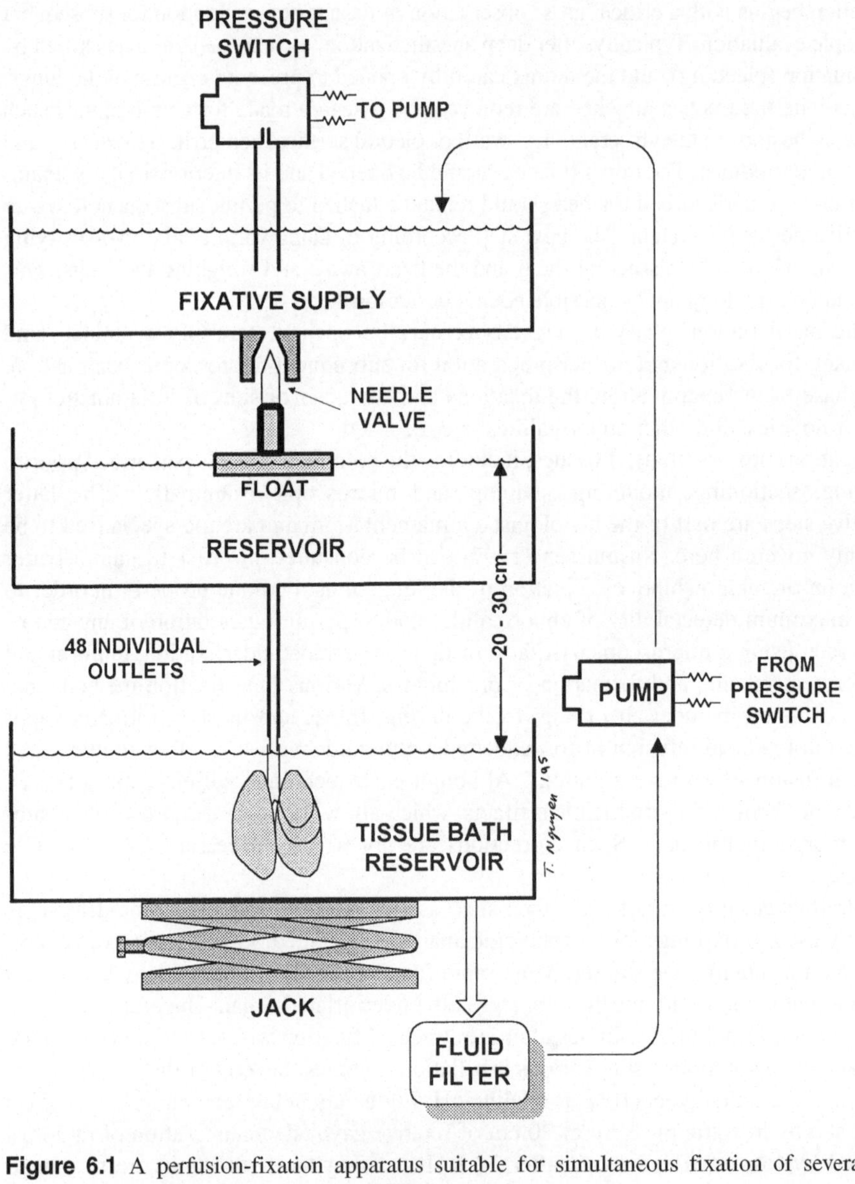

Figure 6.1 A perfusion-fixation apparatus suitable for simultaneous fixation of several lungs. *Source*: Redrawn from McClure et al. (1982).

The lungs sit in the lower reservoir which is fed by seepage of fixative out of the lungs. A fluid-level actuated pump and filter system returns fixative from the lower reservoir to the upper one. An apparatus such as this should be used inside of a ventilated hood to prevent exposure of laboratory personnel to fixative vapors.

Sampling of the fixed lung is complicated by two factors. First, the lung is a heterogeneous organ that varies considerably in structure with depth into its interior. Second, many morphologic responses to inhaled materials occur with a spotty distribution within the lung. These two factors lead to the necessity for both acquiring a relatively large number of distributed samples for detailed study, as well as the need to obtain large sections for survey examination.

The trachea should be sampled at several sites to include upper and lower regions as well as portions of the front and rear walls, which are structurally distinct. The main bifurcation of the trachea (carina) should be sampled when particle inhalation has occurred as this area is likely to be a site of heavy deposition. Bronchi can be sectioned longitudinally, provided care is taken in orienting the lung lobes. Thus, views of these airways at various generational levels can be provided.

Unless a whole lung (right or left) is needed for other types of studies (e.g., biochemical), one should obtain histologic samples from the parenchyma of both sides. Sampling should include several sites, proximal (near the lobar origin), intermediate, and distal (near the tips or edges of lobes).

In addition to microscopic evaluation of fixed, stained sections, a variety of other techniques are available for examining morphologic injury:

- histochemistry and cytochemistry, in which special stains are used to mark enzymes, intracellular components, and a variety of tissues and biochemicals;
- autoradiography for tracing cell transformations including replication, differentiation, and death, as well as metabolic pathways;
- freeze-fracture methods for obtaining samples for transmission and scanning microscopy;
- tracer methods such as use of instilled or injected horseradish peroxidase, hemoglobin, or ferritin, for visualizing vascular or epithelial permeability.

The above techniques are briefly discussed with references in the aforementioned papers by Dungworth et al. (1976, 1995) and Rasmussen (1997). Such morphologic techniques are often essential components of a thorough evaluation of inhaled materials. Therefore, the input and collaboration of persons experienced and trained in morphologic techniques are useful.

MORPHOMETRY

Quantitative anatomical measurement techniques (morphometry) for the lung have become generally available through the publications of Weibel and associates (1963a,b, 2007). These methods, aimed at correlating structure with function, include tissue preparation, sampling, and theoretical and practical aspects of morphometry of lung airways and tissues.

Morphometric measurements made on sections of fixed lungs provide quantitative, statistically testable parameters that can be used to document lung damage or relate structural changes to alterations in lung function. Morphometric techniques in general are too varied and involved for comprehensive detailed description here. However, estimation of pulmonary lung surface area, using the mean linear intercept method, will be summarized as an example. (An automated method has been described by Levine et al., 1970.)

A line of known length, L (e.g., 1 mm), is placed randomly on lung sections and the number of alveolar septa, M, touching the line is recorded. This process is repeated N times and the mean linear intercept, L_M, computed by the following formula.

$$L_M = \frac{N \bullet L}{\Sigma M_i}, \; i = 1 \text{ to } N \qquad \text{(eq. 6.1)}$$

If N is sufficiently large, and measurements are taken at random locations throughout the lungs, then the internal alveolar surface area, S, is calculated using the following equation:

$$S = \frac{4 \bullet V_L}{L_M} \qquad \text{(eq. 6.2)}$$

where V_L is the inflated volume of the alveolar portion of the lung. In practice, the "line" used to tally alveolar crossings consists of two line segments (Fig. 6.2) perpendicular to one another. This eliminates bias due to tissue dimensional distortion that occurs during the cutting of sections.

Determination of the lung's alveolar volume, V_L, can be made on the fresh or fixed organ. If a fixed lung is used, a correction for shrinkage may be necessary. The total lung volume can be determined by the amount of fluid it displaces when completely immersed, or alternatively by measurement of the buoyant force on the immersed lung as described by Scherle (1970). According to Weibel (1963a,b) for human lungs, the total lung volume multiplied by 0.9 yields the alveolar volume.

When making morphometric measurements on laboratory animals, one should use well-matched controls, for example, litter-mates. The control lungs should be fixed, sectioned, and measured simultaneously with the exposed specimens, and at all steps the

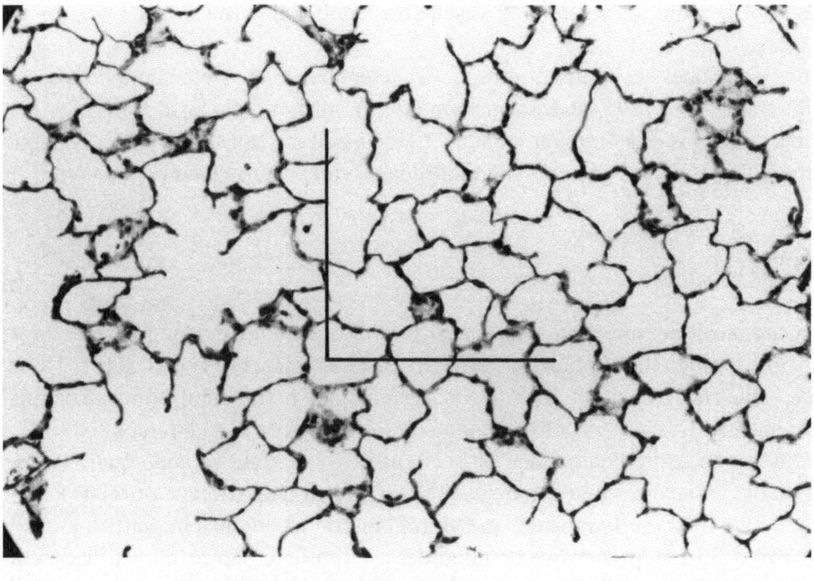

Figure 6.2 A standard length line superimposed on a lung tissue slice for morphometric measurement of mean linear intercept.

hands-on work should be performed blind, that is, without knowledge of which treatment has been given to any specimen.

When properly performed, morphometric measurements can be among the most precise and sensitive tools available to the inhalation toxicologist. Other parameters treatable by morphometric methods include alveolar wall thickness, capillary diameters, alveolar numbers; also volumes, sizes, and numbers of tubular airways (Haefeli-Bleuer and Weibel, 1988; Hyde et al., 1994, 2007).

PULMONARY FUNCTION

Function testing provides information that can often be directly related to an adverse health effect. The most obvious aspect of lung function is gas exchange; however, the respiratory tract is functionally active in a variety of ways not directly involving gas exchange (Heinemann and Fishman, 1969; Gardner et al., 1977, 1988; Parent et al., 1991; McClellan and Henderson, 1995; Gardner, 2006). Nonrespiratory functions of this organ include immunologic, metabolic, excretory, sensory, and trapping of blood emboli, to list only a few. In fact, the present state of understanding of lung function indicates that many new functions await discovery.

Reviews of lung function testing from the point of view of the inhalation toxicologist have been published (Wilson et al., 1976; Murphy, 1994; Mauderly, 1995; Frampton and Utell, 1999; Rennard and Spurzen, 2006). The reviews cover pulmonary function testing in humans, unanesthetized, large laboratory animals and anesthetized, large and small laboratory animals. The authors stress the need for adequate pre-exposure testing so that changes due to inhalation exposure can be accurately determined. Also, a sequential approach to pulmonary function testing is offered. In this approach, screening tests, which are relatively rapid and inexpensive, are performed first, and on the basis of these tests more elaborate confirmatory testing is performed. Sample pulmonary function tests are listed in Table 6.1.

In addition to those tests listed in the table, stress testing is often useful. In such tests laboratory animals or human subjects can be exercised on treadmills (Mautz, 2003), wheels, cycle ergometers, or by swimming. Other stress tests involve voluntary maximum breathing maneuvers and breathing through an added dead space, such as an open-ended tube.

In addition to pre-exposure measurements, the use of sham exposed subjects is recommended in order to evaluate the influence of handling, measurement, and exposure procedures on pulmonary function. Pulmonary function can change as a consequence of excitement, inactivity, environmental odors, noise, and other factors.

Because most pulmonary function tests are nondestructive, they are useful for comparative studies in which the responses of laboratory animals are related to those of humans. Function testing may be performed on essentially all mammals, and a great number of tests have been developed for (or adapted to) large, medium-sized, and small laboratory animals. Examples of techniques for ponies are given in papers by Garner et al. (1971) and Mauderly (1974a). These authors point to the manageability of the unanesthetized pony as well as its cardiopulmonary similarities to humans.

The dog has been widely used in pulmonary function studies. As with the pony, the unanesthetized dog is manageable and adaptable to a variety of function measurements (Dubin and Westcott, 1969; Mauderly, 1972, 1974b; Banchero et al., 1979; Haskins et al., 2005; Talavera et al., 2006). The long history of the dog in inhalation studies has led to a valuable database on normal values, and an extensive array of testing methods.

Table 6.1 Pulmonary Function Tests for Use in Inhalation Studies

Category	Test	Physiological interpretation	Limitations
Ventilatory exchange	Respiratory rate	Frequency of breathing	Very few
	Tidal volume	Volume of breathing	Very few
	Minute ventilation	Total inspired or expired volume per minute	Very few
Static lung volumes	Total lung capacity	Elasticity of lungs and thorax, muscle strength	Requires maximal effort or external forcing pressure
	Vital capacity	Elasticity of lungs and thorax, muscle strength	As above
	Residual volume	Elasticity of lungs and thorax, muscle strength	As above
	Expiratory reserve volume	Expiratory force, diaphragm position	As above
	Functional residual capacity	Elasticity of lungs and thorax	Almost none
	Inspiratory capacity	Inspiratory force, diaphragm position	See total lung capacity
Respiratory mechanics	Static lung and thoracic cage compliance	Stiffness of respiratory system	Requires cooperation, or anesthesia plus external forcing
	Static lung compliance	Stiffness of the lungs quasistatic	As above
	Static thoracic cage compliance	Stiffness of chest wall	As above
	Dynamic lung compliance	Stiffness of lung at specified frequency	Usually requires cooperation, or anesthesia plus external forcing
	Static volume pressure curves of saline-filled excised lungs	Tissue distensibility	Problems of leakage and nonuniform filling plus excision
	Spirometry forced, expired volume vs. time	Overall mechanical function	Usually requires cooperation, or anesthesia plus external forcing
	Flow-volume maximum expiratory flow volume curves	As above	As above
	Flow-volume inspiratory maximum inspiratory flow volume curves	As above	As above
	Lung and thoracic cage flow resistance	Changes in total respiratory system	Specificity of interpretation is limited
	Total lung flow resistance	Flow resistance of airways and lung tissue	Requires cooperation, or anesthesia plus external forcing
	Airway flow resistance	Flow resistance of airways	Requires panting

Continued

Table 6.1 Pulmonary Function Tests for Use in Inhalation Studies—cont'd

Category	Test	Physiological interpretation	Limitations
	Small airways flow resistance	Partition of flow resistance to airways usually of 2–3 mm diameter or less	Requires invasive procedure
	Work of breathing	Work of moving lungs, thoracic wall, or both	Results uncertain without anesthesia and paralysis of respiratory muscles
Distribution of ventilation	Closing volume	Closure of dependent airways	Requires cooperation, or anesthesia and forced breathing
	Single breath nitrogen washout	Distribution of ventilation	As above
	Multibreath nitrogen washout (or ^{133}Xe)	As above	As above
	Regional pulmonary function, ^{133}Xe technique	Topographical distribution of ventilation	As above
Pulmonary circulation	Cardiovascular pressures	Detects hyper- or hypotension in vascular system	Animals should be studied under similar conditions and activity levels
	Cardiovascular volumes, flows resistance, and work	Cardiovascular performance	As above
	Distribution of perfusion	Regional distribution of blood in the lung	Useful only in larger animals, requires radioisotopes
	Right to left pulmonary vascular shunt during O_2 breathing	Percent of the cardiac blood output bypassing ventilated portions of the lung	Requires very accurate O_2 measurements in mixed venous blood and alveolar air
	Matching of ventilation and perfusion	Regional distribution of lung ventilation relative to perfusion	Useful on larger animals, uses radioisotopes
	Histamine, fibrinopeptide B, bradykinin analysis	Analysis of concentrations of vasoactive agents	Requires availability of surgical and analytical techniques
	Edema evaluation by various methods, radioisotopic, wet/dry weight ratios, and gas equilibration	Estimate of extravascular fluid accumulation in the lung	Various methods require animal sacrifice, radioisotopes, or indirect evaluations
Regional ventilation/perfusion matching	Several methods available	Evaluation of ventilation and perfusion in lung subunits	Various methods require arterial blood and alveolar gas measurements, radioisotopes, and cooperation or anesthesia

Continued

Table 6.1 Pulmonary Function Tests for Use in Inhalation Studies—cont'd

Category	Test	Physiological interpretation	Limitations
Diffusion	Several methods available	Evaluates alveolar membrane thickness and pulmonary capillary status	Usually requires cooperation or anesthesia
Blood gases	Several methods available	Total alveolar ventilation	Requires access to appropriate blood vessels

Source: Excerpted from Wilson et al. (1976).

With respect to small laboratory animals, a variety of pulmonary function techniques have been successfully applied to mice, rats, hamsters, guinea pigs, and rabbits. Chief difficulties involved in the use of such animals are associated with their small size and lack of ability to cooperate. These animals have relatively fast respiratory rates and small lung volumes, both of which can tax the limits of common measurement systems. Measurements have been made on small animals under various conditions of consciousness by several investigators (Thomas and Morgan, 1970; Amdur et al., 1978a,b; Costa et al., 1991; Mautz, 1997). A compact review of published respiratory measurements in these species by Likens and Mauderly (1979) summarizes the techniques available, the types of measurements, and normal values that had been previously reported. Schaper (1993) reviewed the relationship between changes in respiratory rates of mice and occupational air concentration exposure limits for 89 chemicals. A high correlation ($R^2 = 0.78$) was seen, validating the model.

OTHER ENDPOINTS

Pulmonary Defense

The lung must constantly maintain a variety of active and efficient defense mechanisms. Such defense mechanisms include changes in breathing pattern, mucus secretion, bronchoconstriction, cough, foreign body engulfment, killing of viable organisms, and physical clearance of deposited materials by mucociliary action. Quantitative assessments may be made for each of these defense mechanisms. Two examples, the inactivation of infectious agents and the physical clearance of deposited materials, will be briefly discussed.

Studies of clearance, killing, or inactivation of infectious organisms in animals exposed to airborne pollutants have a long history (Green and Goldstein, 1966; Coffin et al., 1968; Gardner et al., 1977; Goldstein et al., 1978). Goldstein et al. (1978) looked at the effect of ozone (2.5 ppm for 5 h) on the ability of rat alveolar macrophages to inactivate inhaled *Staphylococcus aureus.* In this study, groups of 24 chronic, respiratory-disease-free Sprague Dawley rats were infected by aerosolized *S. aureus.* Eight rats were killed immediately for assay of viable bacteria in the lungs and measurement of lysosomal (capable of cell digestion) enzymes in pulmonary macrophages. Half of the remaining animals were exposed to clean air. Exposed rats were then killed after the exposure period for histochemical and histologic determination of the locations and viabilities of bacteria, and for enzyme activity determinations. An additional 15 control rats were exposed to *S. aureus,* but not ozone, and the rates of bacterial ingestion by macrophages were measured 0, 2, and

4 h after infection. In the bacterial inhalation phase, which lasted 20 min, each rat inhaled more than 10^5 bacteria. The fraction of bacteria engulfed by macrophages of control animals increased with time from about 35% at 0 h to about 87% and 92% at 2 and 4 h, respectively. Animals undergoing 5 h of clean air exposure before being killed had cleared about 70% of the deposited bacteria from the lungs. In contrast, ozone-exposed animals exhibited negative clearance (bacterial proliferation), having more than twice as many bacteria present at killing than were originally deposited. Also, ozone-exposed animals had a significantly smaller percentage of bacteria engulfed by macrophages than did animals exposed to clean air. Ozone exposure also was associated with an absence of enzyme activity in those macrophages that contained bacterial microcolonies. Enzyme activity was not depressed in those macrophages without ingested bacteria. Thus, enzyme impairment was associated with inability to kill inhaled bacteria and ozone exposure led to such enzyme impairment. Recently, Gilmour et al. (1993) reviewed the mechanisms by which ozone can suppress bacterial defenses, emphasizing the effect on alveolar macrophages. Cohen (2006) reviewed the extensive related topic, pulmonary immunotoxicology.

Another type of study, in which the physical clearance of inhaled inert tracer particles is compared in control and pollutant-exposed subjects has been pursued by several investigators. As an example of this type of study, Kenoyer et al. (1981) investigated the effect of ozone exposure on the early and late clearance rates of radiolabeled tracer particles inhaled by rats. Groups of up to 30 rats were exposed, nose only, to an aerosol of radiolabeled, monodisperse, polystyrene latex microspheres for 20 min prior to exposure to either clean air or ozone. After 4 h of chamber exposure, feces were collected periodically for 2 days and the thorax regions were gamma-ray counted, in vivo, for up to 17 days. The fecal radioactivity excretion curves were analyzed to determine a half-time for the early clearance phase. The thoracic activity data were analyzed to determine the rate of clearance of material remaining after 2 days. Ozone was found to delay short-term clearance after a 4 h exposure to 0.8 ppm or greater, and to accelerate longer term clearance at the same exposure levels. Presumably, inactivation of cilia contributed to the early particle clearance impairment, and an influx of macrophages probably produced the acceleration of material cleared between day 2 and day 17. At killing 30 days later, ozone-exposed animals had less radiolabel in their lungs than did control animals. Such studies in which tracer particle clearance is quantitated provide insight into the defense mechanisms that rid the lungs of particulate materials (Albert et al., 1974; Schlesinger et al., 1979, 1997; Newton, 1995).

Lung Development

Birth and the transition to air breathing are associated with major physiological and anatomical changes in the mammalian respiratory system, but subsequent events in the continuing differentiation and development of the lung are equally significant to normal lung function. The pattern of lung development postnatally appears to be similar in many mammalian species including humans, dogs, rats, mice, cats, and rabbits. Although most investigators conclude that the full number of tubular airways are present at birth, there is typically a significant increase in the number of alveoli throughout the period of early maturation after birth. Lung development was reviewed in Chapter 2.

The question of possible interference with the sequence of alveolar development by environmental, disease, or toxic factors has been addressed (Emery, 1970; Thurlbeck, 1975, 1977; Burri and Weibel, 1977; Reid, 1977; Harding et al., 2004). In summary, development of a full complement of alveoli may be disturbed by infection, congenital diaphragmatic hernia, childhood onset kyphoscoliosis, and hyperoxia. Few studies have been conducted on the potential effects of common airborne pollutants on postnatal lung development. Bartlett et al. (1974) applied morphometric techniques to the lungs of 3- to 4-week-old rats that were continuously exposed to 0.2 ppm ozone for 30 days. Although exposed animals had a significant increase in lung distensibility over controls, the lung weight, alveolar number, and histologic appearance were not altered significantly. Freeman et al. (1972) reported that continuous, lifetime exposure of rats to 10–20 ppm NO_2, which were initially 1 month old, led to loss of alveoli and alveolar surface. Greater volumes of fixative were required to fill the NO_2-exposed lungs to 25 cm pressure. However, the effects of direct destruction of already-formed alveolar tissue by this gas obscured any potential interference with the sequence of alveolar development. Phalen et al. (1986) exposed groups of 6-week-old Beagle dogs to clean air, 1 ppm or 2 ppm ozone, at 4 h/day for 5 days. Seven animals from each exposure were euthanized at age 12 weeks for morphometric analyses. A small increase in alveolar mean linear intercept was seen in the 1-ppm ozone group. The small effect was unlikely to have any physiological significance.

Airborne oxidants such as ozone and NO_2 have marked effects on the transitional zone between terminal bronchioles and alveolar ducts as well as on alveolar septae. For this reason, these and similar gases should be examined more closely with respect to potential interference with postnatal lung development. Documentation of the effects of ozone, for example, on this distal airway region is abundant. A study of Zitnik et al. (1978) illustrates the effects of relatively low levels of this gas. Mice were exposed to 0.5 ppm ozone continuously for up to 35 days. In a 7-day exposure sacrifice group, minimal changes were seen in bronchiolar epithelium, but the "proximal alveoli of alveolar ducts contained accumulations of alveolar macrophages, and interalveolar sepatae were thickened by accumulations of mononuclear cells . . . ." It is clear that ozone's damage to the lung includes the same anatomical region in which new alveoli are developing in the young mammal. Additional studies on the effects of inhaled substances on alveolar development are needed.

Behavior

Inhaled materials can influence behavior in several ways, avoidance of exposure being the most obvious protective response. Studies of air pollutants in the laboratory and in the outdoor environment indicate that behavioral modifications seen in humans include increased or decreased aggression (Rotton et al., 1979), alterations in interpersonal attraction (Rotton et al., 1978), and deficits in performance of some tasks (Evans and Jacobs, 1981). The mechanisms involved in modification of task performance included effects on attention, coordination, memory, problem solving, and capacity for physical work.

The quantitative methods for study of behavior in inhalation toxicology come from the field of *behavioral pharmacology*, which involves the study of the behavioral actions of drugs. A major difference between toxicology and pharmacology is the focus on harm rather than benefit. For an early reference covering the principles of behavioral pharmacology, see Thompson and Schuster (1968). The relationship between physiological and behavioral pharmacological studies is demonstrated in Table 6.2. The principles of experimentation in the physiological realm and the behavioral one are identical. If one replaces the test drug in this example with a test airborne pollutant, the application to inhalation toxicology studies is apparent. A key feature of a behavioral study is the use of quantitation when recording responses. The temptation to merely use qualtitative descriptions of behavior should be resisted.

An alternative approach to human behavioral studies comes from the field of human performance. A symposium proceedings (Folinsbee et al., 1978) contains several papers on

Table 6.2 Comparison of Investigational Procedures Used in Physiological and Behavioral Pharmacology

Experimental phase	Physiological experiment: effect of test drug on systolic blood pressure	Behavioral experiment: effect of drug on lever pressing
Preparation of the subject	Anesthetization Surgical placement of carotid artery catheter	Food deprivation Conditioning of lever pressing activity
Establishment of control baseline	Assessment of blood pressure's stability in experimental condition	Assessment of stability of lever pressing rate under reinforcement schedule
Solvent (drug carrier) control testing	Assessment of the effects of the drug solvent on blood pressure	Assessment of the effects of the drug solvent on lever pressing rate
Effects of *x* mg of drug per kg body weight	Measurement of extent and duration of blood pressure change	Measurement of extent and duration of level pressing rate change
Reestablishment of control baseline	Observation of blood pressure for its return to control levels	Observation of lever pressing for its return to control levels

Source: Adapted from Thompson and Schuster (1968).

topics including heat stress, air pollution, work physiology, cold stress, and altitude. Outcome measures used in such studies include running performance, endurance, work capacity, thermoregulation, adaptation, heart rate, oxygen consumption, etc. Elucidating mechanisms of adaptation to stress, a major thrust in human performance studies, is also useful in toxicology.

An example of a behavioral animal inhalation study is provided by Stinson and Loosli (1979). In this study, young adult, specific-pathogen-free mice, in individual exercise-wheel furnished cages, were allowed to exercise spontaneously. A revolution counter, read twice daily, recorded the number of revolutions, in either direction, of each animal's wheel. Filtered air was used to acclimate the animals to the exercise-wheel cages for 2 weeks. Half (20) of the acclimatized animals were then transferred to another exposure chamber containing air pollutants for a 2-week exposure. After exposure, the two groups exchanged chambers for another 2 weeks. Finally, the second group in the exposure chamber was placed in the filtered air chamber for 2 weeks. Wheel revolution counts between the hours of 5 p.m. and 8 a.m.—the period of controlled darkness and thus high activity—were analyzed. Each mouse's daily revolution count was divided by the mean daily count during the last week of acclimation, and the resultant number called an "activity index." The data for exposure to 0.3 ppm ozone, for 1 ppm NO_2, for 2 ppm SO_2, and for all three combined are shown in Figure 7.5, Chapter 7. All of the studied atmospheres produced an initial drop in activity followed by either a partial or complete recovery to the pre-exposure level during the pollutant exposure period. This behavioral study involved quantitation as well as a useful design for studying combinations of inhaled materials.

Weiss and Rahill (1995) reviewed behavioral measures suitable for use in inhalation studies. Among the sophisticated array of behavioral tests in animals are reduced effort, responses to stimuli, memory modification, complex task performance, reaction time, and coordination. Each test has mechanistic interpretations that shed light on the organ systems that are affected. Behavioral studies, although not currently as popular as physiological or anatomical ones, may become increasingly important in the settling of environmental and workplace air standards.

Biochemical

Normal Lung Biochemistry

Recognition of the lung as an important metabolic organ rather than simply one for gas exchange is an important scientific advance. The lung is the most richly perfused organ, other than the heart. The metabolic functions of the lung include substrate utilization, surfactant synthesis, xenobiotic (foreign substances) metabolism, and endogenous (internally produced) metabolism. Specific cell types in the lung participate to varying degrees in these functions, and each provides a variety of potential quantitative endpoints for use in inhalation studies. The endpoints selected for a given study will depend upon which are suitable for the suspected actions of the studied material.

Useful reviews on the normal biochemical functions of the mammalian lung have been published (Dahl, 1995; Fisher, 1995; Bogdanffy and Keller, 1999). *The Biochemical Basis of Pulmonary Function* (Crystal, 1976), contains several contributed chapters on lung biochemical topics. Two other edited books (Parent, 1991; Harding et al., 2004) cover many aspects of lung biochemistry, during development and aging, and comparative mammalian topics. Measuring changes in lung biochemistry is an important aspect of many inhalation studies.

Substrate utilization refers to the metabolism of substances that are used by the lung, or any tissue, for its own requirements. Glucose appears to be the major oxidizable substrate used by the lung for its energy requirements. According to Fisher (1976) about half of the glucose carbon atoms used by the lung are converted into lactate and pyruvate, about one-fourth are oxidized to CO_2, and most of the remainder are incorporated into various tissue components, including proteins, nucleic acids, polysaccharides, and lipids. Energy derived from substrate metabolism contributes to several functions of the lung including maintenance of bronchial ciliary movement, phagocytic cell activity, glandular secretion, bronchial muscular constriction, secretion of surfactant, and replication of the various cell populations. One expects impairment of all of these functions in the absence of active oxidative metabolism.

Surfactant metabolism is an essential function of the lung. The formation and secretion of surfactant is recognized as an important synthetic activity of alveolar type II cells. Surfactant, a lipoprotein-lining material, lowers the surface tension of alveoli allowing them to resist collapse under the action of surface and elastic forces (Pattle, 1965). A deficiency of pulmonary surfactant is related to an important disease state, that is, respiratory distress syndrome *(hyaline membrane disease)*, in the newborn. According to Stewart et al. (1979) a disruption in surfactant metabolism is a relatively nonspecific and sensitive response to many toxicants.

The lung serves as a major organ for transformation and detoxification of foreign materials presented to it via the airways or blood circulation. Pathways of this xenobiotic metabolism include hydroxylation (transfer of oxygen from one molecule to another) by cytochrome P-450–linked reactions (Dahl, 1995; Bogdanffy and Keller, 1999).

The lung synthesizes and transforms a variety of hormones and other agents that profoundly influence homeostasis. Lungs rapidly clear serotonin and norepinephrine (which serve as neurotransmitters and vasoconstrictors) from circulating blood. Similarly, the lung can convert angiotensin I to angiotensin II, resulting in the formation of a potent vasoconstrictive compound capable of increasing blood pressure. On the other hand, lysis of bradykinin results in a loss of the potent vasodilation and bronchoconstrictive activity of this polypeptide. The lung is also an important site for synthesis, storage, and release of prostaglandins. In addition to having regulatory effects on virtually every organ system of the body, prostaglandins appear to be capable of producing strong constriction and dilation of the blood vessels of the lung (Hyman et al., 1979).

Lung Lavage

Lavage, that is, washing the interior of the lungs with saline, is a method for recovery of lung fluids and free cells. Techniques for using lung lavage in detecting lung damage have been described by Henderson (2005). Analysis of fluid and recovered cells can be used to detect early inflammatory and other lung responses.

The value of lavage is seen in a study of concentrated ambient particles in dogs (Clarke et al., 2000). The exposures, using paired dogs in chambers were 6 h/day for 3 days. Cell counts and protein levels were measured in lavaged fluids, along with measurements in blood samples. Although the effects of air pollutants were subtle, the authors noted increased macrophages and neutrophils in the lavages, and changes in several circulating cell counts.

Detoxification, Activation

Because the lung is metabolically active, the biochemical responses to disease, injury, or other insult are varied. Most papers on the lung's biochemical responses to inhaled pollutants focus on the direct effects of individual pollutants. The lung has two important biochemical roles in inhalation toxicology: detoxification and activation.

For example, a mechanism of detoxification of lipid-soluble toxicants involves the attachment of a polar, reactive chemical group, which then combines with another chemical manufactured in the body to form a water-soluble conjugation product that is efficiently excreted (Dauterman, 1980). On the other hand, reactions such as methylation (addition of a methyl group) that generally produce transformed compounds of lower water solubility, but sometimes lower toxicity, can also be classed as detoxification reactions, according to Dauterman.

Biochemical transformation of inhaled materials can also result in the creation of forms having increased toxicity. A review by Boyd (1982) on metabolic activation of lung toxicants describes three prototype mechanisms of lung injury involving metabolic activation of foreign blood-borne substances. Mechanism I involves the activation of an "inert" substance to a lung toxicant *in situ* (in the lung itself). Mechanism II involves activation of the potential toxicant in the liver, followed by transport via blood to the lung where the critical tissue damage occurs. In mechanism III, the initial compound is alternately oxidized and reduced cyclically, leading to the depletion of important energy stores and/or the formation of reactive oxygen-containing compounds. In this third mechanism, both the loss of an energy store and the formation of reactive molecules can lead to lung injury. Presumably, elevated oxygen levels in lung tissue enhance the potential for injury by mechanism III. Such prototype mechanisms provide a useful framework in which specific toxicants can be investigated.

Work on the biochemical aspects of air pollutants such as ozone, nitrogen dioxide, and sulfuric acid particles has led to important insights into the possible mechanisms by which they adversely affect whole animals. Coffin and Stokinger (1977) reviewed the postulated biochemical mechanisms for ozone toxicity. Among these mechanisms are: (1) release of histamine due to nonspecific tissue injury; (2) oxidation of important sulfhydryl group–bearing compounds; (3) oxidation of polyunsaturated cell membrane lipids; (4) formation of active free radicals; and (5) formation of compounds that have neurochemical activity. The relative importance of these varied mechanisms in injury to whole animals is not yet clear. However, such mechanisms can be related in specific cases to events such as protection against injury (as by agents such as antioxidants, antihistamines, and free-radical scavengers), production of lung edema, and even development of tolerance upon repeated exposure (by induction of increased levels of protective biochemicals).

Several other examples of specific biochemical events following exposure to inhaled materials could be given. However, the major points in this section can now be made. The main importance of biochemical studies lies in their potential for identification of the mechanisms that underlie biologically important phenomena in inhalation toxicity. An understanding of biochemical events is also essential to the selection and validation of laboratory models. Further, biochemical knowledge is often a part of the foundation upon which prevention, and treatment, of injury and disease is based.

Extrapulmonary Responses

Although inhalation toxicologists have primarily focused on respiratory tract injury, the fact that the lungs are centers for the systemic distribution of inhaled materials should not be overlooked. The specialty *inhalation toxicology* covers all of the adverse effects associated with inhaled materials. Every organ and tissue in the body, including the embryo and fetus can be affected by inhaled substances. For rapidly dissolving substances, inhalation can mimic intravenous injections, but with less risk of blood clotting, infection, and direct vascular damage. Even slowly dissolving, inhaled substances can enter the blood stream, lymphatic system, and/or gastrointestinal tract, and achieve wide distribution. The number of possible extrapulmonary effects of inhaled aerosols and gases is enormous. For an appreciation, skim through a medical physiology review book (Ganong, 1999) with inhalation in mind. Viau and Robinson (2006) reviewed the testing of inhaled pharmaceuticals for reproductive toxicity. In such studies, rodents, rabbits, and dogs are commonly exposed nose-only, or in chambers. Among the endpoints used are: fertility (at least 30 endpoints are used in rats); early embryonic development; pre- and postnatal health, teratology; and pediatric measures (using dogs). Although such studies, which are required for drug development, are expensive, the potential cost of not performing them may be much greater.

A dedicated issue of the journal *Inhalation Toxicology* (vol. 16, nos. 6–7, 2004) contains 16 peer-reviewed papers on "non-pulmonary effects of inhaled particulate matter." Gong et al. (2004) examined heart rate variability in asthmatics and normals exposed to concentrated air pollutants, and Urch et al. (2004) investigated arterial vasoconstriction in subjects exposed to such pollutants. Vedal et al. (2004) looked at cardiac arrhythmias in patients environmentally exposed to natural air pollutants. And Oberdorster et al. (2004) examined the translocation of inhaled, ultrafine carbon particles to the brains of rats. These are only samples of the extrapolumonary research papers in the dedicated issue. Such research is on the leading edge of inhalation toxicology.

CONTROLS

Controls are generally of two types: untreated *(procedural controls)*; or treated with a material with known effects *(positive controls)*. The untreated, that is sham-exposed subjects, are necessary in order to ascertain the degree of change produced by an exposure regimen. Control group subjects should be closely matched to the treated groups. For experimental animals, this means that they should come from a single population and have the same treatment, except for administration of the test material, as the experimental group. Here, "same treatment" implies identical housing, handling, placement in apparatus, evaluations, anesthesia, and even in some cases administration of an "inert" substitute study material. Thus, a procedural control should go through all of the experimental procedures, without actual exposure to the studied material. Ideally, personnel handling animals or making measurements on them should also be "blind" with respect to whether the animal is in a control or experimental group.

In some cases, controls are selected from a separate batch of animals or kept in a housing area *(cage controls)* rather than put into the exposure apparatus. In such cases, the investigator runs the risk of lack of acceptance of the findings, especially when the effects of treatment are small with respect to normal variability. Thus, cage controls cannot be recommended.

Positive controls, that is, subjects exposed to a material with known documented effects, are usually necessary to initially validate an experimental procedure. Once it has been demonstrated by positive controls that the experimental methods are capable of reliably detecting the sought response, such controls are only periodically necessary. Positive controls are also useful for comparing a known toxicant to one that is unknown, or previously unstudied, using the methods at hand. Such positive controls give a good indication of the relative toxicity of a studied material.

The quality of control experiments is one measure of the skill of the investigator as well as being a factor in the peer review of grant applications and scientific manuscripts.

BATTERIES OF ENDPOINTS

Whenever possible, multiple endpoints should be used in toxicity studies. The reason is that use of a single endpoint can lead to improper conclusions regarding the potential hazard of a material. It is possible, for example, for a material at a given concentration and exposure duration to have no measurable influence on pulmonary function and still impair important functions, such as defenses against infection. It is not uncommon for investigators to describe a "no effect level," that is, a threshold dose, without a reminder that this conclusion does not necessarily apply to unstudied effects. Such reporting may be understood by researchers but can mislead others.

Selecting endpoints in a toxicological study is a matter requiring sound judgment and experience. For inhaled materials, prior knowledge of the areas of deposition within the respiratory system is of great value in selecting endpoints. A relatively water-soluble gas such as sulfur dioxide is expected to be largely absorbed within the upper airways, nose, mouth, and trachea. Therefore, endpoints that might respond to inhalation of this gas alone include: (1) upper airway cellular morphology, replication rates, or population kinetics; (2) mucus production rate; (3) upper airway tracer particle clearance rates; (4) susceptibility to upper respiratory infections; (5) reflex changes in breathing pattern; (6) bronchoconstiction; and (7) avoidance behavior. On the other hand, an exposure to sulfur dioxide alone is not as likely to alter the following: (1) alveolar cell morphology, replication rates, or population kinetics; (2) deep lung clearance rates; or (3) susceptibility to lower respiratory tract infections. The foregoing example is not intended to imply that toxicologists should not test for unexpected events, they should. However, if one has limited funds, or is in need of a responsive endpoint (e.g., when studying the influence of a cofactor on the toxicity of a primary agent), a limited set of reliable, appropriate endpoints is accepted practice.

When a previously unstudied material or combination is used in an inhalation study, one may wish to consider a balanced battery of endpoints. "Balanced" refers to the inclusion of endpoints that cover several important pheomena. Failure to pursue a balanced set of endpoints increases the probability of inadvertently missing important responses. A timely example of a balanced approach to toxicity testing relates to the recent *nanomaterials* revolution. Nanomaterials are those having one or more small dimensions (<100 nm) and have significant applications in medicine, dentistry, engineering, chemistry, physics, transportation, recreation, and consumer products. The European Centre for Ecotoxicology and Toxicology of Chemicals (ECETOC), held a workshop in 2005 to consider testing strategies to evaluate the safety of nanomaterials (Warheit et al., 2007). In addition to recognizing the need for improved characterization of nanomaterials, several safety tests were recommended. The tests included: (1) identification of appropriate *dose metrics* (size, shape, dissolution characteristics, surface area, etc.); (2) cellular uptake; (3) inflammation; (4) cytotoxicity; (5) histopathology; (6) extrapulmonary effects; (7) deposition patterns in the respiratory tract; (8) skin penetration; and (9) susceptible subpopulations. In a recent review, Stern and McNeil (2008) reviewed research on nanotechnology-related materials. Citing over 160 references, the authors highlighted the importance of surface characteristics, and the potentially significant pulmonary toxicity of some nanomaterials. It appears that a 'material-specific' approach to evaluating emerging nanomaterials is required. Evaluating the potential adverse effects of promising new materials, using realistic exposure scenarios, represents an important and challenging endeavor (Tsuji et al., 2006). For a legal perspective on regulating nanotechnology, see Lin (2007).

7

Experimental Designs

INTRODUCTION

By tradition, the design of scientific experiments has been the responsibility of the principal investigator and colleagues selected by him or her. This task, which requires proper training, experience, and insight, is one of the key elements in scientific investigation. The principal investigator is also responsible for the proper conduct of the study, evaluation of the data, and dissemination of the results. A published study will add to or detract from the investigator's status as a scientist. There are no hard rules to follow when designing scientific studies. A good design may support or contradict a hypothesis, or lead to new insights. Experiments that accomplish this can utilize or ignore any of several techniques including statistical analysis, matched controls, quantitation, repetition, etc. Whether or not a design is good or bad depends upon the quality of its fruit; the degree to which it contributes to useful knowledge.

In the sections that follow, some basic elements of experimental design will be presented. These include statistical considerations, acute studies, dose–response data, repeated exposures, chronic studies, studies of carcinogenesis, mutagenesis, and teratogenesis, and studies of atmospheres that contain multiple components.

BASIC STATISTICAL CONSIDERATIONS

Two Types of Statistics

Statistics, in our context, refers to the scientific use of computational techniques for gaining new insights or for making decisions in the presence of uncertainty. Two broad categories of statistics, "descriptive" and "inferential," exist. *Descriptive statistics* are used to summarize information using means, modes, medians, ranges, standard deviations, histograms, graphs, etc. *Inferential statistics*, historically linked to gambling strategies by way of the work of Blaise Pascal (1623 to 1662), are strongly based upon probability theory. The scientist involved in inhalation studies may need to make a decision as to whether a given exposure has, or has not, produced an effect (i.e., hypothesis testing). At sufficiently high levels of exposure, this decision is usually straightforward. On the other hand, variability may obscure the changes produced by an exposure, and statistical computations become an indispensable tool.

The remainder of this section will deal with relatively elementary types of experiment in which one group is exposed to a test material and compared to a control group that has not been exposed to the material. The question then becomes: Do the collected data sets on the two groups differ? An important assumption is that the groups are comparable and differ substantially only in their exposure to the test material. This assumption must be true in order to validly apply statistical methods. In reality, strict comparability is not possible: the groups are separate and therefore not identical. Even comparing data taken pre- and postexposure on a single subject does not solve the problem. Postexposure, the subject is older and also has a different history. A group of subjects randomly assigned to one of two chambers, one having clean air and one having polluted air, are also different in ways other than those due to individual differences or exposure to the pollutant. No two chambers are truly identical in all respects. Thus, in reality, an experimenter must make the groups and treatments as identical as is practical. Certainly, random assignment can be important. A common procedure involves paring subjects by some relevant characteristic (e.g., body weight or sex) prior to random assignment to a group. In any case, the experimenter must be convinced, and convincing to others, that the pollutant exposure is the only difference in the two groups that could reasonably lead to a claimed result. Skill in accomplishing this is acquired by attention to detail, training, and experience.

Type 1 and Type 2 Errors in Hypothesis Testing

Consideration of the errors that can be made in deciding whether or not an exposure has produced an effect in a measured parameter x must be preceded by a mathematical statement of the question. This statement is either a "null" or an "alternate" hypothesis. The null hypothesis, H_o, is:

$$x_e = x_c, \text{ or } x_e - x_c = 0 \qquad \text{(eq. 7.1)}$$

where x_c, is the value (usually a group mean) for the unexposed or control group and x_e the corresponding value for the exposed group. The null hypothesis is a statement that the exposure does not produce a change in the measure of the effect. The *alternate hypothesis*, H_a, states that a change was produced, that is,

$$x_e \neq x_c, \text{ or } x_e - x_c \neq 0 \qquad \text{(eq. 7.2)}$$

In case the exposure can only change the value of x in one direction (increase or decrease), the alternate hypothesis would be,

$$x_e > x_c, \text{ or } x_e - x_c > 0 \text{ (for an increase)} \qquad \text{(eq. 7.3)}$$

$$\text{or, } x_e < x_c, \text{ or } x_e - x_c < 0 \text{ (for a decrease)} \qquad \text{(eq. 7.4)}$$

The investigator must decide whether the null hypothesis should be rejected as false and the alternate hypothesis accepted, or vice versa.

As in gambling, one is not certain which hypothesis is true. Therefore, an estimate of the error, or probability of being wrong in making a decision as to which hypothesis to accept, is computed. Two types of error may occur. It may happen that the null hypothesis is *true* and we conclude that it is *false*. In this case the mistake is a *type 1 error*. The probability of making this error is termed the *significance* level, α (sometimes the p-value), of the statistical test. Alternatively, it may happen that the null hypothesis is *false* and we fail to reject it and conclude that it is *true*. Should this occur, we have made a *type 2 error*.

The probability of making this error is often given the symbol β. The *power* of the statistical test, also a probability, is 1-β. Thus, a powerful test is one in which the probability of making a type 2 error is small.

Clearly, an investigator prefers that their experimental design is such that the type 1 and type 2 errors involved in the statistical test of the null hypothesis are small. That is, one prefers that the significance level is small and the power large. If this is the case, one is on firm statistical ground in deciding to declare whether the exposure has or has not produced an effect.

Computed values of the significance and the power are obtained from the experimental data. In general, the larger the number of subjects or number of repeated measurements made on each subject, the smaller the probability of making type 1 and type 2 errors. Practical considerations, in reality, limit the number of measurements that can be made. Thus, one ideally should decide what values for the probability of a type 1 and type 2 error are acceptable. Figure 7.1 shows a *truth table* that demonstrates the various characteristics of a statistical significance test.

Some Tests of Significance

Tests of significance are mathematical manipulations of data that yield advice on whether to reject or not reject a null hypothesis; for example, whether one should claim that an exposure has produced an effect or make no such claim. Each investigator should, at their own discretion, set the level of significance (p-value) required in order to confidently reject the null hypothesis. Once this level is set, one is, in effect, deciding on an acceptable frequency for making a wrong choice in claiming an effect. For example, one may decide that a 10% error in claiming an effect is acceptable. In this case, one has selected a level of significance, or type 1 error rate, of 0.1, and roughly 9 out of 10 times that investigator will not claim an effect when in fact there is no effect. Persons very fearful of making a false claim of effectiveness can set the type 1 error rate at, say, 1% or 0.01, but they may find that in a great number of cases they can make no claim. That is, they are permitted to claim an effect only in those cases where the experimental data are unambiguous enough that the type 1 error is only 1%. In reality very few data sets are this overwhelming, and, in fact, the investigator may be criticized for being too cautious. As a compromise, a p-value of 0.05 is often used.

The specific significance test chosen will depend upon the experimental design and the nature of the measurements. Data which can have a wide range (theoretically infinite) of values can be subjected to a variety of statistical tests of significance. A simple but

DECISION	TRUE SITUATION	
	H_o	H_a
Reject H_o	α, p: significance Type 1 Error	(1−β): power Correct Decision
Fail to Reject H_o	(1−α): Correct Decision	β: Type 2 Error

Figure 7.1 Truth table demonstrating the power and type 1 and type 2 errors in a significance test.

popular test, *Student's t-test*, can be used to compare mean values for two populations. The assumptions necessary for rigorous application of this test include random selections of groups from normally distributed populations, and approximately equal variances for the two groups. The statistic, *t*, for comparing two sets of data is computed from the mean values of experimental, $\bar{x}_e$, and control, $\bar{x}_c$, groups, the numbers of animals, n_e and n_c, and the pooled standard deviation of the two groups, s_p, using the relationship:

$$t = \frac{\bar{x}_e - \bar{x}_c}{s_p \left(\frac{1}{n_e} + \frac{1}{n_c} \right)^{1/2}} \qquad \text{(eq. 7.5)}$$

where s_p is calculated using the standard deviations of the two groups, s_e, and s_c.

$$s_p^2 = \frac{(n_e - 1)\, s_e^2 + (n_c - 1)\, s_c^2}{n_e + n_c - 2} \qquad \text{(eq. 7.6)}$$

Roughly speaking, *t* is the difference between the group means divided by the standard deviation of the difference in means. One can use this *t*-value to estimate the probability that the two means are drawn from a single population, that is, the type 1 error for stating that the groups differ.

The critical values for *t* for a given type of error are tabulated as a function of the number of degrees of freedom of the data, ν (Table 7.1). For equal group sizes, n,

$$\nu = 2(n - 1) \qquad \text{(eq. 7.7)}$$

and for unequal group sizes:

$$\nu = n_e + n_c - 2 \qquad \text{(eq. 7.8)}$$

If, for example, one selects a value of 0.05, or 5%, for the type 1 error and the number of degrees of freedom are 18, the critical *t* value is 2.10. Thus, the null hypothesis of no difference in two group means will be rejected if *t* is equal to or greater than 2.10. Roughly speaking, if the difference between the two group means is at least 2.10 times the standard deviation of the difference in the means, then we conclude the mean values differ, with a 5% uncertainty in this conclusion, provided the null hypothesis is true.

Some data are not infinitely valued and thus are not usually described by typical theoretical distributions. Distribution-free tests are used for data, including certain ranked or categorical data. Such tests are commonly called *nonparametric*. For example, the endpoint may be death of a subject; it lives or dies. There are only two values, say + 1 (lived) or 0 (died), and no other values are possible. As an example, let's say 50% of the untreated animals are known to live for a designated time and 2 of 10 of the exposed animals lived for that time. Does the exposure increase the probability of death during the designated postexposure period? Further, let's decide upon a significance level of 0.05. A simple, nonparametric test, the sign test, may be applied to the data. Table 7.2 gives critical values for the sign test as a function of the number of subjects, n, and the number of nondeaths, k. In this table, α is the level of significance. One sees that 2 survivors in 10 are not associated with an α less than 0.05. In fact, to meet this level of significance one would have to see only 1 survivor. It is clear that in this case the sign test has not been very powerful.

Table 7.1 Critical Values of t for a Given Level of Significance (α) and Number of Degrees of Freedom (ν) (Two-Tailed Test)[a]

	α			
ν	0.30	0.20	0.10	0.05
2	1.39	1.89	2.92	4.30
4	1.19	1.53	2.13	2.78
6	1.13	1.44	1.92	2.45
8	1.11	1.40	1.86	2.31
10	1.09	1.37	1.81	2.23
12	1.08	1.36	1.78	2.18
14	1.08	1.35	1.76	2.15
16	1.07	1.34	1.75	2.12
18	1.07	1.33	1.73	2.10
20	1.06	1.33	1.73	2.09
30	1.06	1.31	1.70	2.04
40	1.05	1.30	1.68	2.02
60	1.05	1.30	1.67	2.00
120	1.04	1.29	1.66	1.98
∞	1.04	1.28	1.65	1.96

[a] *Note*: the sign (+ or –) of t is ignored. *Source*: Adapted from Wine (1964).

That is, we might fear that if the exposure did produce excess deaths we might fail to detect it. The situation can be improved by enrolling more animals in the study. The study is repeated with 50 animals and, say, 15 survive. In this case, 15 is less than the critical value of 18 and the test advises us that we have less than a 5% chance of being wrong in claiming an effect on mortality.

Additional tests of significance are available to the investigator. The appropriate test will depend upon the nature of the experimental design and the data to be tested. Several general works on statistics are available: Table 7.3 gives a small selection.

The reader should be warned that for all but the simplest experimental designs, the advice of a professional *applied statistician* is recommended.

Table 7.2 Critical Values of k for the Sign Test

	α		
n	0.05	0.10	0.25
10	1	1	2
12	2	2	3
14	2	3	4
16	3	4	5
18	4	5	6
20	5	5	6
30	9	10	11
40	13	14	15
50	17	18	20

Source: Adapted from Wine (1964).

Table 7.3 Selected Statistical References[a]

Reference	Description
Fisher, R.A., Sir, *The Design of Experiments, 9th Ed.*, Macmillan Publishing Co., New York, 1971	The logic of the scientific method and mathematical arguments are discussed in this classic work on the basic statistical principles which underlie experimentation
Gad, S.C., *Statistics and Experimental Design for Toxicologists, 3rd Ed.*, CRC Press, Boca Raton, FL, 1998	A basic overview of statistical concepts, and methods employed by toxicologist
Norman, G.R. and Streiner, D.L., *Biostatistics: The Bare Essentials*, Mosby, St. Louis, MO, 1994	An elementary, yet fairly complete soft-cover text with problems and answers. Expect at least two jokes per page
Snedecor, G.W. and Cochran, W.G., *Statistical Methods, 8th or 9th Ed.*, The Iowa State University Press, Ames, IA, 1980 or 1989	Standard reference for users of statistical methods. Many of the examples and exercises are agricultural or biological in nature. The later editions were published after the original authors deaths by D.F. Cox

[a] Table prepared with the assistance of statistician T.T. Kurasaki.

Group Size

In general, the greater the number of members in the groups tested, the lower the type 1 and type 2 errors will be. Using the *t*-test as an example, we will see that the necessary group size for detection of a given magnitude of effect at a given level of significance can be calculated from a knowledge of the population standard deviation. The definition of *t* for two groups of equal size, n, and equal standard deviations, s, is:

$$t = \frac{\overline{x}_e - \overline{x}_c}{(2s^2/n)^{1/2}} \quad \text{(eq. 7.9)}$$

And, the critical value of *t* for a type 1 error of α is:

$$t_\alpha \le \frac{\overline{x}_e - \overline{x}_c}{(2s^2/n)^{1/2}} \quad \text{(eq. 7.10)}$$

Solving for n, one has:

$$n \ge 2\left(\frac{st_\alpha}{\overline{x}_e - \overline{x}_c}\right)^2 \quad \text{(eq. 7.11)}$$

This simple inequality can be used to estimate the minimum group size required for the detection of a given difference in group means at a desired level of significance. To demonstrate the use of this inequality, let's set a value of 0.05 for α. Assume that $\overline{x}_c = 10$, $s = 2.5$, and that we wish to detect a 10% change in $\overline{x}_c$, that is, $\overline{x}_e$ is either 9 or less, or 11 or greater. In order to set a value for t_α, we must make a guess about the number of degrees of freedom. Because this is essentially what we are trying to calculate, the accuracy of our initial guess will determine how many iterative calculations we will have to make before we are close enough to our final estimate of n to stop. As an initial value, assume n is 10 which gives a ν of 18. From Table 7.1, $t_{0.05} = 2.10$. Substituting values into the inequality we have:

$$n \geq 2 \left[\frac{(2.5)\,(2.10)}{1} \right]^2 = 55 \qquad \text{(eq. 7.12)}$$

The calculated value of n, is larger than our guess of 10. As a next guess for n we can try 50 and recalculate a new n. This calculation gives,

$$n \geq 2 \left[\frac{(2.5)\,(1.99)}{1} \right]^2 = 49.5 \qquad \text{(eq. 7.13)}$$

This new value for n is very close to our guess and we can end our calculation. A reasonable group size for the experiment is 50 subjects. An experiment using 50 subjects in each group is therefore expected to provide us with the ability to detect a 10% change, positive or negative, in the mean value, provided the standard deviation of the group is not significantly altered by the treatment. This calculation, made for instructional purposes, is typically performed by statistical software which is discussed in the books by Gad and by Norman and Streiner (Table 7.3).

This simple example has counterparts for more complex tests of significance, which are beyond the scope of this book.

What of our type 2 error? This estimate is complex, even for our simple example, but tables exist for estimating a type 2 error for a *t*-test. In our example, consulting such a table provides a value for β of 0.53. Thus, there is a 53% probability for the type 2 error. In other words, there is a 53% chance that were the null hypothesis in fact false, we would fail to reject it. This means that the treatment could have had an effect which we might fail to detect. This event, called a false-negative result, could be due to either the treatment producing a real (but possibly not biologically significant) change in the group mean but less than 10%, or simply a failure to detect a change of 10% or greater due to the random nature of variation in the data that went into calculating the group means. Estimation of type 2 errors is usually a task for the professional statistician or a statistical software program. Such an estimate should be made early, rather than after completion of the experiment.

The Role of the Statistician

The involvement of a statistician in an inhalation study can take the form of advisory work or collaboration. He or she most effectively serves the project when they have an understanding of the biological and medical issues, and a knowledge of the project activities. The statistician will need to acquire an accurate understanding of the nature of the data and potential sources of variability and bias. For a discussion of the general problems associated with statistical consultancy, the reader is referred to papers by Sprent (1970), and by Kenett and Thyregod (2006).

Throughout the course of a study, there are functions which the statistician can perform. In the initial stages, an experimental design will be planned and the statistician's responsibility can be to devise a mathematical formulation of the problem and a statistical plan. It may be possible in this planning phase to discover the kinds of results and confounding factors that might be forthcoming. As data are being collected, careful and continued statistical monitoring are essential. Refinements in the acquisition of data might be suggested in order to reduce errors, to make data collection more efficient, or to account for changes in study design that are suggested by early data. Finally, the analysis and interpretation of data will fall within the domain of the statistician. An attempt should be made to provide clear and convincing results stated in a language understandable not only to the investigator, but also to other interested parties. In addition, any limitations or important qualifications in the results should be clearly described.

EXAMPLES OF COMMON DESIGNS

Acute Exposures/Dose–Response Relationships

Acute exposures are usually measured in minutes or hours, times that are short in relation to the life spans of cells, tissue repair processes, infectious disease episodes, fertility cycles, gestational periods, and mammalian life spans. Thus, an acute exposure largely comes and goes while the subject is in a given state of health and development. Klaassen and Doull (1980) define an acute inhalation exposure as "... continuous exposure for less than 24 h ... ." Others consider acute exposures as lasting up to a few weeks. No quantitative definition can be given. An acute exposure experiment may be conducted for purely scientific purposes, for example, to see how a system responds to a stimulus that is briefly applied, or in order to mimic some "real world" exposure. Acute exposures are important for several reasons: the metabolism, development of injury, and repair are observable without the confounding effects of reinjury, adaptation, or sensitization. Further, acute exposures serve to simulate a variety of important occupational, environmental, and other human exposures, including accidents, medical exposures, air pollution episodes, military and police actions, and self-inflicted or substance abuse activities.

When elevated concentrations are used, one must be aware that events such as avoidance, edema formation, suffocation, and bronchial constriction may be present that would not occur at lower concentrations. Numerous examples of acute toxicity tests are found in books, such as those edited by Derelanko and Hollinger (1995) and by Massaro (1997).

One interesting facet of inhalation toxicology that bears on the design of acute studies is the area of *dose-rate effects*. If one considers total dose as the product of exposure concentration (C), minute ventilation (V_m), exposure time (T), and fractional uptake in the subject (F), one can formulate simple models that relate this dose to a given response (R). A simple model relating dose to response is:

$$R = X\,f(D) = X\,(k_1\,C^a \cdot k_2\,V_m^b \cdot k_3\,T^c \cdot k_4\,F^d) \qquad \text{(eq. 7.14)}$$

where X is a factor relating the units of response (percentage change or another quantitative measure) to the units of dose. If the superscripts are all one or zero, a linear relationship will exist between dose and response. The product of C and V_m is the exposure dose rate in units, for example, of mg/min, or rate of presentation of airborne material to the subject. If the response is independent of the dose rate, then brief high-concentration (h) exposures are equivalent to prolonged low-level exposures (l), provided that the total dose delivered to the subject remains constant, that is:

$$(C \cdot V_m \cdot T \cdot F)_h = (C \cdot V_m \cdot T \cdot F)_l = \text{constant} \qquad \text{(eq. 7.15)}$$

Studies involving ozone performed by the author and associates (Phalen et al., 1978b; Reischl, Mautz, 1982) have shown the importance of dose rate to response. In these studies, one in rats, one in dogs, the response was proportional to the concentration to the second power, that is:

$$R_{rat} = KC^2T, \text{ and } R_{dog} = K(CTV_m)^2 \qquad \text{(eq. 7.16)}$$

Thus, for a given dose, more rapid delivery produced a greater response than did slower delivery. A mechanism that can produce such an effect is overloading the rate of production of a protective biochemical. The major point is that one cannot assume a priori that acute exposures to high concentrations are equivalent to prolonged exposures at lower concentrations. More subtly, if a given dose can produce different responses, depending upon the dose rate, the concept of a dose–response relationship is greatly complicated.

One usual objective of an acute study is the development of a *dose–response* characteristic curve. The importance ascribed to this relationship is seen in a quotation from Klaassen and Doull (1980): "The characteristics of exposure and the spectrum of effects come together in a correlative relationship customarily referred to as the dose–response relationship. This relationship is the most fundamental and pervasive concept in toxicology."

The relationship between a given response and the dose is often used to establish that the suspected material, and not some spurious factor, produces the observed effect. With due caution necessitated by the earlier paragraphs, we will turn to a closer examination of the dose–response relationship. Assuming that there are no dose rate or similar complicating effects, a relationship between dose and response, such as that in Figure 7.2 can be examined.

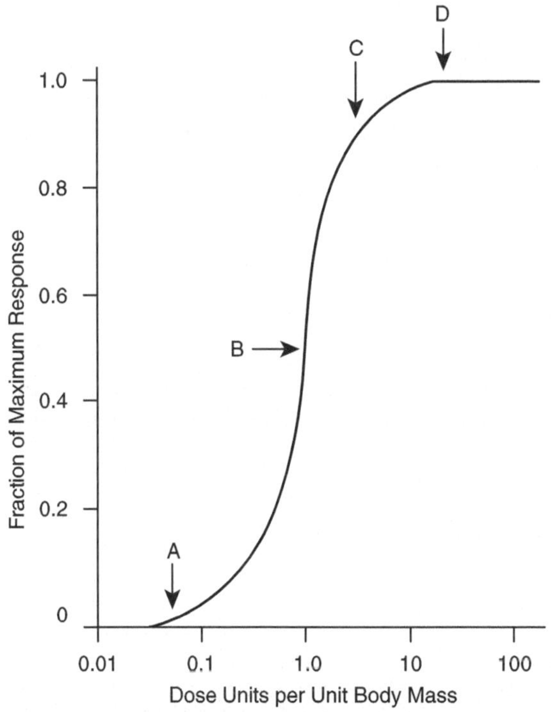

Figure 7.2 A hypothetical dose–response relationship. Regions of interest include: A, the threshold; B, the point of maximum sensitivity; C, the region of saturation of response; and D, the maximum response.

The curve could be that of an *individual subject* or the mean response for a *group of subjects*. Implied is some *maximum biological response*, beyond which increases in the dose are ineffective. Also, no uncertainty is drawn on the curve. Random errors in response and dose measurements, as well as biological variability, will complicate the dose–response curve in any real experiment.

Several points on the curve are of interest. Below point A no response is seen; point A thus represents the *threshold* or *no-effect dose*. At B, the slope of the dose–response curve is maximum. That is, a given small variation in dose about point B produces a maximum change in response. Thus, B is the point of greatest *sensitivity* for response. When designing a study to examine the effect of some ancillary factor or costress (e.g., subject age, activity, nutritional state, or ambient temperature or humidity), one might initially fix the dose at B because the subject's response is very sensitive to changes in controlled variables. At point C, the response is becoming saturated and at point D, a maximum effect is achieved. In some cases, for example, when percent lethality is the endpoint, *saturation* obviously must occur. In other cases, such as when respiratory rate is the endpoint, one also expects saturation to occur at some level. In fact, because unbounded phenomena do not usually occur in biological systems, point D can be expected to occur in most experiments.

Numerous substances, for example, trace nutrients including metals, have beneficial effects at low levels of exposure but adverse effects at higher levels. With respect to dose response, this phenomenon has been called: *biphasic response*; *low-dose stimulation*; *hormesis*; *stimulatory-inhibitory*; J- or U-*shaped response*; etc. Calabrese (2006) provided a review of *hormesis*, and the implications for risk assessment. A potential conflict in establishing exposure criteria exists when low doses are beneficial for some groups and/or individuals, but harmful to others.

Repeated Exposures/Dose Fractionation

Many human environmental exposures involve repeated inhalation. This is obvious in the occupational setting but also true for outdoor air pollution exposures and domestic indoor exposures. In fact, nearly all human exposures will have some cyclic element because of the combination of diurnal habits and differences in the quality of air found in commonly occupied settings. Such cyclic exposures involve fractionated doses. Laboratory *exposure schedules* may be designed to simulate realistic dose fractionation patterns. Although no standard schedules are universally accepted, Table 7.4 gives some examples of laboratory study schedules designed to simulate various human dose patterns.

Fractionated dose patterns are extremely interesting, yet poorly explored, in inhalation toxicology. In such exposures, the interplay of phenomena such as adaptation, sensitization, development of injury, and interrupted repair can be varied and largely unpredictable.

Table 7.4 Various Human Exposure Situations Involving Fractionated Doses, and Some Corresponding Laboratory Exposure Schedules

Human exposure	Possible exposure schedule
Occupational	7–10 h/day, 5 days/week, 50 weeks/year, 20–40 years
Domicile	10–20 h/day, 7 days/week, 50–52 weeks/year, 75 years
Vehicular	1/4 to 1 h, 2–3 times/day, 5–7 days/week, 50–52 weeks/year, 50–75 years
Environmental	2–24 h/day, 7 days/week, 52 weeks/year, 75 years
Episodic environmental	2–6 h/day, 2–20 days, 1 to several times/year, 75 years

The importance of the details of a *dose fractionation* schedule is seen in an early study involving ozone. Farrell et al. (1979) exposed 14 human volunteers to 0.4 ppm ozone 3 h/day for 5 consecutive days; control values were determined by a similar exposure regimen to clean air during the previous week. Forced vital capacity and specific airway conductance were measured in each subject after each 3-h exposure. The authors summarized their findings in the following way: "The forced vital capacity was significantly lower than the control value on the first 3 days of exposure to ozone, but there was no significant difference on the fourth or fifth days. ... All subjects were symptomatic on the first and second days of exposure to ozone. Symptoms resolved thereafter, with only one subject remaining symptomatic on the final day of exposure to ozone." *Tolerance* was seen to override the effects of repeated exposure.

Other examples yielding similar results in laboratory animals had been previously described, indicating that such tolerance is not limited to humans. Stokinger (1965), in a review on ozone toxicity, wrote: "The edemagenic response and its associated phenomena of enzyme and metabolite alterations are markedly reduced by simple interruption of the O_3 administration with exposure to air; intervals as brief as 15–20 minutes in repeated 30-minute exposures to O_3 reduced edema and mortality in laboratory animals."

The use of laboratory-derived dose–response curves for acute exposures in the experimental sequence leading to the establishment of "safe" concentration limits for human populations is discussed from the traditional toxicologic viewpoint by Klaassen and Doull (1980), and Salsburg (1981), and from the Russian point of view by Filov et al. (1979).

Chronic Exposures/Carcinogenesis, Mutagenesis, and Teratogenesis

Chronic Exposures

A chronic exposure takes place over a significant fraction of the normal life span of the subject. Definitions vary but, in general, chronic studies involve exposures lasting about 1 year or more. In such exposures, it is generally believed that there is the greatest potential for the manifestation of disease. For many exposure materials, defense mechanisms, including the development of tolerance, may be depleted and nonrepairable injury which is below the limits of detection after brief exposures may accumulate and result in loss of structure and function. Even when adaptation occurs, it is likely that the costs to the subject of such adaptive shifts will become recognized in chronic exposures. Also, material introduced via the lung can build up in other tissues and produce effects at sites distant from their point of entry. For this, and a variety of other reasons, one must be aware of the likelihood of significant extrapulmonary health effects.

Several practical and theoretical factors complicate the performance and interpretation of chronic inhalation studies. Practical difficulties include: maintaining the reliability of exposure equipment and procedures; maintaining constancy of biological endpoint assays; and handling larger numbers of subjects. These practical difficulties underscore the necessity for careful planning that includes: the availability of back-up equipment; availability of repair services; procedures for checking and maintaining long-term quality control; and adequate veterinary and routine animal support. The practical demands of chronic studies should not be underestimated.

The theoretical difficulties in performing and interpreting chronic studies are many. In a prolonged study, subjects develop and age, and the growth and developmental sequences along with natural aging pathologies are intimately intertwined with the effects of study materials. Several substances exhibit their effects only after years of exposure, and sometimes decades after exposure is terminated: Asbestos and some other carcinogens are notable for such behavior. Thus, rodents, with life spans measured in terms of a few years, are not necessarily adequate for examining the chronic disease potential of many agents. Nonhuman primates, or other long-lived species, can lead to study costs over 100-fold greater than those for rodent studies. The issue is complicated by the lack of comparative mammalian information on normal developmental, maturational, and senescent sequences (Harding et al., 2004). As a result, much of what we know about the chronic disease potential of inhaled materials has come from human experiences.

Discussions of life span and developmental considerations in the selection of animal models for chronic inhalation studies are found in treatments by Page (1977), Stara and Kello (1979), and Harding et al. (2004). Specific issues in the use of rodents in chronic inhalation studies are discussed by Fox (1977), and Boorman (1981). An especially helpful paper by Arnold and colleagues (1977) covers such topics as personnel needs and training, required daily and weekly observations on animals, intensive care, and clinical laboratory testing recommendations.

Chronic studies usually involve one or a combination of several endpoints including changes in life span, various toxicities, cancer production, and assessments of teratogenic, reproductive, and mutagenic properties of inhaled materials. Each type of endpoint carries special requirements with respect to selection of animal models, doses used, exposure schedules, and monitoring and follow-up procedures. *The CRC Handbook of Toxicology* (Derelanko and Hollinger, 1995) covers many topics related to chronic studies which have only been briefly discussed here.

Carcinogenicity

Chemical carcinogens are distinct from most other toxicants in that they typically have effects that are delayed and persistent, can often be more effective in divided doses than in single doses, and have distinct mechanisms with respect to the subjects' macromolecules, especially genetic elements (Pitot and Dragan, 2003). Carcinogens may produce malignant (life-threatening) tumors, and this tumorgenicity can manifest itself by: (1) an increased incidence of those types of tumors found in control subjects; (2) the occurrence of tumors at earlier times than in controls; (3) the production of types of tumors that are not seen in controls; and (4) the increased multiplicity of tumors per subject. Further, agents may be directly carcinogenic or can act as *cocarcinogens* or *promoters* for other carcinogens. The known mechanisms of carcinogenesis are many and varied, often involving metabolic activation of noncarcinogenic elements, and interactions with viral agents, dietary elements, and immunologic and hormonal factors.

Chronic studies of carcinogens typically involve rats, mice, or hamsters, with exposure beginning at an early age. Under meticulous care, exposures can continue over the entire lifetime of these species, but 21-month to 2-year exposures are also routinely performed. The number of animals enrolled in a study depends upon the spontaneous occurrence of cancers, the desired detection limit for increase in this level, as well as the study duration and other experimental design considerations. The advice of professional statisticians experienced in carcinogen studies is typically required before the study is begun.

Several references, including the review by Pitot and Dragan (2003), discuss carcinogenesis, carcinogens, and toxicological techniques. *Inhalation Carcinogenesis*, by Hanna et al. (1970) contains nearly 30 separate papers covering topics including the relationship between inhalation exposures and carcinogenesis, inhalation exposure techniques, cellular and physiological injury patterns, studies with carcinogens, and issues in future program planning. *Morphology of Experimental Respiratory Carcinogenesis,* by Nettesheim et al. (1970) contains over 25 contributed papers on lung structure and function, human lung tumors, classification of lung tumors in domestic and experimental animals, and current experimentation. Hahn (1995) reviewed inhalation carcinogenesis for chemical and radiological agents, including basic biological principles and rodent models.

Mutagenicity

Mutagens have the ability to alter genetic material in cell nuclei in such a way that subsequent cell division or fusion leads to transmission of the altered genetic information. Such mutations can lead to the production of cancer, death of an embryo from various causes, or the production of congenital abnormalities (birth defects). In addition to these specific consequences of genetic mutations, an increase in mutation frequency in the human gene pool can have effects on future generations. Preston and Hoffmann (2003) presented a good overview of genetic toxicology, including mechanisms and available assays, both in vitro and in vivo. Unfortunately, modern testing systems are not yet equal to the task of accurately identifying those compounds and combinations that are mutagenic to people by the routes of exposure that occur in the human environment.

Many currently used tests for mutagenicity are conducted in vitro, or on bacterial or isolated cell systems where the probability of false results, both positive and negative, is great (see Pitot and Dragan, 2003 for a review). Such tests must be supplemented by whole animal studies if reasonable judgments on the control and/or introduction of medications and other substances are to be made: This point was clearly summarized by Butterworth (1981). Bacterial and cell culture mutagenicity assays can indicate a potential problem, but do not take into account important factors such as the ability of the chemical, or its active metabolites, to reach the germ cells in intact subjects.

The use of whole animal systems in short-term testing has been reviewed by Preston and Hoffman (2003). Such systems include the incorporation of classical cytogenetic methods using assays on preparations made from blood and bone marrow samples. Such methods have the advantages of permitting repeated tests on a laboratory animal subject, as well as allowing one to monitor humans who are inadvertently exposed to suspected mutagenic agents. A disadvantage is that other tissues that may be adversely affected are not sampled. Butterworth (1981) points out four relevant practical considerations when such samples are used for counting chromosome anomalies in laboratory studies.

- Experienced, properly trained personnel must be used to read the slides.
- For statistical validity, at least 50–100 cells must be examined per subject and the sampling should include at least 5 animals of each sex at each dose level.
- Proper controls must be run because accidental factors, such as viral infections, can lead to chromosomal alterations. Positive and solvent controls must be included.
- Because aberrant cell survival times can be short, samples should be drawn within 48 h of exposure, specifically at 6, 12, and 18 h.

Further, prudent investigative technique would require that several other criteria be met. These include the use of more than one species, the use of both a maximum-tolerated (with respect to lethality) dose and one lower dose, dosing via the route expected for human exposure (oral, dermal, or inhalation), examination of control and dosed specimens in a randomized blind fashion, and verification of delivery of the desired dose when inhalation is the route of exposure.

The recently available large number of *transgenic* rodent models promises to revolutionize mutagen testing. A relatively comprehensive review (280 pp.) by Lambert et al. (2005) covers the transgenic rodent assays for mutagenesis, their applicability and limitations, recommendations for performing such assays, and their use in a "regulatory context." The authors point out that all routes of administration can be used, but that many discrepancies exist (with respect to informing regulatory needs) as a result of "nongenotoxic mechanism of action," "inappropriate administration," or "inadequate study design." This review is an important resource for anyone involved in the use of transgenic rodent models for informing risk assessments.

Teratogenicity

Teratogens are agents that cause congenital (birth) defects of structure, function, or metabolism. One of the foundations of teratogenic studies is embryology, which is concerned with the sequence of cellular proliferation, migration, transformation, and the formation of organs during intrauterine development. This process, depicted in Figure 7.3 for humans, is subject to irreversible disruption at critical periods (shown by shading in the figure). Thus, the specific teratogenic effects of an agent are determined by the dose, and time of exposure during gestation. Similar developmental sequences exist for nonhuman mammals, and information on periods of organogenesis is available (Andersen and Goldman, 1970; Altman and Dittmer, 1972; MacKenzie and Hoar, 1995). For lung development, see Harding et al. (2004). According to Harbison (1980), chemical teratogens have several characteristics: (1) the effects are dose related with teratogenic effects occurring between levels that have no adverse effects (a lower threshold dose) and levels that are lethal to the developing organism; (2) they may or may not be mutagens, and if mutagenic, the congenital abnormality may be passed on to future generations; and (3) teratogens are often selective

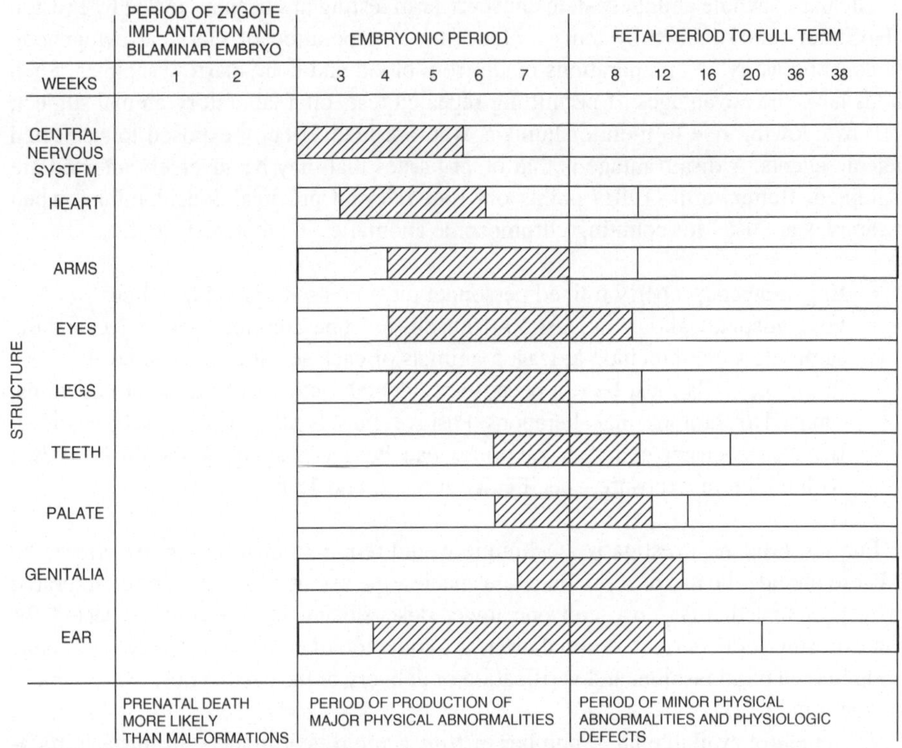

Figure 7.3 Summary of the prenatal human developmental sequence. Critical periods for producing birth defects are shaded. *Source*: Redrawn after Harbison (1980).

in their action on developing tissues. Many classes of teratogens have been identified in human and laboratory animals, but great differences in species sensitivity are known to exist (Harbison, 1980).

Because teratogenic testing on a broad scale has been intensified, experimental designs are currently evolving. Caution must be applied in examining teratogens (Schwetz, 1981); Schwetz provides several, still relevant, important points.

- The probability is that any agent administered at a high enough dose to embryos at the proper stage of development will cause disturbances in development (also known as *Karnofsky's law*).
- The cost of conduct of a teratogenic study is small with respect to the possible consequences of not doing the study.
- Exposure must occur during the critical period of organ development.
- At least two species should be used for each study, rat, rabbit, mouse, hamster, monkey, dog, cat, and sheep are common choices.
- At least 20 pregnant females in each of two dose groups should be used.
- The highest tolerated dose (with respect to survival) and at least one lower dose should be used.
- Animals should be exposed via the same route (oral, dermal, or inhalation) through which humans would be expected to be exposed.
- The number of controls should equal or exceed the number of dosed animals so that the background level of congenital abnormalities can be precisely defined.
- Animals must be strictly randomized into control and dosed groups; controls should be given treatment identical, except for dosing, to the exposed groups, including placement into chambers.
- In inhalation studies, the actual delivery of dose should be verified.

Reproductive effects of the material under study can greatly complicate teratogenic studies. For example, alterations in fertility and the spontaneous abortion rate can obscure detection of teratogenic effects.

MULTICOMPONENT ATMOSPHERES

Combinations or mixtures of agents may have effects that are not predictable from knowledge of the effects of each component acting alone. Modification of biological effects can be especially strong in particle–gas combinations. Such interactive effects can derive from many phenomena including:

- *Physical and chemical interactions occurring before inhalation.* Examples include adsorption of gases on particle surfaces, dissolution of gases in liquid particles, and chemical reactions that lead to changes in particle size, hygroscopicity, acidity, irritancy, solubility, carcinogenicity, toxicity, etc.
- *Biological interactions occurring within the subject.* For example, one agent may stimulate or deplete a natural chemical or physiological defense against another agent. Another agent may alter the rate and/or depth of breathing and thereby increase or decrease the dose due to a coexisting agent.
- *Other types of biological interactions may occur that lead to modification of dose delivery or response.* For example, a mixture may change the level of perception of exposure and thus stimulate or inhibit avoidance behavior such as rodents

hiding noses in fur or cage corners. Also, a combination can lead to changes in level of physical activity that change the ventilation rate.

- *Chamber artifacts may occur in exposures to mixtures*. For example, one component of a mixture may react with excreta and produce or destroy chemical species (such as ammonia) that interact with the atmosphere or the subjects.

Studies of simple or complex mixtures can be exceptionally complicated depending on the study material and exposure methods, so it is difficult to provide general guidance in the design of such experiments. However, four types of study design can be described, other designs being possible depending upon the specific study objectives. The four designs may, for convenience, be called *combinatorial, nested* (or subtractive), *supplementary*, and *sequential.*

To illustrate a *combinatorial design*, consider a study in which the combination of three air pollutants (A, B, and C) are being investigated. For further simplification we will keep each pollutant at one concentration; which could be some expected maximum environmental level. We wish to know how each pollutant acts alone and whether or not any pairs or all three together exhibit interactions such as a *positive synergism* (greater than additive effect), a *negative synergism* (antagonism or less than additive effect), or simple *additivity* in their effects when combined. Clearly, one can answer these questions by performing eight separate exposures, looking at the effects of all combinations of the pollutants, including the sham exposure case where none is present. Thus, in the combinatorial design the eight study atmospheres would consist of:

A	Alone,
B	Alone,
C	Alone,
A + B	Paired,
A + C	Paired,
B + C	Paired,
A + B + C	All combined, or
O	Clean-air sham exposure.

The nested or *subtractive design* involves comparison of the effects of a full combination with that of the combination minus each ingredient taken away one at a time. Such a design is applicable to a mixture of pollutants when the issue is the relative importance of each ingredient in contributing to an overall effect. Such an experiment can provide information on which ingredients should be given priority for selective control or elimination from the environment. This is a practical issue in the urban setting where industry, power generation, transportation, and home heating each contribute specific pollutants to the breathing air. If one considers a mixture of pollutants A, B, C, and D, the nested design would be applied to test all four ingredients together and all possible "nests" or combinations of any three pollutants. Thus, in this example the following six separate atmospheres would be generated, subjects exposed, and comparisons made.

A + B + C + D	Full combination
B + C + D	Nest with A subtracted
A + C + D	Nest with B subtracted
A + B + D	Nest with C subtracted
A + B + C	Nest with D subtracted
O	Clean-air sham exposure

The *supplementary design* (Figure 7.4) is only efficient when two-pollutant combinations are tested. This design is good for determining positive and negative synergism or simple additivity of the combination. In a simple form, an initial total dose is established and the two pollutants A and B are mixed in various ratios, their sum always adding to the same total. The process is repeated as necessary at higher or lower total doses for each pollutant. In the example given, each response curve is generated by measuring the response seen in five separate exposures; inclusion of more points on the curve provides better definition, but requires more effort.

A *sequential design* is used when one wishes to test the effects of administration of agents in various sequences. For two air pollutants, A and B, one compares subjects first exposed to A followed by an exposure to B, to subjects exposed to B first followed by A. For completeness, one may wish also to include groups exposed to each agent followed by an exposure to clean air. Also, one may wish to expose a group to A and B simultaneously. As for each of the previous study designs, clean-air exposed controls are required. The atmospheres studied under a sequential study design might then be as follows:

A	Followed by B
B	Followed by A
A	Followed by O (clean air)
B	Followed by O (clean air)
O	Followed by O
A + B	Simultaneously, followed by O

An example of an inhalation study with complex atmospheres using a behavioral endpoint was described in Stinson and Loosli (1979). The study is an example of the combinatorial design without the binary combinations. Young white mice were housed in individual cages having a small rest area and an exercise-wheel with a revolution counter.

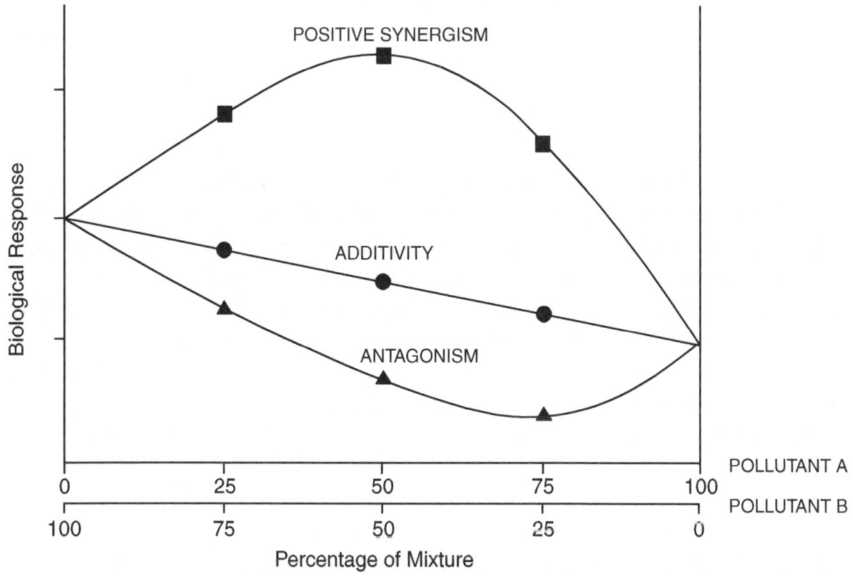

Figure 7.4 An example of the supplementary design for investigating the effects of mixtures of two pollutants. In the *top* curve, a positive synergism is shown, the *middle* curve represents simple additivity, and the *bottom* curve antagonism.

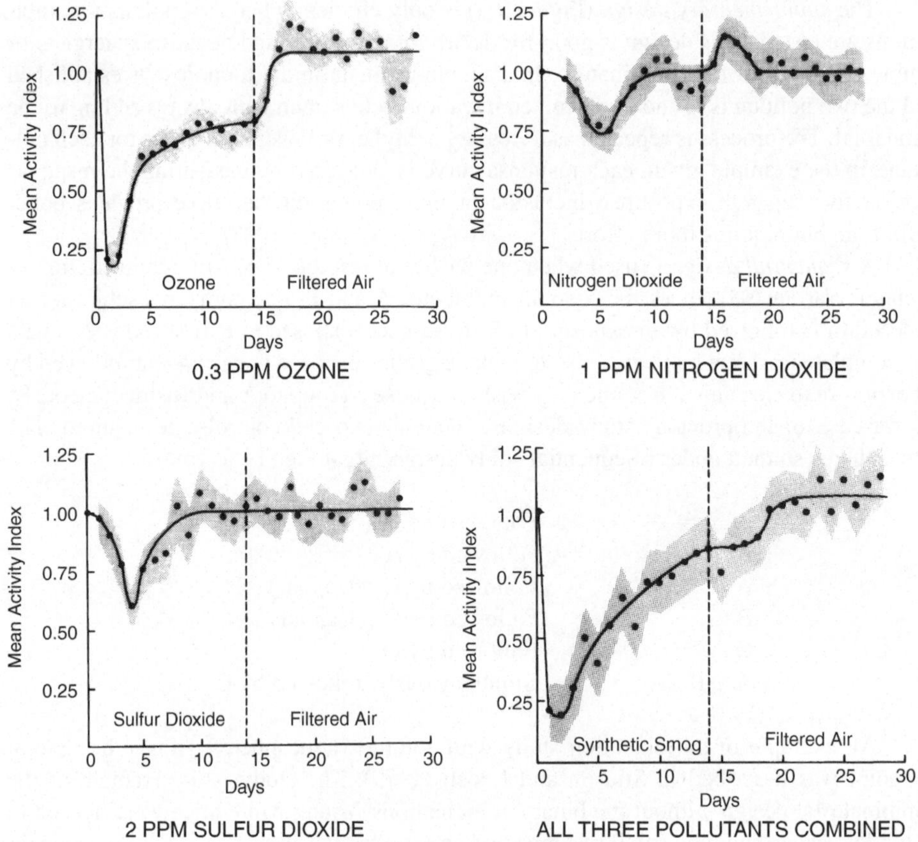

Figure 7.5 A combinatorial design study using a behavioral endpoint (spontaneous running activity) in which the gases ozone, nitrogen dioxide, and sulfur dioxide were studied individually and combined. The effect of the combination is apparently dominated by ozone. *Source*: Adapted from Stinson and Loosli (1979).

In this environment, animals were continuously exposed to either clean filtered air, or 0.3 ppm ozone, or 1 ppm nitrogen dioxide, or 2 ppm sulfur dioxide, or the combination of the three pollutant gases ("synthetic smog"). Groups of 20 were exposed for 2 weeks, preceded and followed by 2 weeks in clean air, and the number of revolutions run by each mouse between 5 p.m. and 8 a.m. were recorded. The ratio of activity measured during exposure to average daily activity during the last week of pre-exposure was defined as the activity index. The resultant data are shown in Figure 7.5. Each pollutant gas had its characteristic effect on spontaneous running activity, namely a depression followed by partial or full recovery toward the end of exposure. A most interesting finding was that the effects of the trinary combination are nearly identical to that of ozone alone. A reasonable conclusion is that the gases tested, under the conditions examined, did not produce a synergistic effect. It is interesting to note that the authors did not find any clinical or histopathologic indications of toxicity in the tested mice. The level of ozone studied was similar to that for which exercising human subjects show pulmonary function changes and report symptoms such as cough, throat tickle, and congestion (Savin and Adams, 1979).

8 Facilities and Support Considerations

INTRODUCTION

High quality inhalation studies can only be conducted in a specialized setting with a great deal of technical support. An inhalation study requires not only the exposure apparatus, but also analytical support, atmosphere generation and characterization systems, an excellent vivarium (if animals are used), support for the measurement of biological effects, data analysis capabilities, space for personnel, and additional support for fabrication, repair, maintenance, and calibration. The requirements are more complex for studies that involve chronic exposure, large numbers of animals, complex atmospheres, or hazardous exposure material. Further, the buildings have stringent air-conditioning requirements, should not be in high outdoor pollution areas, and should be isolated from nonlaboratory human traffic. Today, serious attention must be given to security systems in order to protect the research. When facilities-related compromises are made due to cost or other constraints, heroic efforts may be required in order to generate meaningful data.

The cost of inhalation facilities and the necessary support can be enormous, depending on the nature of the studies. In 1982, the author's laboratory added two new inhalation chambers with an aerosol equilibration line and throughput air purification equipment at a cost of about \$50,000 (1982 dollars) not including personnel costs during design, assembly, and checkout. Further, this cost did not include the building modifications or pollutant generation and characterization equipment. A small (4,000 ft^2) acute inhalation laboratory with four chambers and on-site animal housing will cost a minimum of about 1 million dollars including an inexpensive building. This modest facility may require a staff of at least two doctoral-level investigators, half a dozen expert technicians, animal technologists, general laboratory helpers, and an administrative support person.

Several facilities-related references are available, including: *A Guide to Laboratory Design* (Everett and Hughes, 1981); *Laboratory Design Guide* (Griffin and Arch, 1998); *Guidelines for Laboratory Design: Health and Safety Considerations* (DiBerardinis et al., 2001); *Biosafety in Microbiological and Biomedical Laboratories (BMBL), 5th Ed.* (CDC/NIOSH, 2007); *Guidance for Protecting Building Environments from Airborne Chemical, Biological, or Radiological Attacks* (CDC/NIOSH, 2002); *Guide for the Care and Use of Laboratory Animals* (ILAR, 1996); *Marks' Standard Handbook for Mechanical Engineers* (Avallone and Baumeister, 1996); and some recent papers (Kim et al. 2005; Vogelweid et al. 2005). These references cover several topics including basic design features, fire protection, ventilation and fume extraction, laminar flow rooms, storage of hazardous items, earthquake protection, and special requirements for work with radioisotopes and hazardous microbes.

FACILITIES COMPONENTS

Laboratory Buildings

The building should have sufficient room for efficient day-to-day operations and spatial isolation of incompatible activities such as animal care and chemical analyses. Virtually any laboratory-type construction will do as long as the floor is solid enough to support heavy equipment, and the air supplies for laboratory, office, and the vivarium are separate. For the protection of equipment, animals, and hazardous chemicals, security and fire alarm (or extinguishing) systems are recommended. Figure 8.1 depicts the author's facility, which comprises about 8,000 ft^2 on a remote corner of a university campus. This facility is close to the minimum size necessary for self-contained inhalation studies using several species of laboratory animals.

Exposure Systems

The throughput air supplied to an exposure system is cleaned and conditioned as described in Chapter 3. The equipment used is best located either outdoors or in a room separate from other functions, as air cleaning and moving equipment is typically noisy and a source of significant quantities of heat. This noise and heat interferes with control of the exposure as well as the tranquility of subjects and personnel. For efficiency, the outlet of an air purification system should be close to the exposure apparatus to which it is connected.

Chamber installations typically require high ceilings, a strong disinfectable floor, unobstructed surrounding space, good room air temperature control, and large floor drains. If air-moving pumps are placed on a roof above the chambers, their mountings should prevent excessive vibration: Spring-supported platforms are suitable.

Access to all sides of a chamber is recommended, and it is wise to leave plenty of space between adjacent chambers. The space surrounding a chamber will be needed for placement of monitoring, generating, and recording equipment, carts, and movement of laboratory personnel. Thus, about 10 m^2 (100 ft^2) of surrounding floor space should be allowed for a 1 m^3 volume chamber.

The floor in an animal exposure chamber room must be easy to clean, resistant to spilled chemicals, and have good water drainage toward large floor drains, 3 in (7.6 cm) or greater in diameter. Unavoidably, this floor will become soiled with animal urine and feces, will become wet during chamber cleaning, and will periodically receive a variety of spilled lubricants, exposure materials, and other agents. Cement protected by a hard, durable coating such as epoxy-based paint or rubberized plastic (Dex-O-Tex®, Dex-O-Tex, Rancho Dominguez, California, U.S.A.) is suitable. This nonslip, fireproof, waterproof, abrasion- and corrosion-resistant product, developed for use on exterior ship decks, has given several years of excellent service in the chamber and animal housing rooms of the author's laboratory.

Air-conditioning for a chamber room must be of high quality because the temperature inside of a chamber will be influenced by that of the surrounding room air (Bernstein and Drew, 1980). One must avoid large spatial or temporal temperature gradients within the room. Such gradients are caused by locating inlet air-conditioner ducts on one side of the room and return ducts on the other, and by air-conditioning systems that cycle on and

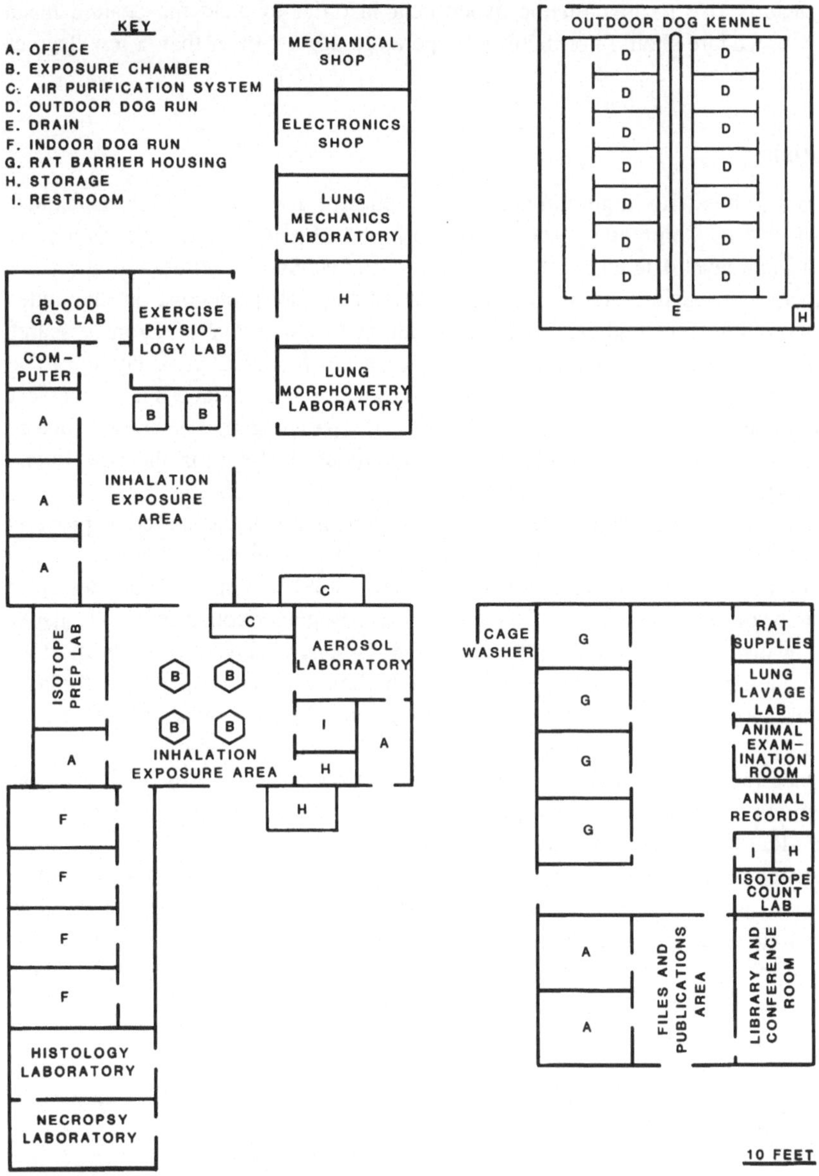

Figure 8.1 The Air Pollution Health Effects Laboratory on the University of California, Irvine Campus. This 8000 ft^2 facility supported inhalation research using rodents and dogs. Research staff included four at the doctoral level, four with master's degrees, and six technicians.

off during an exposure. In order to promote thermal uniformity, quiet stirring fans can be placed within the room to provide air circulation. Equipment that generates significant amounts of heat can also alter the internal temperature of a chamber. Such equipment should not be near chambers unless the surrounding air is well mixed with the rest of the room. During an exposure, the heat load from equipment, lights, animals, and laboratory personnel can produce a steady temperature increase unless the room air-conditioning system is of adequate capacity. If one is stuck with an undersized unit, the room temperature can be lowered below the desired mean temperature before the exposure and then allowed to steadily rise throughout the experiment in order to yield the desired mean value. Such a procedure should not involve temperature drifts greater than a few degrees centigrade.

Animal Housing

Institutional vivariums with large numbers and varieties of animals are often inadequate for inhalation studies. Commonly, rodents housed in such facilities will have respiratory tract infections that invalidate all but the most coarse studies. Some institutional vivariums will have air-barrier housing and meticulous isolation and care protocols, but the added expense often prohibits such high quality care. In addition to respiratory infections, research animals may be exposed to contaminants (airborne and in food, water, or bedding) that modify their response to experimental atmospheres. Thus, in most instances it is necessary for the inhalation toxicologist to establish and maintain separate animal housing dedicated to their studies. Equipment and supplies for animal studies are found in the *Lab Animal Buyer's Guide* (Lab Animal, 2007).

Such housing must be well planned, scrupulously maintained, and staffed by properly qualified and trained animal technicians. In order to protect the health and environmental exposure history of the animals, many of the same principles that apply to the design of exposure systems must be followed in the vivarium. Thus, proper air-conditioning, cleanable, well-drained walls and floors, and adequate space for day-to-day operations must be provided. In addition, a light–dark cycle, a balanced high purity diet, proper room for large animal exercise, and stringent monitoring, cleaning, and disinfection schedules are all necessary.

The space required for housing various animal species can be estimated using the recommendations for space per animal in the *Guide for the Care and Use of Laboratory Animals,* (ILAR, 1996). An excerpt from this source is shown in Table 8.1. The rooms or outdoor areas in which animals are caged should have adequate space for carts; movement of cages; placement of items such as balances and other equipment; cage and dish washing; food preparation; clean caging, drugs, cleaning and disinfecting supplies; and boots, lab coats, coveralls, gloves, etc. At a typical facility, animal cages may occupy only about 60% of the vivarium space.

Table 8.1 Space Recommendations for the Housing of Laboratory Animals

Animals	Weight	Floor area/animal	Cage height
Mice	< 10 g	39 cm^2	12.7 cm
	10–15 g	52 cm^2	12.7 cm
	16–25 g	77 cm^2	12.7 cm
	> 25 g	97 cm^2	12.7 cm
Rats	< 100 g	110 cm^2	17.8 cm
	100–200 g	148 cm^2	17.8 cm
	201–300 g	187 cm^2	17.8 cm
	301–400 g	258 cm^2	17.8 cm
Dogs	15 kg	0.72 m^2	[a]
	15–30 kg	1.08 m^2	[a]
	> 30 kg	2.16 m^2	[a]

a Cage height (floor to top) must allow occupant to stand in a "comfortable position."
Source: Data from the *Guide for the Care and Use of Laboratory Animals* (ILAR, 1996).

The air in an animal housing area should be of proper temperature and humidity, copiously supplied, and cleaned free of viable and nonviable airborne contaminants (both particulate and gaseous). These areas should have about 15 to 20 air changes per hour, with recirculated air limited to less than 75% of the total throughput. The temperature and humidity should be maintained within the ideal limits for the species used (20–23°C for most common laboratory animals and greater than 20% relative humidity all year round). The air supply should have filters and scrubbers for removal of particles such as dander, dirt, viruses and bacteria, and of gases such as ammonia, hydrocarbons, and outdoor air pollutants (e.g., ozone, sulfur dioxide, and oxides of nitrogen). This air cleaning is also needed to protect vivarium personnel (Kacergis et al., 1996). The main component of the system can be a modified commercially available heating and cooling unit. The cooling coils should operate continuously while the heaters will be used only when the thermostat calls for heat. The air passing over the cooling coils can be lowered to a temperature of about 10°C to prevent excess humidity. Air purification (see Chapter 3 for details) can be achieved by charcoal and chemical filtration using buffered potassium permanganate impregnated on activated alumina pellets. These pellets are an effective "odor oxidant" which can remove a wide range of polar, inorganic, and organic molecules through absorption, adsorption, and oxidation. The purification unit may precede or follow the air-conditioning unit. Replenishment air should enter the air stream before the air-conditioning unit so that outside air will be cleaned before entering the animal rooms. Particulate matter is conveniently removed in two stages. First, one may use a "30% gross filter" which will remove large particles and protect downstream units. The second stage filter, such as an HEPA type, will operate at nearly 100% efficiency for particles 0.3 μm in diameter.

In addition to adequate room air-conditioning, some animals (rats, for example) may require additional protection from airborne infectious organisms. Two methods are used to provide this additional protection: filter-topped cages, and laminar flow air-barrier isolators. Filter tops, available for covering plastic rodent boxes, can reduce the intrusion of particles. Such tops are usually used in housing conditions that involve animals living in contact with bedding material, so they do not protect animals from contaminants generated in cages (such as ammonia and bedding dust). Nonetheless, filter tops provide an economical additional level of protection over conventional open caging.

Excellent protection of rodents from airborne infections is afforded by air-barrier housing systems. In these systems, a gentle flow of recirculated purified air is maintained over and around the cages. Two types of systems, vertical air flow and horizontal air flow, are used. For an upright cage rack, the horizontal flow system (Fig. 8.2) insures that each animal is surrounded by fresh clean air. In a vertical air flow system, animals in lower cages may receive air from those above, allowing for possible spread of airborne infections. In both types of barrier systems, the air is ideally passed through a high efficiency filter, such as the HEPA type described in Chapter 3. Air-barrier housing systems are currently available commercially, or they may be custom designed. An additional feature in the custom unit in Figure 8.2 is the inclusion of a pollutant gas trap in addition to the particle filter.

Walls and flooring must be disinfected in the animal housing area. This necessitates use of tough impervious coatings and construction which is free of cracks and inaccessible reaches. Adequately sloped floors leading to drains are important to maintaining a clean environment. A floor pitch of at least 0.25 in/yd (about 0.6 cm/m) is recommended for good drainage. In heavy use areas such as dog kennels, large floor drains are recommended. Heavy duty disposal units beneath the drain openings can aid in removing animal wastes and thus reduce future maintenance costs. Recommendations regarding animal housing are contained in the *Guide for the Care and Use of Laboratory Animals* (ILAR, 1996). Such recommendations are periodically updated and should be carefully followed.

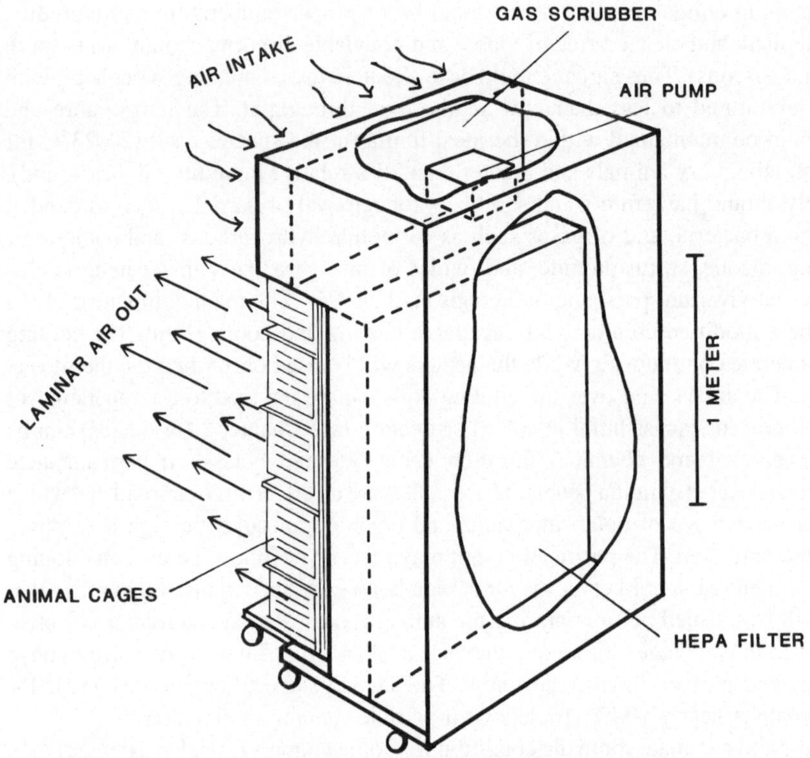

Figure 8.2 Custom-designed horizontal flow laminar air-barrier module for isolation of a rack of 20 rodent cages. Note gas- and particle-cleaning capabilities for treating the air supply.

Design of animal housing facilities should be guided by people who are experienced in the care and use of research animals and who are familiar with the research program.

What has been presented applies to ordinary inhalation studies. When strong pathogens, toxins, or carcinogens are studied, additional safeguards will be required: The Centers for Disease Control and Prevention (CDC) and The National Institutes for Health (NIH) have several relevant guidelines (e.g., CDC/NIOSH, 2007). Further, more stringent requirements, including sterilization of food and water, and sanitary procedures, are indicated for the protection of highly disease-susceptible animals, especially when they are held for long periods.

Necropsy

When research animals are used in the laboratory, necropsy facilities should be available. The purposes of a necropsy are to determine and document cause of death and to obtain tissues for further study. To these ends, weighing scales, a standing-height dissection table, a refrigerator with freezer, a large sink (with a disposal unit), a camera with stand, good lighting, and surgical tools are recommended. One should also have a disinfectable desk or writing table. The ventilation in this area should be generous, as open containers of fixative solutions will be present.

Some specialized equipment may be useful in the necropsy area, such as a perfusion–fixation apparatus (especially appropriate for lung tissue), equipment for injecting tracer dyes, a centrifuge, and a quick-freezing apparatus. A computer will allow for on-the-spot data reduction.

If live animals are to be brought to the necropsy area, equipment and drugs for humane euthanasia should be present along with comfortable temporary holding cages complete with food and water.

Excellent housekeeping with periodic disinfection of all surfaces, including walls, floor, light switches, and door handles is essential in order to prevent this area from becoming a source of infection.

Necropsy forms, carefully filled out and signed by the person performing the necropies, should be designed and should become a part of the research records. Animal number, species, sex, age, weight, cause of death, date, general animal condition and behavior, organ weights, organ gross appearance, location and identification numbers of tissue samples, protocol number, and general comments will be useful. If the autopsy is performed blind (without knowledge of exposure history), space on the form describing the experimental treatment should be filled in soon after all necropsy operations are completed. Approved, up-to-date protocols must be readily available in the necropsy room.

Data Handling

It is not difficult to generate scientific data at a rate that exceeds the capacity for reduction and analysis. This is especially true when data collection is automated, as by the interfacing of equipment with computers. Most successful researchers soon learn that raw data should be quickly digested and analyzed: Putting data into final form as soon as possible after they are collected is advisable. Only by such a quick turnover of information can one provide timely, often essential, feedback into an ongoing study. Also, if serious data gaps are detected soon enough, the protocols can be amended so that needed data can be gathered. Sadly, problems with data sets are often detected after the experiment when all hope for remedy is long past. The most important consideration regarding experimental data is the design of the study itself. Care must be taken to acquire all essential data without developing a large amount of extraneous information.

Modern data handling equipment includes a computer. The computer is ideally suited for real-time interactions with equipment, calculations on large data sets, and curve-fitting and iterative computational routines. The computer can standardize operations and remove human error. On the other hand, use of the computer has some costs. Writing and debugging programs can require an inordinate amount of time. Some programs evolve, becoming more and more complex and less and less fathomable to human beings. Programs can become so complex, that hand-check calculations are impossible. Further, the computer performs only specified operations and can overlook interesting, unanticipated information or relationships in data sets that would be obvious to an observant person. In short, dependence on mechanical data handling routines introduces a distance between the investigator and the data. This distance can save time, but can also prevent one from gaining new insights into the phenomena under study.

Another consideration in selecting a computer is the availability of software. Without ready-made software, one shoulders the burden of writing programs. Useful software includes curve-fitting routines, statistical analysis packages, and data output programs for generating histograms, frequency distributions, and various plots.

When selecting laboratory computing equipment, one is advised to seek expert advice. Weighing the cost, power, flexibility, and future operating personnel time commitment is an important part of the planning process. Once the data handling system has been selected, it should be given an appropriate home. Heat, dust, moisture, and pests (such as insects and wild mice) are enemies from which the system must be protected. Further, one should be able to sit at a terminal in a quiet location, free of human traffic, in a comfortable chair, and with plenty of elbow room, in order to develop an intimate one-to-one relationship with one's scientific data. Good lighting and air-conditioning are important in the room selected for a computer. A carpet and "acoustic" ceiling can be helpful in reducing distracting noises from cooling fans and printers.

If the computing machine is interfaced to laboratory equipment, long transmission lines can result in poor signal transmission due to 60-cycle interference (from lighting and other electrical equipment), as can improper electrical grounding practices. Experience and expertise are required to establish and maintain such automated capabilities.

Analytical Support

Analytical support functions primarily include equipment calibration, and physical and chemical analyses of atmospheric and biological samples. Additionally, reagents and standards used in the laboratory may require analysis of strength, purity, and other properties. The analytical laboratory must be convenient, so that timely analyses can be performed. The most important item in the analytical chemical laboratory is the analytical chemist. Many needed chemical analytical methods will require some adaptation, or will be either unprecedented, very difficult to perform, or buried in the esoteric (not in English or more than 20 years old) scientific literature. Similarly, physical characterizations are often best performed by a physicist, geologist, physical chemist, or materials scientist.

A list of even the basic equipment for chemical and physical analyses would be quite lengthy and include such general items as balances, ovens, vacuum dessicators, hoods, temperature controllers, pH and conductivity meters, microscopes, volumeters, centrifuges, flow controllers, pressure regulators, viscometers, pyrometers, hydrometers, refractometers, condensers, distillers, spectrophotometers, shakers, stirrers, pumps, timers, manometers, impactors, impingers, bubblers, radiation detectors, etc. One can visualize the potential expense and space requirements for analytical functions in an inhalation laboratory. For sophisticated sensitive analyses of many materials, one must consider such specialized, expensive equipment as the mass spectrometer, X-ray diffractometer, vacuum microbalance, atomic absorption spectrophotometer, gas chromatograph, permeation tube apparatus, fluorescence spectrophotometer, and transmission and scanning electron microscopes. Clearly, there is almost no limit to the possible sophistication with respect to analytical capabilities. Resources that can help in identifying analytical instruments include the *Lab Animal Buyers Guide* (Lab Animal, 2007), and the *NIOSH Manual of Analytical Methods* (NIOSH, 2006).

One must tailor the analytical support space, equipment, and personnel to the research program, and study materials of interest. Specific information on chemical analytical techniques and alternative analytical methods for the elements and many compounds may be found in the *NIOSH Manual of Analytical Methods* (NIOSH, 2006), which is updated periodically, and describes analytical methods for a large variety of chemicals in air, water, and tissues.

It is difficult to provide clear guidance on the amount of in-house space required for analytical support. The author's inhalation laboratory, which is engaged in inhalation exposures of several laboratory animal species to urban air pollutants singly and in combinations, has analytical support space that is approximately equal in size to the total space devoted to inhalation exposure chambers. Thus, such a facility having six chambers in rooms of about 600 ft^2 of floor space should have about 400 to 600 ft^2 devoted to analytical functions. This range is probably realistic for most inhalation studies

(Note that English units are used here, as they are still used in building construction in the United States.)

Analytical space should have good lighting, piped-in air and vacuum, and plenty of lab bench surface area to allow for custom analytical setups and organization and manipulation of samples. Some analytical equipment, sensitive balances, and microscopes for example, are compromised by vibration. Vibration from air-conditioning systems, elevators, human and vehicular traffic, and large motors can lead to difficulty in use of these instruments. Vibration isolators and heavy bench tops can greatly reduce, though not completely eliminate, vibration problems. Vibration isolation, or more specifically, attenuation of transmitted vibration, is achieved by lowering the natural vibrational frequency of the protected system, W_{inst}, being isolated, so that this frequency is much lower than the external vibrating frequency, W_{ext},. Because the external vibrations are not easily controlled, the protected system (weighing balance, microscope, etc.) is often modified using a vibration isolator in order to make the frequency ratio, R, as small as is practical.

$$R = \frac{W_{inst}}{W_{ext}} \qquad \text{(eq. 8.1)}$$

In the case of a laboratory instrument (Fig. 8.3), a platform support of mass M suspended by a set of springs, pads, or air cushion with spring constant K is commonly used to decrease the instrument system's natural frequency. The natural frequency of the instrument system (assuming the instrument mass is negligible with respect to the support's mass) is:

$$W_{inst} = C\frac{K^{1/2}}{M} \qquad \text{(eq. 8.2)}$$

where C is constant. Thus, one attempts to make the support's mass M large and/or the spring constant K small in order to protect from vibration.

Toxicity Testing

The specific biological endpoints used in an inhalation study will dictate the space, equipment, and personnel requirements. If adequate space is available, each separate type of endpoint can be housed in its own area. Thus, supplies, equipment, and work space can be organized to maximize the efficiency of the measurements. Equipment, supplies, and laboratory arrangement will depend upon the specific types of measurements made in the

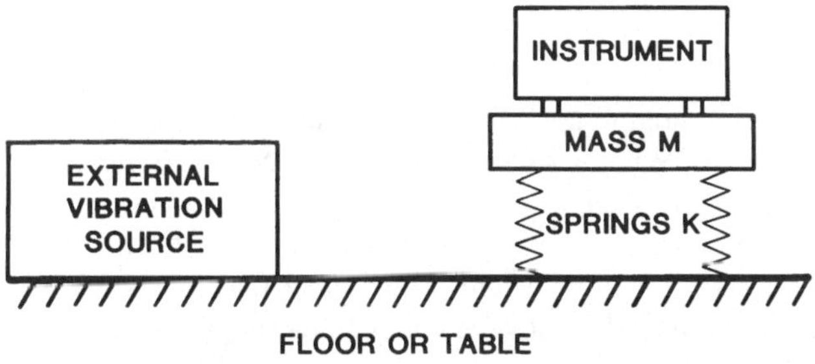

Figure 8.3 A vibration isolator having a support slab of mass M suspended by a set of springs or pads with a spring constant, K.

Table 8.2 Common Types of Biological Endpoints Used in Inhalation Studies and Common Requirements for Each

Histology: Requires apparatus for necropsy, tissue fixation, embedding, sectioning, staining, evaluation, and ventilation (fume hood) to remove vapors
Pulmonary physiology: Utilizes spirometers, recorders, plethysmographs, gas analyzers, ventilators, holding cages, restraining systems, custom masks, and sedation capabilities. Special studies may require treadmills, radioisotopes, and surgery support
Pulmonary biochemistry: Requires apparatus for isolation of biochemicals, and organic chemical quantitative analysis, radioisotope methodology, and techniques for rapid freezing and lypholization. Isolated perfused lung preparations require additional apparatus
Infectivity: Requires incubators, isolators, and high-containment exposure system with special caging. Microbiological assay and radioisotope techniques may be required
Particle deposition and clearance: Requires well-defined, possibly radiolabeled aerosols, and inhalation exposure apparatus. Particle sizing and counting techniques and methods for following particle clearance may be required. Rodents may need holding cages that prevent coprophagy
Morphometry: Requires infusion fixation equipment plus apparatus required for histologic studies. Precise measurement of organ volumes and dimensions require specialized equipment. Automated analysis requires computer hardware and software
Behavior: Requires computer hardware and software for recording of behavioral events, specialized instrumented caging including exposure capability, and environmental noise-masking equipment

laboratory area. Table 8.2 lists a few of the common types of endpoints used in toxicity studies and indicates some of the special requirements for each. The decision as to how many and which endpoints will be used is addressed in Chapters 6 and 7.

The cost of an inhalation exposure itself is generally large, and inclusion of several endpoints may be required to justify the overall research effort. Once an exposure atmosphere is established, it may be relatively easy to expose large numbers of animals that can then be shared among various endpoint measurement protocols. Use of multiple endpoints also allows for correlation of the separate effects observed. For example, Figure 8.4 shows the relationship between inhaled particle clearance data observed with inhaled radiolabeled tracer particles and the magnitude of histologic lung lesions seen in groups of rats exposed to various levels of ozone. Exposed rats were assigned to one or the other endpoint. From such data one can begin to suggest mechanisms for the disturbance in clearance. Similarly, histologic, morphometric, and biochemical studies can be used to shed light on the events seen in behavioral or physiological assessments. Thus, one is advised to include multiple quantitative endpoints in an inhlation study.

Shop Support

Experienced investigators visiting a laboratory are quick to notice the quality and accessibility of the research team's shop tools. When a soldered connection comes loose, a pipe bursts, or a thread strips, research activities can grind to a halt. Also, when a new apparatus is needed or modifications to an existing setup required, delays can result. Most large research institutions have central shop facilities where professional machinists, electricians, plumbers, and carpenters are found. Alternatively, commercial shops may be available in a nearby community. Certainly, large or complex jobs will usually be done by such professionals. However, the numerous small, urgent, or especially creative unplannable tasks that are essential to a quality research program are often best done by the research

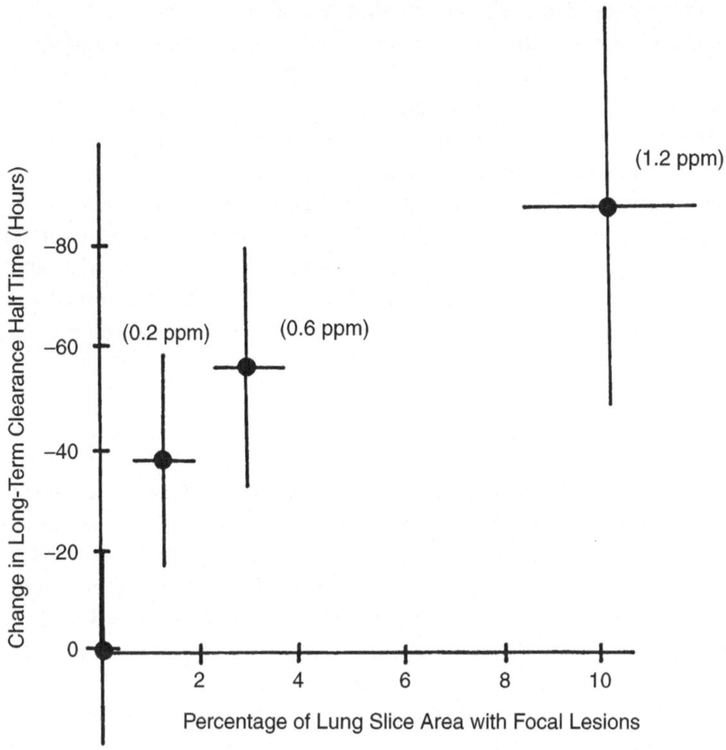

Figure 8.4 Observed correlation between histologically observed lung lesions and alterations in clearance of inhaled tracer particles in groups of rats exposed to ozone.

team itself. For this purpose, a shop area and equipment should be available to scientific personnel. The size and sophistication of this shop will depend upon the research program, the skills of the research staff, the nearness and cooperativeness of professional shops, and the available space and money. A mature research program will usually evolve an adequate shop. When establishing a new program one should anticipate the future need for shop space, supplies, and equipment. The shop should be under the control of a safety-conscious supervisor.

Inhalation studies are associated with a variety of mechanical hardware items for plumbing, ducting, enclosing, sampling, cleaning, pumping, etc. Further, complex, varied, and often fragile instrumentation is required, along with specialized apparatus such as exposure masks, radioisotope shielding, scrubbers, diluters, etc. The materials most frequently utilized, including tubing (plastic, glass, copper, stainless steel), pipes (copper, iron, plastic), metal sheet and foil (iron, stainless steel), and other items such as plywood, lumber, plastic sheets, wire screen, gaskets, mold-making liquids, etc., should be kept on hand.

Essential basic tools include: hammers; screwdrivers; adjustable, box, pipe, and specialty wrenches; pliers; metal shears; pocket knife; carpenter's square; soldering iron; hack saw; coping saw; files; brace and bits; wire cutters; first-aid kit; gloves; and safety goggles. Other essentials include a good vise, sabre saw, electric drill with hole saw attachments, power sander, miter box and saw, stopper borers, socket wrench set, torque wrench, wet and dry shop vacuum cleaner, circular saw, tubing cutter, glass-cutting tool, heat gun, chisel set, clamps, and a plane. Complex items requiring training of personnel in terms of proper use and safety are also useful and include a drill press, band saw, welding and brazing equipment, belt sander, and a small lathe. More sophisticated equipment generally requires operation by a skilled professional machinist.

Checkout, calibration, and minor repair of electrical equipment often must also be done by research personnel. For this work, one needs a high quality large range volt-ohm-ammeter, a microammeter, and possibly an oscilloscope.

Supplies must be maintained for the shop functions and will include nuts, bolts, nails. screws, connectors, a variety of types of tape (electrical, metallic, thread-sealing, duct, masking, strapping), lubricants, adhesives, liquid sealing agents, wire, solder, and basic building and fabricating materials. As one might also expect, good initial organization and meticulous work habits are necessary to prevent the shop area from degenerating into a hopeless mess.

Conference, Library, and Office Space

Adequate communication among the research staff and with visitors cannot be taken for granted. Space specifically assigned for these purposes should be included in facilities planning. Provision for a conference room is an important way of both providing a place for discussions, and for encouraging them. This room should be large enough for the entire staff and a small group of visitors to meet at once, and should be furnished with a chalk/dry-eraser board and table with chairs. One can estimate the size of a conference room for a given number, N, of people. Assume each person requires about 12 ft^2 for a chair, legroom, and writing surface, that 4 to 6 ft along one wall are needed for use of the chalkboard, and that 2 to 3 ft of corridor width is required along walls and between seated groups. Depending upon the arrangement of tables, for an approximately square room, the number of square feet S_c as a function of N is approximately:

$$S_c = 12N + 50N^{1/2}, \ (4 \leq N \leq 40) \qquad \text{(eq. 8.3)}$$

For larger groups, theater-type seating, with all facing the front, is necessary. Assuming a theater-type chair with writing arm requires 8 ft^2 and surrounding space adequate for aisles and a lecturer is allowed, one has:

$$S_c = 8N + 50N^{1/2}, \ (N > 40) \qquad \text{(eq. 8.4)}$$

Library space should be adequate to hold commonly used books and those journals that must be on hand rather than in a distant central library. Reading and working room for use by about 20% of the staff at any one time should be allowed. For a research group of up to 40 persons assuming 4 ft^2 of bookcase space per person, the library size S_l, in square feet should be about:

$$S_l = 7N + 10N^{1/2}, \ (N < 40) \qquad \text{(eq. 8.5)}$$

Office space requirements are dictated by a policy decision on who does and does not require an office. Professional staff investigators and high level technicians usually should have offices separate from laboratory space. Single occupancy offices are typically about 70 to 100 ft^2. For double occupancy, about 120 to 150 ft^2 are adequate. Perhaps the most common complaint registered regarding office space is the lack of a window: inclusion of windows in office spaces is far cheaper at the time of construction than afterward.

Table 8.3 Important Factors in Selecting a Location for an Inhalation Studies Laboratory

1. *Space*: Parking, animal quarters, hazardous material storage, general storage, and possible needs for expansion should be considered.
2. *Ambient air quality*: Urban, industrial, and agricultural activities all degrade the air quality.
3. *Resources*: Scientific, technical, and medical libraries, supplemental laboratory and medical services are important resources.
4. *Utilities*: Inhalation studies tend to have large needs with respect to power, water, sewer, and waste disposal.
5. *Security*: Radioisotopes, chemicals, and animal subjects must be adequately isolated from intentional or accidental unauthorized access.
6. *Accident*: Fire, explosion, and leakage must be considered from point of view of both the needed emergency response and contamination of the surrounding environment.

FACILITIES LOCATION

In many instances, the location of an inhalation laboratory is predetermined and no choice is available. In other instances, choices are available as to where the inhalation facilities will be moved or built. So many factors are important to consider in choosing a location that no simple advice can be given. However, several factors that directly impact on the research are listed in Table 8.3.

Adequate space is an essential requirement in selecting a location. Inhalation studies suffer when cramped because many needed functions interfere with one another. In addition, possible outdoor housing of air purification equipment, compressed gas tanks, and large animal quarters should be considered.

The cost of cleaning ambient air to laboratory standards is dependent upon the level of local air pollution. Urban air, with pollutants from traffic, home heating, and power plants can be difficult to clean. Similarly, areas of high industrial concentration may have problematic air. Agricultural areas are likewise associated with poor ambient air quality due to pesticide and herbicide spraying practices and waste burning activities. Even relatively unspoiled regions can have high levels of locally generated air contaminants such as mineral dusts, spores and pollens, methane, and gaseous sulfur compounds. Local source types and strengths, along with local meteorological conditions will determine the local air quality. A survey of the local air quality should be made in advance so that necessary air-cleaning and air-conditioning equipment, and their costs, can be identified.

The question of the altitude at which inhalation or aerosol studies are conducted often arises within the scientific community. The reason for this is that several important factors that affect animals, aerosol particles, and instruments will vary with altitude. Examples include air pressure, air density, oxygen and other gas partial pressures, molecular mean free-path, speed of sound, viscosity of air, and thermal conductivity of air. Graphs and formulae for the physical parameters as a function of geometric altitude are found in reports published jointly by the National Oceanic and Atmospheric Administration, the National Aeronautics and Space Administration, and the U.S. Air Force (1976). Although scant information is available on the effects of variations in altitude on aerosol sizing instruments, a paper by Yeh et al. (1981) from Albuquerque (altitude about 1500 m above sea level, barometric pressure = 0.82 atm) points out the potential importance to particle size analysis. Using an electrical aerosol analyzer to size monodisperse, aluminosilicate particles (0.1 μm diameter), the instrument gave a reading for count median diameter of 0.50 μm; a 50% error. By correcting the instrument's data reduction constants for laboratory altitude, the new count median diameter was 0.077 mm, which is significantly closer to the expected value.

The effects of reduced atmospheric pressure on animals is not only a function of the pressure itself, but also the duration of residence. Acclimatization involves sustained increases in ventilation, diffusing capacity, and the oxygen-carrying capacity of the blood due to increased hemoglobin concentration (Grover, 1978; Guyton, 1991). Although altitudes up to about 2,400 m are considered safe for normal individuals, physiological effects are seen at much lower altitudes (Lambertsen, 1961). At about 1,500 m, the oxygen partial pressure in air is about 80% of the value at sea level and the arterial oxygen content is reduced to about 95% of the sea level value. Buskirk (1969) reviewed the literature on the effects of altitude on maximum work capacity. This review indicates that from sea level to about 1,500 m there is no consistent change in maximum oxygen consumption, but as altitude increases further there is a linear decrease in this parameter. Although no firm value can be given for the maximum altitude at which generalizable inhalation studies can be conducted, one might consider 0 to 1,500 m as the approximate range for normal cardio-pulmonary and cognitive function.

The remaining items in Table 8.3 (resources, utilities, security, and accident planning) are important factors in selecting a laboratory location. Inadequate initial attention to any of them is likely to lead to inefficiency, hardship, and/or public opposition.

9
Animal Models

INTRODUCTION

General Considerations

An *animal model* is an intact preparation (usually mammalian) having an identified scientific utility. The selection of animal species and strains for inhalation studies must be based on objective criteria. Although many practical factors such as available space, cost, public pressure, and difficulties in handling will influence the selection, these considerations should not be primary. Criteria by which animal models are selected often include the following.

- Similarity to humans with respect to anatomy, physiology, biochemistry, and susceptibility to injury or disease.
- Life span that is appropriate for the study.
- Possession of unique qualities, such as sensitivity to certain toxicants, or increased measurement accuracy.
- Existence of an appropriate database on the model.
- Existence of proven procedures and apparatus pertaining to the model.
- Ease of extrapolation or interpretation of data.

A given model is usually strong in some criteria and weak in others, making the selection process a compromise. For in-depth information on the most commonly used animals in toxicology (mouse, rat, hamster guinea pig, rabbit, ferret, dog, minipig, and primate, see *Animal Models in Toxicology, 2nd Ed.* (Gad, 2007a). Although this thorough (933 pp) book is not oriented toward inhalation exposure, it covers the history, husbandry, study designs, dosing methods, pathology (by organ system), metabolism, model selection and scaling, clinical pathology, laws and regulations relating to animal research, and commercial sources of laboratory animals. Two other large monographs (Derelanko and Hollinger, 1995; Massaro, 1997) extensively cover toxicology models, methods, and regulations: They also have substantial material on inhalation toxicology. For discussions of species considerations for inhalation studies, see Cantor (1989), Pauluhn (1994b), and Koenig and Luchtel (1997). Parent's (1991) book, *Comparative Biology of the Normal Lung*, is a valuable resource for inhalation toxicologists.

Models of Human Diseases

Models of human diseases generally fall into two categories: natural and contrived. An impressive report, *Naturally Occurring Animal Models of Human Disease: A Bibliography* by Bustad et al. (1975) contains nearly 900 citations. Only naturally occurring diseases are included, with most infections omitted. Organization is by organ system—alimentary, cardiovascular, ear, endocrine, eye, hematopoietic, musculoskeletal, nervous, reproductive, respiratory, skin, and urinary—and by disease class. Table 9.1 lists the respiratory system disease entries. An earlier review by Cornelius (1969), citing over 300 references, covers about 300 animal disease models. Animal models of human lung diseases cited include pulmonary emphysema (cattle and horse), atelectasis (horse), pulmonary adenomatosis (cattle), pneumonia (dog), and induced lung tumors (mouse). A concise, but important, review on large animal models of human disease by Lewis and Carraway (1992), lists 12 respiratory diseases (including cancer, emphysema, asthma, infections, and hyaline membrane disease). The animals include cats, dogs, sheep, goats, swine, cattle, horses, and nonhuman primates.

Many common human respiratory viruses also infect laboratory animals. A review by Clyde (1980) covers models of respiratory syncytial virus (ferret, cotton rat, chimpanzee, and calf), parainfluenza type 3 (hamster and guinea pig), adenovirus (hamster, dog, rabbit, and piglet), and rhinovirus, which is responsible for common colds (cattle, chimpanzee, and gibbon). The ferret contracts human influenza, and it is an important model in virus and vaccine research (Maher and DeStefano, 2004). In recent years, several new animal infectivity models have been introduced: The extensive review by Conn et al. (2000) covers models of bacterial and viral pneumonia, and influenza, and discusses the strengths and weaknesses of dozens of models. The models include mice, rats, hamsters, guinea pigs, rabbits, ferrets, calves, lambs, chimpanzees, and nonhuman primates. Another review by Bakker-Woudenberg (2003) covers experimental designs, species selection, exposure techniques, and infection monitoring methods. The review includes viral, bacterial, fungal, and parasitic pulmonary infections. Patterson and Carrion (2005) discuss the increasing demand for nonhuman primates in infectious disease research. This demand is largely driven by pressure to develop vaccines for emerging infections and bioterror threats: other animal models do not permit the needed rapid development and approval of vaccines. These papers are valuable starting points for designing studies of pulmonary infections. Many of the

Table 9.1 Naturally Occurring Respiratory Disease Models

Animal model	Species
Allergic alveolitis	Cattle
Bacterial pneumonia	Subhuman primates
Diffuse fibrosing alveolitis	Cattle
Emphysema	Cattle, horses
Hypersensitivity pneumonitis	Cattle, horses
Influenza	Subhuman primates
Lung injury	Subhuman primates
Pulmonary alveolitis	Rats, cattle, rabbits
Respiratory distress syndrome	Cattle
Tuberculosis	Subhuman primates
Lung tumors	Subhuman primates, rodents

Source: Adapted from Bustad et al. (1975).

aforementioned animal models, although invaluable, do not accurately replicate the course of disease in humans (Fehrenbach, 2002; Bakker-Woudenberg, 2003; Shapiro, 2007).

Human emphysema can be classified into four types: panlobular, centrilobular, paraceptal, and irregular. Each type involves different anatomical locations, but they all involve the permanent destruction of walls of airways beyond the terminal bronchiole. Among the laboratory models for emphysema is the blotchy mouse which has a genetic defect in connective tissue synthesis, the rat and guinea pig exposed to high (over 75 ppm) levels of NO_2, dogs and rabbits exposed for long periods to cigarette smoke, and a variety of animals after intentional exposure to proteolytic enzymes. The blotchy mouse model presents an opportunity for study of biochemical and cellular events leading to emphysema. Such animal models are primary for testing theories of the events leading to human diseases. Transgenic and other animal models, although valuable, suffer from not faithfully replicating the complex events seen in the human disease (Fehrenbach, 2002; Shapiro, 2007).

Port and colleagues (1977) compared normal and emphysematous lungs from nine mammalian species, including the human, using light and scanning electron microscopy. The studies were undertaken in recognition of the need for laboratory animal models to understand the roles of "environmental, chemical, and microbiologic factors" in the development of emphysema. Included in the study were emphysematous lungs, from rats and mice exposed by inhalation to NO_2 and from hamsters exposed intratracheally to papain (a proteolytic enzyme): The human and horse lungs had mild emphysema. Because alveolar wall pores (*pores of Kohn*) are believed by some to progressively enlarge in emphysema, the average number of pores per alveolus was quantified in each species (Table 9.2). Among their conclusions were that alveolar pore enlargement was involved in human and horse emphysema but not in NO_2, or papain-induced emphysema, and the recommendation that alveolar pore number be used as one guide in the selection of animal models for studies of human emphysema. The special value of the horse in emphysema research is due to its similarity to humans in gross and subgross anatomy (in pleural and septal structure), as pointed out by Tyler and Julian (1991).

Kilburn (1976) discussed models for studying the biological effects of air pollutants (Table 9.3). He reviewed the known effects on humans of air pollutants, ascribing these effects to two causes: impairment of lung defenses; and airway responses such as narrowing that may be protective but lead to altered pulmonary function. These effects may eventually

Table 9.2 Number of Alveolar Pores (Pores Of Kohn) in Normal Lungs of Various Mammalian Species as Determined Using Scanning Electron Microscopy

Species	No. alveoli examined	No. pores per alveolus	Age of subjects
Rat	123	1.28	270 days
Mouse	226	3.08	200 days
Hamster	218	3.64	180 days
Rabbit	122	4.70	9 months
Human[a] (nonsmoker)	150	6.14	73 years
Dog	136	7.20	28 months
Guinea pig	119	7.93	2 years
Monkey	54	8.50	2 years
Horse[a]	151	27.13	?

[a] "Normal" portions of emphysematous lungs were selected for quantitative study.
Source: Adapted from Port et al. (1977).

Table 9.3 Animal Models for Human Diseases Caused by Air Pollution

Human diseases	Spontaneous models	Contrived models
Asthma	Dogs with IgE antibody	Guinea pig: anaphylaxis
Bronchitis	Rats with mycoplasma	Rats: SO_2 exposure
Bronchiolitis	None	Rats and guinea pigs: NO_2 or O_3 exposure
Cancer	Mice	Hamster: benzo(a)pyrene with ferric oxide particles
Diffuse alveolitis	Sheep	Guinea pig and monkey: *Aspergillus*
Emphysema	Horses, rabbits	Dog: PMN leukocytes; rat: Cd exposure
Fibrosis	Kimberly horse disease	Dog: silica exposure; rat: paraquat exposure
Pulmonary edema	Cattle with brisket disease	Dog: α-hapthol thiourea exposure: cow: 3-methylindole exposure

Source: Adapted from Kilburn (1976).

lead to several disease states including bronchitis, bronchiolitis, asthma, diffuse alveolitis, bronchogenic cancer, emphysema, fibrosis, and pulmonary edema.

More recently, emphasis has been placed on introducing new compromised animal models into inhalation studies of particulate air pollution (Nikula and Green, 2000; Kodavanti and Costa, 2001: Costa and Kodavanti, 2003). An entire issue of *Inhalation Toxicology* (12(9), 2000) has extensive reviews covering models of pulmonary infection (Conn et al., 2000), asthma (Bice et al., 2000), age and susceptibility (Mauderly, 2000), and cardiac disease (Muggenberg et al., 2000). These thorough reviews by well-known researchers collectively cite over 700 references. These papers underscore several points: no single model is adequate for understanding a given human disease; animal model development is still in an immature state; and any particular study must be designed with both the strengths and limitations of the model clearly in mind.

Bland (2005) reviewed animal research on neonatal "new bronchopulmonary dysplasia (BPD)." Premature baboons and lambs on ventilators while being surfactant treated were called "authentic models." Such models have significantly improved the treatment of ventilated newborns suffering from BPD.

Stuart (1976) discussed the characteristics of several species commonly used in inhalation studies aimed at evaluating human risks, citing over 150 references. The paper was geared toward evaluation of hazards from ionizing radiation with an emphasis on cancer production. The species discussed include the laboratory mouse, rat, hamster, beagle dog, miniature swine, subhuman primates, and equines (horses, ponies, donkeys, etc.). After discussing some of the uses, advantages, and disadvantages associated with each species, Stuart made several recommendations: use at least three species; use several dose levels; use the same exposure route of exposed human populations; and use animals whose metabolism of the studied agents is similar to that of humans.

Asthma involves bronchial irritation, constriction, and excess secretions. Patterson and Kelly (1974) compared human asthma and similar disease states in guinea pigs, dogs, rhesus monkeys, and some asthma-related in-vitro models. They described two types of human asthma: extrinsic, which can be produced by the reaction of immunoglobulin (Ig)E antibodies with an inhaled antigen; and intrinsic, which is not closely related to antibodies and is of unknown cause. No animal models of intrinsic asthma were well understood at that time, so the review was directed toward the extrinsic (IgE antibody) form. The authors indicated that the dog is known to naturally exhibit human-like asthma. The guinea pig can be artificially immunized so that it develops respiratory obstruction upon inhalation challenge with antigen. Thus, experimental preparations using the guinea pig or its isolated bronchial tree muscles have been useful in studying immunologic and neurotransmitter phenomena in bronchial constriction. The dog spontaneously develops circulating (in blood) antibodies to inhaled ragweed and other pollens. The authors discussed the ways in which canine and human extrinsic asthma are similar and different. They indicated that although an IgE-mediated clinical asthma has not been demonstrated in subhuman primates, it is likely to exist. Rhesus monkeys can be sensitized by administration of human IgE antibodies, resulting in an acute respiratory response upon antigen inhalation, but unlike humans, monkeys require fresh resensitization each time the challenge experiment is repeated. This may be due to the production of antibodies that act to destroy the human IgE. Pulmonary function changes in sensitized dogs or monkeys upon challenge with aerosolized antigen are very similar to those of asthmatic humans. More recently, a sensitized (by ragweed injection) specially bred beagle dog has been proposed for the study of immune mechanisms in asthma, and evaluation of therapies (Redman et al. 2001). The types and limitations of current animal models of human asthma were discussed by Kurucz and Szelenyi (2006) and by Zosky and Sly (2007) from the viewpoint that the human disease is unique and not fully replicated in animal models. Pauhuln (2006) reviewed small animal models of asthma, pointing out the sensitized Brown Norway rat and the guinea pig's roles in testing allergenicity, and in mechanistic studies.

In summary, Table 9.4, serves to introduce the great variety of laboratory animal models available to the inhalation toxicologist.

EXTRAPOLATION FROM LABORATORY ANIMALS TO HUMANS

"There appears to be worldwide agreement concerning the fact that extrapolating laboratory animal toxicity data to man remains a major unresolved problem in toxicology." (Dixon, 1976). Few would contest Dixon's words, but the impetus is so strong that extrapolation of data is a common practice: The reasons include the health, economic, and other pressures to make decisions before a human database is available. Extrapolations are also necessary when tests on humans cannot be done due to lack of appropriate nondestructive endpoints, and for planning definitive human studies.

Extrapolations must always be considered tentative until validated by human epidemiologic, clinical, or other data. Examples of conditions under which tentative extrapolations from experimental animal studies to humans can be justified include the following.

- Several species were studied such that a generalizable mammalian response pattern was identified and the human response was extrapolated on the basis of body size, life span, etc.
- The biochemical or physiological mechanisms of the experimental animal's response are well understood along with the relevant animal and human similarities and differences in biochemistry and physiology.
- The studied agent belongs to a chemical or physical class in which other members have been studied, and animal and human responses directly compared.
- Validated dose-adjustment, toxicokinetic, and metabolic models support extrapolations.

A number of authors have addressed the complex problem of extrapolation of experimental animal inhalation toxicity data to humans (Oberdörster, 2001; Witschi, 1994;

Table 9.4 Selected Laboratory Animal Models used to Study Some Lung Diseases

Human lung disorder	Characteristics of human disorder	Some potentially useful animal models
Allergic alveolitis (hypersensitivity pneumonitis)	Tissue reaction to inhaled antigenic organic aerosols leading to infiltration of tissues by inflammatory cells, edema, fibrosis, and emphysema	Rat, cattle, rabbit, mouse, horse, cow, dog, sheep, guinea pig, and monkey
Bacterial pneumonia	Acute lung infection accompanied by excess secretion, edema, and inflammation	Dog, rodents, and primates
Bronchial asthma	Heightened reactivity of bronchial tree to stimuli leading to bronchial constriction and inflammation. Wheezing, cough, dyspnea, and tenacious sputum may be present	Guinea pig, dog, horse, monkey, sheep, rat, and mouse
Bronchiolitis	Inflammation and hardening of bronchioles (small noncartilageneous bronchi); may be accompanied by tenacious sputum and airway blockage due to ingrowth of connective tissue	Dog, various rodents, and monkey
Bronchitis	Enlargement of mucus secreting glands of bronchi accompanied by airway luminal narrowing by mucous plugs, edema, inflammation, and fibrosis	Monkey, horse, and dog
Cancer	Cellular proliferation of primary or metastatic origin leading to occlusion, compression, or destruction of normal airway structures	Cattle, mouse, hamster, swine, primates, equines, and dog
Emphysema	Destruction and permanent loss of septal walls of alveoli, alveolar ducts, and respiratory bronchioles.	Horse, cattle, dog, rabbit, rat, mouse, guinea pig, and hamster
Fibrosing alveolitis	Irreversible thickening of alveolar septa by formation of fibrotic (scar-like) tissue. May be associated with edema, presence of inflammatory cell response and bronchiolization of alveolar walls, especially in early stages	Horse, rat, dog, hamster, and guinea pig
Influenza	Highly contagious infection of the upper airway potentially spreading to the lung parenchyma, producing bloody engorgement and edema of tissue and infiltration of mononuclear cells and neutrophils	Ferret, horse, rat, cattle, chimpanzee, and mouse

Table 9.4 Selected Laboratory Animal Models used to Study Some Lung Diseases—cont'd

Human lung disorder	Characteristics of human disorder	Some potentially useful animal models
Pneumocniosis	Any of a number of changes in lung structure due to the persistence of inhaled inorganic dust. Particulars vary depending on the disease-producing dust, but wheezing, excess muscus, emphysema, fibrosis, and infection are common complications	Rat, monkey
Simple mechanical injury	Contusion of chest leading to edema, hemorrhage, and collapse of bronchi and bronchioles	Nonhuman primates, sheep
Tuberculosis	Bacterial infection with initial pneumonia-like inflammation leading to formation of tubercles which are encapsulated nodules progressing through stages including fibrosis, calcification, liquefaction, etc. resulting in destruction of normal lung tissue	Nonhuman primates, rabbit, rodents
Viral and bacterial diseases related to bioterrorism	Severe, often rapidly fatal, diseases including anthrax, plague, etc.	Nonhuman primates, ferret, rodents, lambs

Mauderly, 1996b; Kodavanti and Costa, 2001; Gerde, 2005; Jarabek et al., 2005). Dixon (1976) reviewed the general extrapolation problem, emphasizing several points attributed to Weil (1972).

- One or more species should be used that handle the test material similarly to humans. Metabolism, absorption, excretion, storage, and other physiological factors should be considered.
- Several dose levels should be used, unless a large maximum realistic human exposure dose is found to be nontoxic.
- The biologically insignificant (regarding risk) dose should be found and a safety factor above this dose applied for human protection.
- Randomization of dosed and control animals must be done to ensure statistical validity.
- The routes of administration to animals should match those anticipated for humans.

The issue of metabolism of a toxicant includes all significant chemical transformations of the compound. Metabolites may be more, or less, toxic than the original material, and species differences in metabolic pathways are often significant. Parent's edited book, *Comparative Biology of the Normal Lung* (1991) includes 10 chapters on mammalian lung biochemistry. Similarly, Gardner et al.'s, *Toxicology of the Lung, 3rd Ed.* (1999) contains several chapters relating to metabolic and immunologic aspects of humans and animal models. These works provide valuable guidance for selecting animal models.

Individual variations in responses of laboratory animals, which usually are relatively well matched and healthy, may be smaller than those for heterogeneous human populations. A usual means for handling this problem is the assignment of a safety factor, which is some multiple of the lowest effective dose found in the laboratory. One must carefully differentiate extrapolation factors found during the use of animals in preclinical drug trials where the dose is well known, from studies of environmental exposure where doses received by humans are much more variable. Still, information from preclinical trials according to Dixon (1976) shows that when maximum tolerated (no untoward effects) doses (of anticancer drugs) in monkeys and dogs are found, and 1/10 of this dose (on a mg/kg basis) for the more sensitive of these two species is given initially to human volunteers, there is only a 3% risk of exceeding the maximum tolerated dose. If the humans receive 1/3 of the most sensitive of the two species maximum tolerated dose, then about 6% will experience adverse effects. Depending upon the potential importance of the tested drug to future patients and the nature of the untoward effects, such risk levels may be warranted in phase I human clinical trials. Such risk factors are clearly not low enough for chemicals that may produce irreversible debilitating injury. Derelanko and Hollinger's, *CRC Handbook of Toxicology* (1995) has some excellent chapters on U.S. and European approaches to risk evaluation, including the assignment of safety factors.

Krasovskii (1976) discussed the extrapolation of laboratory animal data for the protection of humans exposed to environmental toxicants. The underlying key assumption was that man is a "typical" 70-kg body mass mammal. Krasovskii attempted to derive a "calculated man" using other mammalian data, and the typical human body weight. Using 86 regression equations (i.e., 86 characteristics of mammals) he concluded, "only in the case of four biological parameters did the computed values not agree with the actual values ... the life span indices for man, the relative weight of the brain and the amount of oxygen consumed by the brain exceeded the computed values by 4.7, 8.8 and 11.2 times respectively, and man lagged behind the animals... with respect to the body weight doubling period by 7.7 times." The proposed method also used an extrapolation coefficient, which is the ratio of activity (i.e., quantitative measure of response) in the animal model to that seen in humans for a given dose of the compound. Thus, if an effect (e.g., 10% depression of an enzyme level) occurs at a dose of 1 unit per kilogram body mass in the dog and at a dose of 6 units per kilogram body weight in the human, then the extrapolation coefficient from dog to man is 1/6. For many events (minimum effective dose, response to a given dose, etc.) the plot of logs of the extrapolation coefficients versus logs of body weights can be fit by a straight line. Use of this extrapolation technique requires obtaining information on at least four laboratory species, verifying the log-linear relationship for them, and then extrapolating to humans using a log-log regression on effect versus body weight. Using this method, the average error in extrapolation to humans for a variety of unnamed substances was stated to be not greater than "3 to 4 times." The author suggested that when this method is used, a safety factor of 4 be used to estimate the human equivalent dose. The safety factor decreases the probability of seeing unexpectedly greater effects in humans, or response in humans to some low dose. This method is clearly useful for formulating the extrapolation problem in quantitative terms, but it should not be blindly relied upon in human protection. Krasovskii cautions others in applying his method to anyone but the "average man," as age, gender, and state of health are known factors that can modify the responses to an agent.

Olson et al. (2000) surveyed 150 pharmaceutical compounds and concluded that animal models are of value in predicting significant human toxicities. Rall (1979) addressed the relevance of animal toxicological experiments to humans from the point of view of application of experience gained during drug testing to modern environmental issues.

As evidence for the success of using laboratory animal data for human protection he presented data (Fig. 9.1), which correlates BFD_1 mouse lethality data with maximum tolerated human doses in clinical trials for over 20 compounds. The technique of establishing and exploiting such an empirical relationship between mouse and man is characteristic of what Rall calls the scientific "lumpers." On the other hand, the "splinters" dissect every example, every report of species differences, to try to determine what they can learn, and what different mechanisms are used by the species that respond differently to accomplish the same physiological function. These analytical studies are of immense importance to biomedical science. But when broad guiding principles are considered, he concluded that "the lumpers are the more important." He also stressed the symbiotic relationship between well-conducted laboratory animal experiments and well-conducted human epidemiologic studies in protecting human populations.

More recently, the U.S. Environmental Protection Agency has explored sophisticated, quantitative techniques for extrapolating inhaled particle doses from laboratory animals to humans, taking into account several important biological differences (Jarabek et al., 2005). From the foregoing, it is clear that much progress is being made regarding the extrapolation of laboratory animal data to human populations. It should be equally clear that the general problem is far from being solved.

COMPARATIVE DOSE DISTRIBUTION

Dose and Dose Variability

When subjects are given an agent by any route, the term *dose* is used as a quantitative measure of the amount administered. Thus, the dose of a barbiturate given intravenously

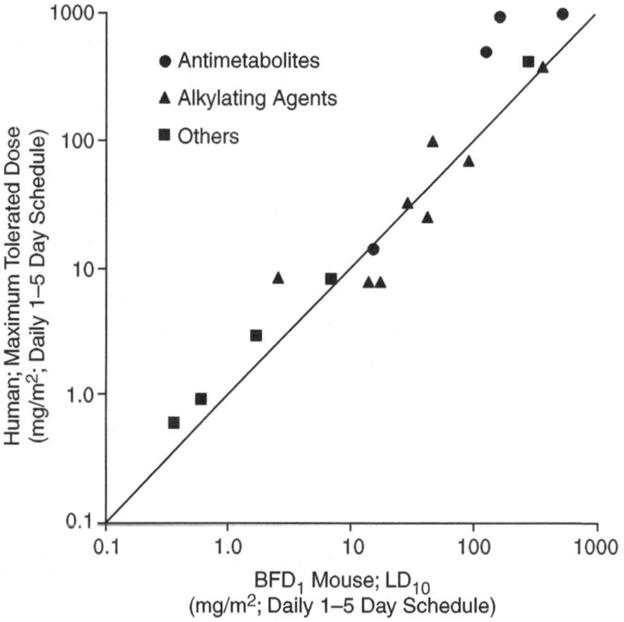

Figure 9.1 Relationship between lethality of anticancer drugs in mice and maximum tolerated doses as determined in human clinical trials. *Source*: Redrawn from Rall, (1979).

to a dog may be 300 mg. Because a given effect, for example, unconsciousness, is produced by different amounts of the drug in different sized dogs, the dose is often expressed in terms of amount of agent per unit of body mass, say 30 mg of barbiturate per kilogram of body mass (30 mg/kg). If the action is principally on the central nervous system, the *effective dose* is that amount of drug present in certain regions of the brain. Reflection will indicate that determination of this effective dose is impossible without destroying the subject, and that the concentration of a drug in a tissue varies with time, climbing to a peak and then declining. The kinetics of drug buildup and elimination in the target tissue will depend upon many factors including the distribution of blood, tissue permeability to the drug, characteristics of other tissues that absorb, store, and release the drug, and rates of metabolism and excretion. Many of these factors are likely to be changed by the drug, as is the sensitivity of the target tissues. Thus, not only is it difficult to theoretically define the dose in the target tissue, but also this dose will change with time, and the instantaneous value may not closely correlate with the effects being seen in the subject. It should also be realized that individuals differ widely in sensitivity and in each of the factors that determine the time-course of concentration of an administered substance in a given tissue. Thus, a population which is given a specific amount of an active chemical will exhibit a distribution of responses.

With respect to airborne particles or gases, the dose to a given tissue depends upon a chain of events. Inhalation will lead to an initial deposition pattern within the respiratory system. Contact of the substance with airways may in itself alter the deposition pattern via stimulating defense mechanisms such as bronchoconstriction, cough, and altered ventilation. Once deposited, the material will immediately enter a dynamic situation. Portions of the material may dissolve in the surrounding fluid. Other portions may be chemically transformed, and portions may be removed from the deposition site. That portion entering the bloodstream will usually behave as if it were injected into a blood vessel. Material that moves directly into tissue spaces, lymphatics, or the gastrointestinal tract may also lead to general distribution throughout the body. Once material is in a tissue, it may alter the structure or function of the tissue and produce effects. The magnitude of the effects depends both upon the dose delivered to the tissue and upon the sensitivity of that tissue. The chain linking inhaled dose and effects is long and complex. Detailed toxicological effects models that incorporate species anatomy and physiology, and chemical characteristics of gases and particles, have been described by Andersen (2003) and Jarabek et al. (2005).

A useful paper, "Variablility in target organ deposition among individuals exposed to toxic substances," was published by Cuddihy et al. (1979). In this paper, a combination of empirical data and theoretical modeling was used to describe the population distribution of organ uptake of inhaled materials. Dogs under laboratory exposure conditions, and free-living humans were compared. In the case of controlled laboratory administration of aerosols to beagle dogs, they found that 2% of the animals received more than 3 times the group mean organ doses. For humans exposed to airborne material in the environment, organ dose variability was greater. A few percent of the exposed individuals received over five times the average population doses to internal organs. Further, the variation in organ doses was much greater when the route of intake was inhalation instead of ingestion. Their findings are summarized in Table 9.5.

Assuming, after Cuddihy et al. (1979), a lognormal population dose distribution, one can examine the effect of the geometric standard deviation on the fraction of the population that is expected to receive a dose in excess of (or below) some multiple of the geometric mean population dose (median dose). In Figure 9.2 (drawn on log probability paper) one can see, for example, that for a reasonable geometric standard deviation of 3 about 2% of the population is expected to receive more than 9 times the median dose. The fraction

Table 9.5 Descriptive Parameters of Population Organ Dose Distributions Resulting from Inhalation and Ingestion of Metals

Population	Exposure	Principal route	Target organ	Geometric mean	Geometric standard deviation
Beagle	Metals	Inhalation	Bone or liver	0.9	1.8
Humans	Ti (soil)	Inhalation	Lung	34 μg/g dw	3.4
Humans	Al (soil)	Inhalation	Lung	30 μg/g dw	3.4
Beagle	Sr	Ingestion	Bone	6.1% Sr[a]	1.3
Humans	Sr (fallout)	Ingestion	Bone	0.13 pCi/g of Ca	1.8
Humans	Pb	Inhalation	Lung	0.17 μg/g ww	1.7

[a] Total amount of radioactive Sr ingested in 250 days.
Abbreviations: dw = dry weight; ww = wet weight.
Source: Adapted from Cuddihy et al. (1979).

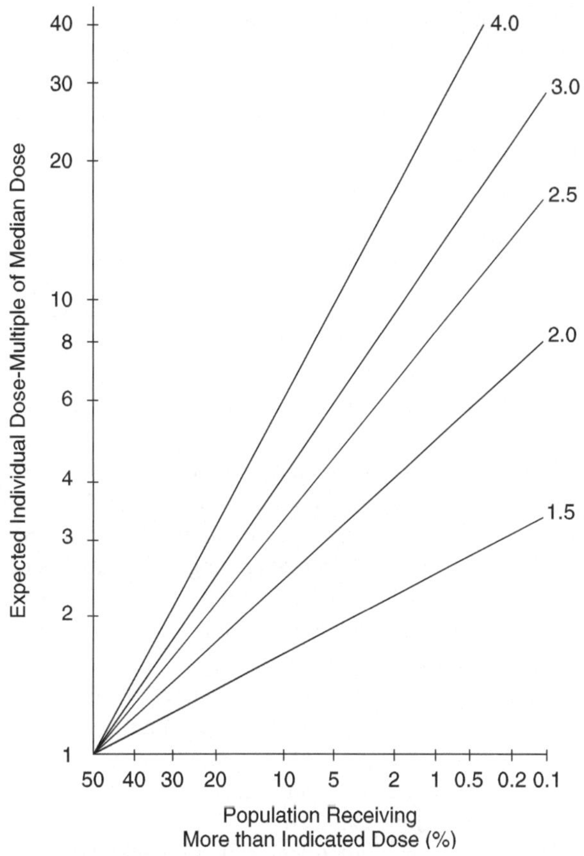

Figure 9.2 Distribution of individual organ doses (geometric standard deviation) in a population exposed by inhalation or ingestion to a material the accumulates in tissue. Lognormal distributions are assumed to be valid.

of the population receiving less than the median dose can be obtained by using the inverse of the median dose. Although one must be careful about trusting extrapolations into the tails of such distributions, it is known that very high doses for small fractions of a population actually occur. Based upon human autopsy data after environmental exposure, some individuals can receive 10 or more times the population median organ doses. If initial inhaled particle deposition doses to sites of high deposition in the tracheobronchial tree are modeled, then factors such as exercise, uneven ventilation, and normal biological variations should be considered. Using this approach, one can identify high-dose individuals that can receive local tissue doses that exceed the average surrounding tissues by a factor of 100 or more (Phalen et al., 2006). Autopsy lungs from nonsmokers support this conclusion, showing particle deposition hot spots that are about 100 times greater than surrounding tissues (Churg and Vedal, 1996).

Comparative Minute Ventilation Per Unit Body Mass

When evaluating an animal model for an inhalation study, one dose-related consideration is the mass-specific minute ventilation, or volume breathed per minute per unit of body mass (or weight) of the animal. Data on various mammals are available (Crosfill and Widdicombe, 1961; Altman and Dittmar, 1974; Mauderly, 1974a,b; Boyd and Mangos, 1981; Lindstdet and Schaeffer 2002). One can estimate the weight-specific ventilation for a mammal by use of relationships derived from the formulas of Guyton (1947). His relationships for tidal volume (V_t), breathing frequency (f), and body weight (w) are:

$$V_t\ (\text{cc}) = 0.0074\ \text{w}\ (\text{g}) \qquad \text{(eq. 9.1)}$$

$$\text{f}\ (\text{min}^{-1}) = 295/\text{w}^{\ 1/4}\ (\text{g}) \qquad \text{(eq. 9.2)}$$

Multiplying V_t and f to get minute ventilation V_m and dividing by body weight one obtains:

$$V_m\ /\text{w}\ (\text{cc/min/g}) = 2.18/\text{w}^{\ 1/4}\ (\text{g}) \qquad \text{(eq. 9.3)}$$

This relationship is plotted in Figure 9.3 along with data points from the references mentioned above. One sees from this figure that not only is the derived relationship reasonably valid, but also that great differences in weight-specific ventilation occur among mammals. Compared to humans, on a per unit body weight basis, dogs ventilate about 3 times greater volumes, and small rodents about 10 times greater volumes of air. Note also that the newborn human has a higher weight-specific ventilation than the adult. This higher ventilation per unit body weight in small animals leads to possible higher lung doses in these animals in relation to adult humans exposed under similar conditions. Conversely, one expects large animals to receive generally lower doses than smaller humans during inhalation exposures under similar conditions. Although Guyton's equations are widely used, more recent versions are available (Bide et al., 2000; Lindstedt and Schaeffer, 2002). The use of ventilation data in dose extrapolation alone can lead to erroneous conclusions, as particle deposition and clearance efficiencies, along with translocation, metabolism, and excretion characteristics, must also be considered (Fiserova-Bergerova, 1995; Jarabek, 1995; Lindstedt and Schaeffer, 2002).

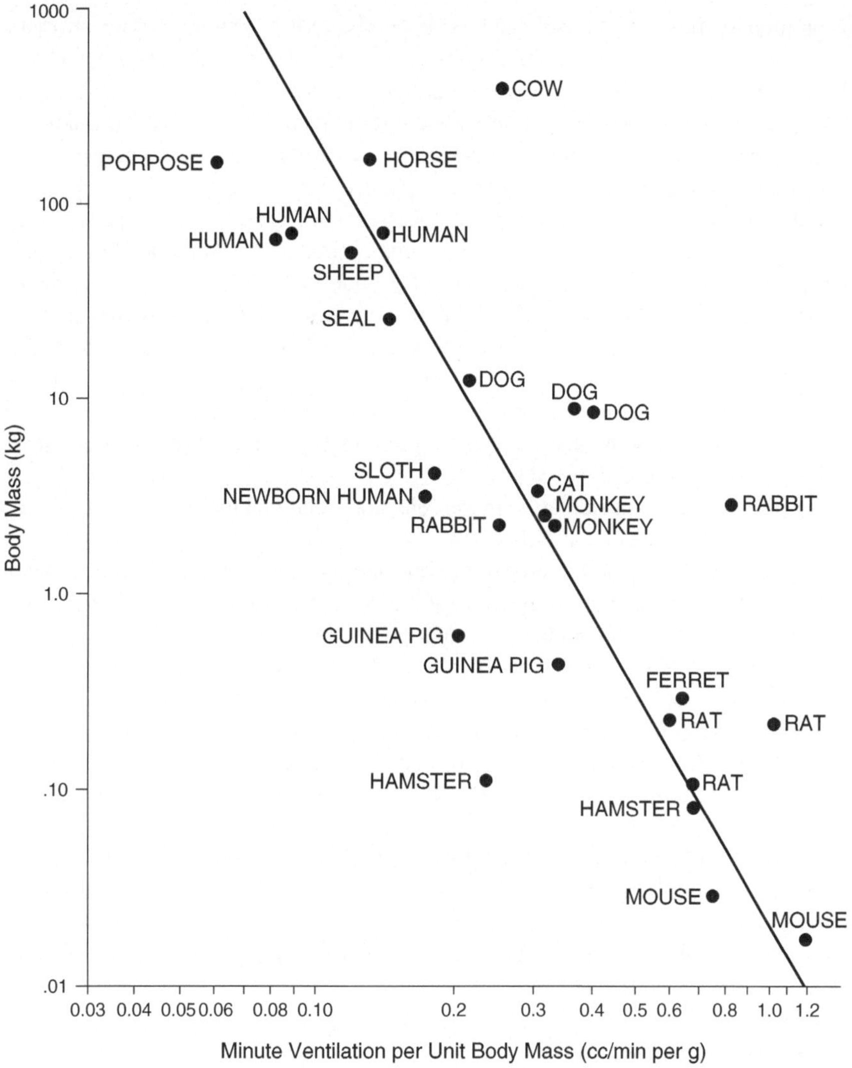

Figure 9.3 The relationship between body weight (log scale) and minute ventilation per unit body weight (log scale). Data points for various mammals were obtained from a variety of published data.

COMPARATIVE PHYSIOLOGY AND ANATOMY

Comparative Pulmonary Function

Several species have been widely used in laboratory inhalation studies: Three in particular (dog, mouse, and rat) have intimately shared our food, our air, and our homes for thousands of years. Other common laboratory species include the hamster, guinea pig, rabbit, ferret, nonhuman primate, and cat. When larger animals are used, the pig, sheep, goat, and horse are frequently chosen. Comparative anatomical and physiological data on the respiratory systems of these animals are available which allow comparison to humans (Parent, 1991; Newton, 1995).

Regarding pulmonary function, Crosfill and Widdicombe (1961) presented data on eight species, viz. mouse, rat, guinea pig, rabbit, monkey, cat, dog, and human. Respiratory data presented included lung weight, lung specific-gravity, functional residual capacity, tidal volume, tidal pressure swing, minute volume, breathing frequency, mean alveolar diameter, lung compliance, chest wall compliance, lung resistance, work of breathing, and derived characteristics including time constant of the lung (compliance times resistance) (Table 9.6). As another example of comparisons of various mammals, Table 9.7 by Boyd and Mangos (1981) includes original measurements on the human, dog, ferret, rabbit, rat, and hamster. For a larger animal, the pony, Table 9.8 shows original data of Mauderly (1974) along with literature values for dogs and humans. More recently, comparative pulmonary physiology in mammals has been reviewed and augmented by several authors (Lai, 1991; Jones and Longworth, 1991; Pauluhn, 1994; Newton, 1995; Schlesinger et al., 1997). It is not possible to use only such data to identify a priori appropriate animal models. Such data do, however, give normal values for various species, and provide some insight regarding pulmonary characteristics that might be measured as endpoints in a toxicological study.

Another physiological consideration in the selection of animal models is the degree to which portions of the lung can be ventilated by *collateral shunts.* Collateral pathways are detected by injecting a gas, liquid, or aerosol into one airway and recovering it from another airway which is not directly downstream from the injection. Several types of collaterals are believed to exist within mammalian lungs: alveolar pores; interconnecting respiratory bronchioles; and accessory bronchiolar alveolar tubes. Collateral ventilatory shunts permit gas exchange to occur between the atmosphere and alveoli, even when the primary air pathway is occluded. Such ventilatory channels can provide reserve in the event of lung injury, and can likewise obscure measurements of degree of occlusion which are based upon assessing gas flow kinetics (such as multibreath gas washout tests). A review on airway obstruction and collateral ventilation published by Macklem (1971) presented information on the collateral ventilation of various species. It appears that the human, dog, cat, and rabbit exhibit significant collateral air flow, but the calf and pig (species with thick and complete interlobular septa) do not. Thus, the former laboratory animal species are more similar to humans with respect to collateral ventilation, but the latter species may be more sensitive for detecting and measuring certain types of lung damage.

Comparative Airway Anatomy

Along with the breathing pattern and physical and chemical properties of an inhaled material, the anatomy of the airspaces will determine how much of and where an air pollutant will initially deposit. Comparing the anatomy of various species helps in the selection of animal models and aids in extrapolation.

A rigorous approach to selection of species as well as extrapolation of findings involve consideration of the mathematical equations that describe deposition of gases and particles in airway-like structures. These equations indicate which anatomical parameters govern uptake, and thus are of interest. Equations for predicting gas uptake by the walls of simple tubes are given by Morgan and Frank (1977). The rate of uptake of a gas during passage through an absorptive tube is a function of the following parameters.

- The area of contact with the gas.
- The difference in partial pressure in the gas phase and the surface liquid phase.
- The ventilation rate.
- The tube diameter and length.
- The gas diffusivity.
- Henry's law constant (inversely proportional to the solubility).

Table 9.6 Mean Values for Selected Respiratory Parameters in Eight Mammalian Species

Animal	Body wt (kg)	Lung wt. (g)	Specific gravity (g/mL)	Tidal vol. (mL)	Minute vol. (L)	Frequency per min(min^{-1})	Alveolar diameter (μm)	Lung compliance per mL of lung (mL/cm H_2O)	Chest wall compilance per mL of lung (mL/cm H_2O)	Lung resistance per mL of lung (cm H_2O/L/s)	Work of breathing per gram body wt. (g cm/min)	Absolute work of breathing (g cm/min)
Mouse	0.032	0.20	0.42	0.18	0.021	109	39	0.11	0.68	395	4.25	66.5
Rat	0.25	1.6	0.51	1.55	0.16	97	59	0.12	0.55	277	1.90	482
Guinea pig	0.69	3.2	0.40	3.7	0.13	42	83	0.15	0.46	480	0.52	272
Rabbit	2.4	9.1	0.43	15.8	0.62	39	94	0.28	0.47	522	0.62	1,502
Monkey	2.45	22.3	0.20	20	0.70	33	89	0.12	0.069	1,028	0.31	817
Cat	3.7	20	0.23	34	0.96	30	133	0.16	0.16	865	0.63	1,857
Dog	12.6	82	0.23	144	3.1	21	74	0.12	0.098	480	0.55	6,720
Man	70	1,065	0.35	400	6.4	16	166	0.065	0.068	2,780	0.43	30,000

Source: Adapted from Crosfill and Widdicombe (1961).

Table 9.7 Mean Values for Selected Respiratory Parameters in Six Mammalian Species

Animal	Body wt. (kg)	Tidal vol. (mL)	Minute vol. (L/min)	Frequency (min^{-1})	Compliance per mL lung[a] (mL/cmH_2O)	Total lung capacity (L)
Hamster	0.122	1.2	0.028	24	0.263	0.007
Rat	0.233	1.4	0.242 (0.16)[b]	115	0.196	0.008
Ferret	0.314	3.98	0.195	44	0.175	0.050
Rabbit	3.14	23.9	0.190 (0.88)[b]	37	0.277	0.111
Dog	9.68	193	3.65	19	0.242	1.26
Human	70	500	6	12	0.083	6.00

[a] At expiratory level; [b] computed from frequency and tidal volume.
Source: Adapted from Boyd and Mangos (1981).

Thus, the major anatomical and physiological parameters that influence gas uptake include airway surface area, airway diameter, airway length, and gas flow rate. Hubal et al. (1996) reviewed gas uptake models for the upper respiratory tract. Tsujino et al. (2005) compared gas uptake simulations in humans, dogs, and rats for common pollutant gasses.

For particle deposition, the mechanisms of diffusion, sedimentation, and impaction are typically used in computational models. The models require the following airway anatomical, physiological, and particle parameters (Yeh and Schum, 1980).

- Airway tube length.
- Airway diameter.
- Airway branch angle.
- Airway angle to gravity.
- Air velocity in an airway.
- Particle diffusion coefficient.
- Particle diameter and density.
- The particle Cunningham slip correction factor.

Thus, airway length, diameter, inclination to gravity, and bifurcation branch angles, along with air flow velocity are the parameters of interest. The usefulness of any given animal model, including humans, is enhanced by having quantitative anatomical data.

An emerging modeling technique, computational fluid dynamics (CFD), models particle and gas uptake in realistic airway structures using scans of airways in living subjects (Kimbell and Subramaniam 2001; Martonen, 2001; Oldham, 2006). The CFD

Table 9.8 Mean Values for Selected Respiratory Parameters in Three Mammalian Species

Animal	Body wt. (kg)	Tidal vol. (L)	Minute vol. (L/min)	Frequency (min^{-1})	Oxygen uptake per unit wt. (mL/min/kg)
Pony	167	1.5	28.6	20	3
Dog	9	0.186	3.7	20	10
Human	74	ND	9.4	ND[a]	4

Abbreviation: ND = not done.
Source: Adapted from Mauderly (1974).

method is potentially very powerful in supporting extrapolations because realistic individual anatomies can be used.

The pioneering morphometric work of Weibel (1963a,b) on the human tracheobranchial tree has stimulated similar quantitative measurements on the airways of several species. Table 9.9 lists some sources of published respiratory tract anatomical data on various mammals. In general, human lungs have more symmetrical tracheobronchial airway branches than any of the common laboratory animals. That is, the diameter ratios of daughter branches of a given parent tube are more near unity in the human. In contrast, mice, rats, hamsters, guinea pigs, rabbits, ferrets, dogs, monkeys, and horses have strongly asymmetrical branching. This difference is evident in casts of the human and dog airways (Fig. 9.4). Similar species differences exist in airway lengths and branch angles (Phalen et al., 1978a; Schlesinger and McFadden, 1981; McBride, 1991; Phillips and Kaye, 1997).

The consequences of anatomical differences on the deposition pattern of inhaled particles are likely to be significant, based upon the dependence of the mechanistic deposition equations upon airway geometry (Landahl, 1950; Phalen et al., 1990). This anatomical effect on particle deposition was demonstrated directly by Schlesinger (1980) by deposition studies in hollow models of the donkey and human major tracheobronchial airways. In the first five generations, similar particle sizes and air flows produced differences in deposition at a given bifurcation by factors of two or more. Schlesinger (1985) and Schlesinger et al. (1997) published extensive information particle, gas, and vapor deposition (and particle clearance) for several common laboratory animals, based on their studies and a review of the literature.

Table 9.9 Quantitative Anatomical Data on Mammalian Respiratory Tract Airways

Species	Region measured	Parameters measured	Reference
Human	Tracheobronchial tree, Parenchyma	Length, diameter	Weibel, 1963a
Human	Tracheobronchial tree	Diameter, branch angles, lengths	Horsfield et al., 1971
Human, dog, rat, hamster	Tracheobronchial tree	Branch angles, parent and daughter length and diameter ratios	Phalen et al., 1978a
Human	Tracheobronchial tree	Length, diameter, branch angles, gravity angles	Yeh and Schum, 1980
Monkey, dog, rat	Nasal-pharyngeal region	Sectional area and perimeter	Schreider and Raabe, 1981
Rabbit, donkey	Major tracheobronchial airways	Parent and daughter tube diameter ratios	Schlesinger, 1980, 1985
Guinea pig	Tracheobronchial tree, parenchyma	Length, diameter	Schreider and Hutchens, 1979
Guinea pig	Tracheobronchial tree	Length, diameter	Kliment et al., 1972
Rat	Tracheobronchial tree	Length, diameter, branch angle, gravity angle	Yeh et al., 1979
Rat	Tracheobronchial tree, parenchyma	Length, diameter	Granito, 1971
Dog	Tracheobronchial tree	Length, diameter	Horsfield et al., 1982
Mouse	Tracheobronchial tree	Length, diameter, branch and gravity angles	Oldham et al., 1994; Oldham and Phalen, 2002

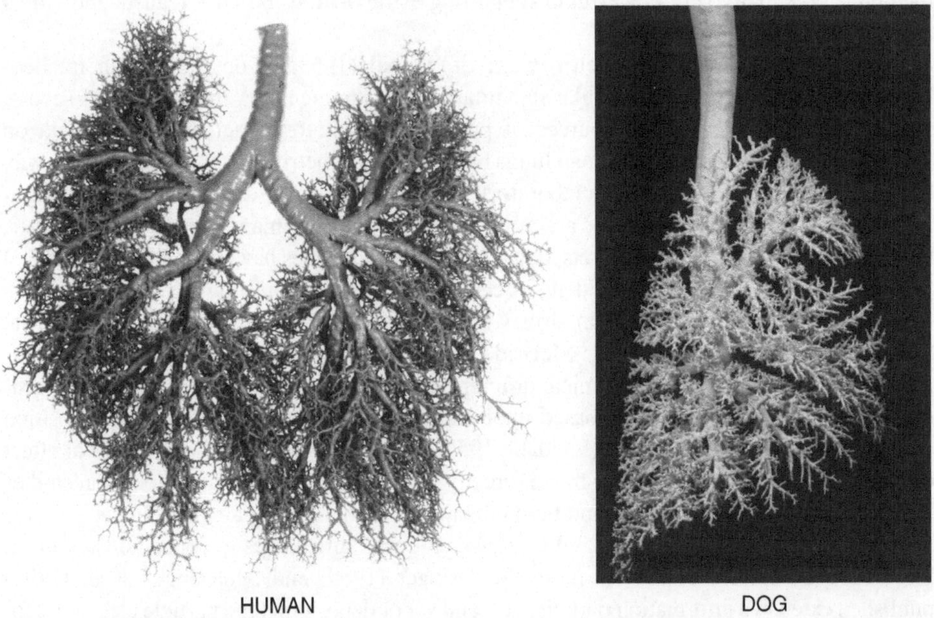

Figure 9.4 Tracheobronchial airway casts of the human and dog showing the more symmetrical branching in the human. The silicone rubber casts prepared in the thorax were hand-trimmed to remove alveolarized portions. Casts were made by the author and associates at the Inhalation Toxicology Research Institute (now, the Lovelace Respiratory Research Institute) in Albuquerque, New Mexico, United States.

A useful way of comparing various species airway anatomies can be borrowed from analytic geometry. As seen in Figure 9.5, the distance between points in space can be calculated from their coordinates. This concept was applied to comparing the airway anatomy of four species by two students of the author, D. Wilner and M. Numamoto. They measured tracheobronchial tree lengths, diameters, and branch angles in the first five generations using two ferret replica airway casts, and then computed three parameters and their standard errors: ratio of daughter pair diameters; ratio of tube length to diameter; and difference (in radians) between each daughter branch angle. Similar calculations were made from published values for other species. These values were then graphed in 3-space (Fig. 9.6). The computed distances (d) between each pair of points showed that the human was relatively distant from each of the other species; d = 1.33 to 2.36. Also, the species that was closest to the human was the ferret, and the species furthest from the human was the rat. Use of distance parameters as a factor in species selection appears promising if appropriate parameters are used and sufficiently large sets of data are obtained.

As was discussed in Chapter 2, experimental data on the deposition of inhaled particles indicate that mammals of similar body size appear to have similar total deposition efficiencies. Greater particle deposition variations are seen among humans than appear to exist, for example, between the "average" human and "average" dog, under similar conditions.

That great differences in nasal anatomy exist among mammals can be seen from the quantitative morphometry of Schreider and Raabe (1981) on the rat, dog, and monkey (Fig. 9.7). Particle deposition equations for small laboratory animals, with a comparison to humans, have been published by Zhang and Yu (1993), and particle deposition measurements in nasal casts of rats are also available (Cheng et al. 1990).

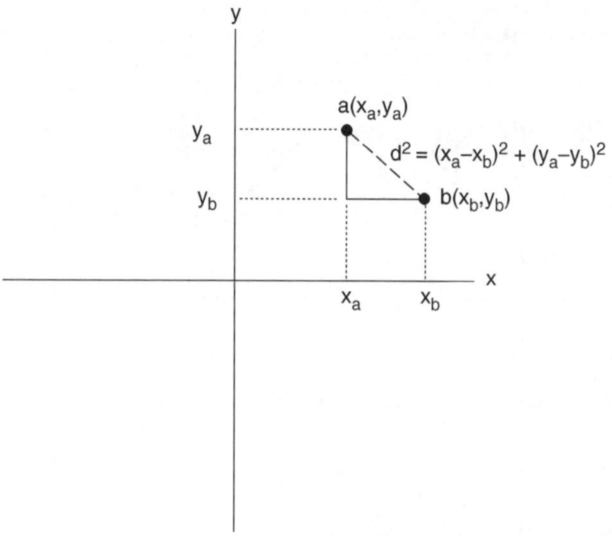

Figure 9.5 The distance between two points, a and b, in a two-space coordinate system is the square root of the sum of the squares of the differences in their orthogonal coordinates.

Several mammalian species were compared with respect to tracheal anatomy by Mawdesley-Thomas et al. (1970). Notable in Table 9.10 from this work is the absence, or small number, of tracheal glands in the rodent, rabbit, guinea pig, hamster, mouse, and rat. The authors concluded that these small rodents were useful for studies involving goblet cells, and that the sheep was a convenient model for study of the bronchial glands.

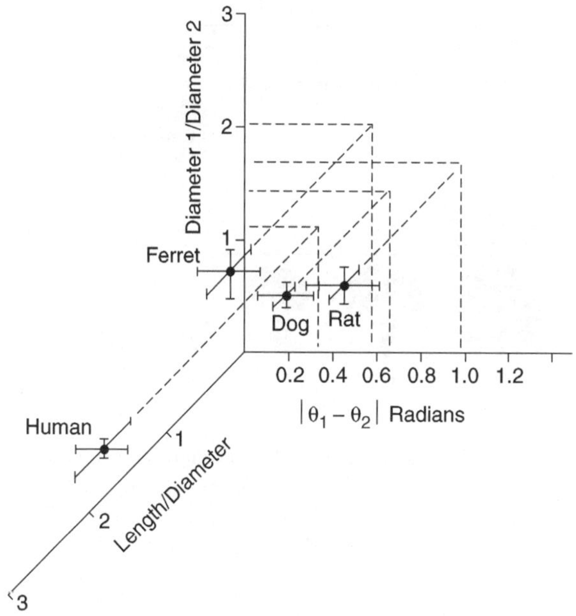

Figure 9.6 Comparative anatomy of tracheobronchial airways (generations 0 through 5) of human, ferret, dog, and rat. The distances between points in 3-space as calculated by the method shown in Fig. 9.5 are human–ferret, 1.33; human–dog, 1.50; human–rat, 2.36; dog–ferret, 0.69; rat–ferret, 0.61.

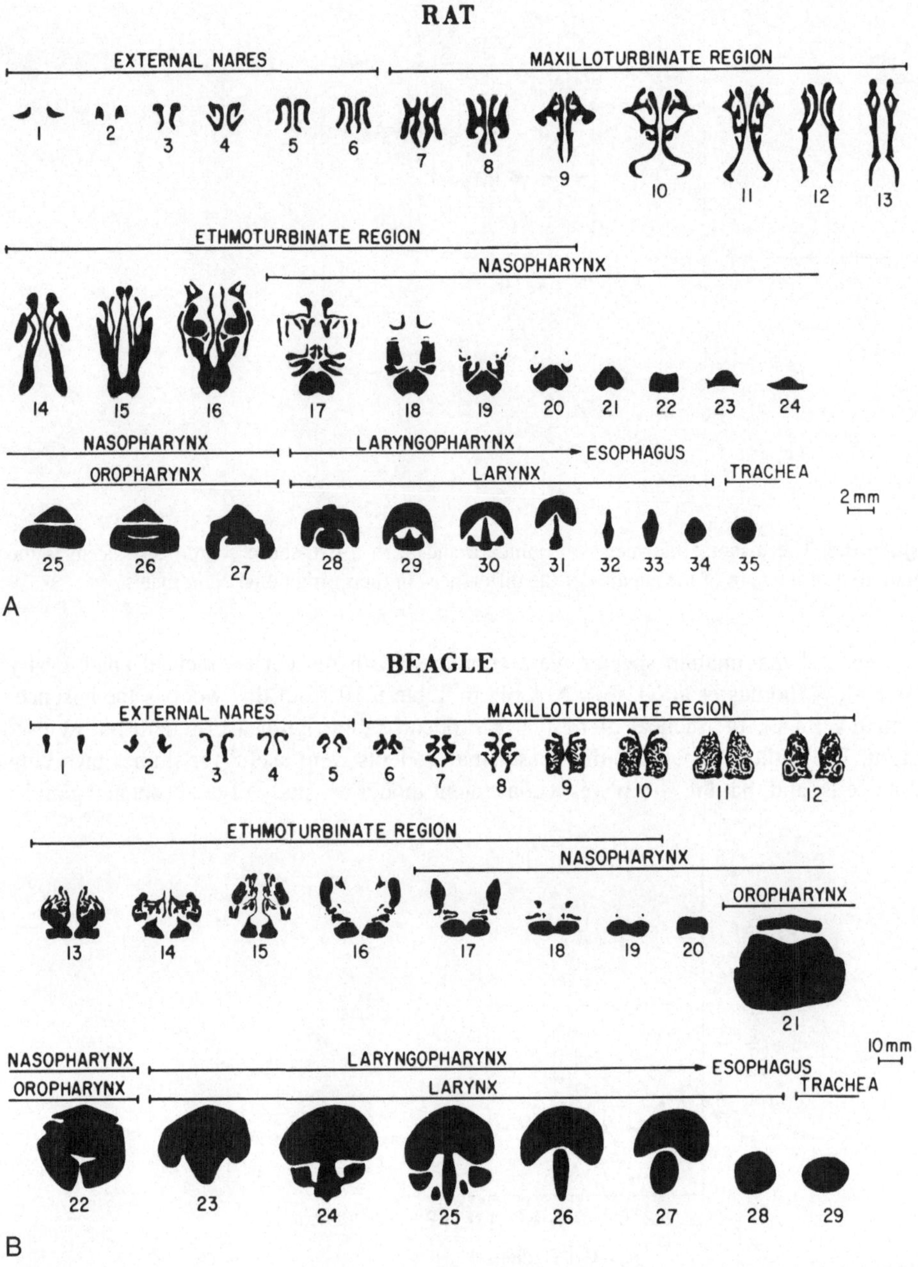

Figure 9.7 Cross-section of the nasopharyngeal airways of the (**A**) rat, (**B**) beagle, and (**C**) rhesus monkey. Areas and perimeters of sections were given by Schreider and Raabe. Figure supplied by J.P. Schreider and O.G. Raabe. *Source*: From Schreider, J.P. and Raabe, O.G. *Anat. Rec.*, 200(2):195–205, of Wiley-Liss, Inc., a subsidiary of John Wiley & Sons, Inc., 1981. Reprinted with permission.

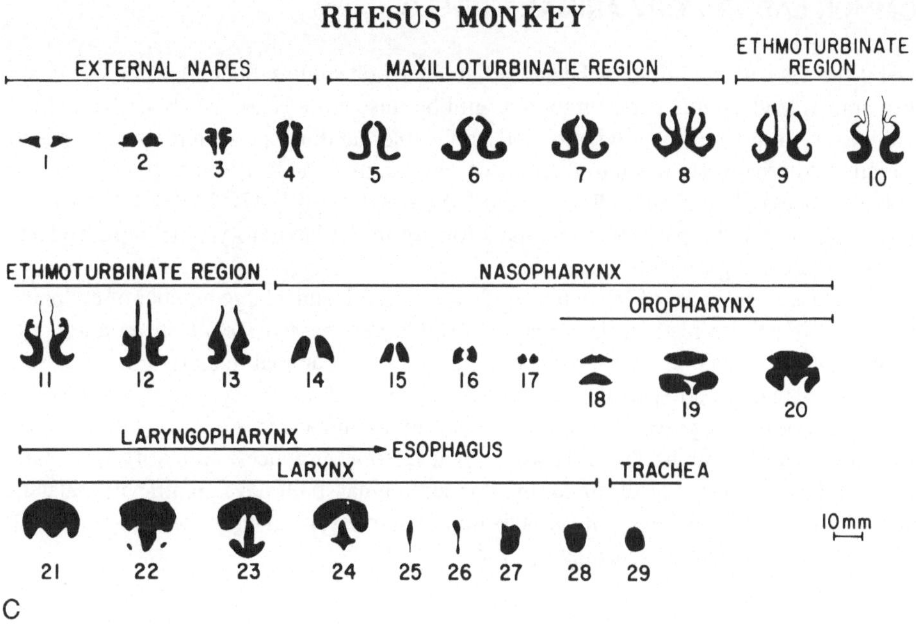

Figure 9.7 Cont'd

Parent's monograph (1991), *Comparative Biology of the Normal Lung*, has 38 chapters on anatomy, physiology, biochemistry, and defense mechanisms. A complementary volume edited by Harding et al. (2004) covers lung development and aging in 28 detailed chapters, and a review by Newton (1995) has 79 tables and figures of importance to animal selection and exposure methods in inhalation studies.

Table 9.10 Anatomical Data: Trachea of Various Species

Species	Length (cm)	Internal diameter (mm)	Goblet cells per cm	Glands
Dog	10	15	350	+++
Baboon	3.5	5	600	+++
Pig	13	10	150	+++
Macaca	6	7	300	++
Mouse	2	2	NR[a]	±
Rabbit	6.5	10	150	–
Guinea pig	3.3	4	600	–
Hamster	1.5	2	6	–
Sheep	24	26	300	+++
Cat	7	7	600	++++
Squirrel monkey	2.5	3	150	+
Rat	3.2	3	8	±
Man	11.0	20	200[b]	+++[b]

[a]NR = not reported, [b] Varies considerably.

Source: Adapted from Mawdesley-Thomas et al. (1970).

COMMON LABORATORY ANIMAL MODELS

Several species have traditionally been used in inhalation studies. Such repeated use of any given animal tends to make it even more useful because more is known about that animal, including its responses to pollutants and applicable handling procedures and exposure apparatus. Alternatively, the introduction of new animal models is usually a slow and involved process. A dedicated 2006 issue of the *ILAR J.* (Vol. 47[4], 2006) contains 11 relevant papers on "Preparation of animals for use in the laboratory," including rodents, rabbits primates, dogs, ferrets and swine.

Because many types of inhalation studies are useful, and a large number of endpoints are used, it is unwise to think in terms of a single *best animal model*. In an attempt to demonstrate the factors that go into selecting or evaluating animal models, a list of advantages, disadvantages and special characteristics for several useful species is presented in Table 9.11. For the purposes of this table, similarity to humans in any given characteristic is considered an advantage. Note, however, that in some instances, similarity to humans may be a disadvantage. For example, the dog and human both have significant collateral ventilatory pathways, making clinical detection of some types of lung damage more difficult than in species with less ventilatory reserve.

Dogs

Dogs are probably the oldest of the domesticated animals: Their bones have been found dating to the stone age (Andersen, 1970). The trend in recent decades has been to use purebred animals from *class A suppliers* (suppliers that raise animals on their premises) of animals instead of "strays" and "pound dogs" that have unknown histories and variable states of health. In the United States, the beagle is a favorite in inhalation studies. The reasons according to Andersen (1970) include: its medium size, moderate length of hair coat, even temperament, adaptability to living in groups, representative conformation for the dog, and the lack of need for cosmetic surgery. The beagle's excellent disposition and personality are assets, because special handling is seldom necessary and a minimum amount of restraint is required for most procedures. However, larger dogs are preferred in some studies, such as those involving surgical research, and epidemiologic investigations.

One must emphasize the necessity for treating dogs, and all experimental subjects, with utmost respect and kindness. This noble, cooperative animal, shaped by centuries of selective breeding, is capable of giving its complete loyalty and trust. Its long history as both servant and companion has provided it with a special place in our lives and hearts. Abuse or harsh treatment of dogs by members of the public or researchers should not be tolerated. Although laboratory animals in general are normally provided with better medical care, nutrition, and general treatment than are pets, it is important that the research community's reputation not be soiled by a few careless or unfeeling individuals. Scientists must protect access to this invaluable research subject by continuing to insist upon the highest standards of care and treatment in research.

Ferrets

The ferrets are carnivores in the family containing weasels, polecats, minks, badgers, skunks, sea-otters, and others. This little (adult male weight about 1 kg) carnivore, first domesticated by the Egyptians (Hahn and Wester, 1969) is available in two strains, Albino and Fitch (dark-brown coat). It can be tamed and handled much like the dog or cat. The greater expense of the dog, coupled with increasing pressure over the use of the dog,

Table 9.11 Information on Several Mammalian Species Commonly Used in Inhalation Studies

Mammal	Advantages	Disadvantages	Special considerations
Dog	Relatively long, 15-year life span. Useful for chronic studies. Convenient size for many procedures. Cooperative temperament, fewer stress artifacts. Nutritional, psychological needs well understood. Normal physiology, morphology well known. Great diversity of natural disease states. Pulmonary aerosol deposition similar to human. Availability of inhalation exposure apparatus designs. Useful as model for immunologic phenomena including asthma. Respiratory bronchiolar airways similar to humans. Useful as model for tracheal mucous movement. Model for pulmonary function measurements. Relatively defined lung growth and senescence. Nasal anatomy well described	Requires spacious housing. Care, feeding is expensive. Favorite subject of antiresearch individuals and groups. Subgross lung type differs from human Nasal anatomy unlike that of human. May be relatively insensitive to some inhaled gases	Has large collateral ventilation. Carotid artery anastomoses in upper respiratory tract participate in brain cooling
Ferret	Used as model for respiratory system viral diseases. Has well-developed respiratory bronchioles, a potential site for small airways disease. Extensive literature available on care, husbandry, and use in biomedical research. Pulmonary function data available. Easy to handle and adaptable to masks, plethysmographs, etc. Airways appear to be more like human than those of dog. Laboratory quality purebred animals are inexpensive compared with dogs. Runs well on treadmill. Litter size is 6 to 10	Has mildly disagreeable scent. Has not been used extensively in past in inhalation toxicology	Has very long trachea in relation to lung size for a mammal. Has large lung volume per unit body weight for a mammal
Nonhuman primates (monkeys mainly)	Nasal anatomy similar to human. Convenient size. Commonly used in inhalation studies. Pulmonary function techniques and exposure apparatus available. Long life span convenient for chronic studies. Respiratory response to antigen challenge is similar to human asthmatic. Chest wall compliance, though lower than that of human, is closer to human than either dog or rat	Husbandry practices not well standardized. High initial cost and lack of general availability. availability. Can transmit serious disease to handlers. Subgross pulmonary anatomy differs from human. Has reputation for being less than cooperative. Targeted by antiresearch community	Several genera exist, having greatly differing characteristics. Wild-captured animals are becoming scarce, resulting in increased use of colony-bred animals

Continued

Table 9.11 Information on Several Mammalian Species Commonly Used in Inhalation Studies—cont'd

Mammal	Advantages	Disadvantages	Special considerations
Horse	Long life span useful for chronic studies. Large size makes many procedures easier. Subgross lung structure similar to human. Has chronic lung diseases similar to those of human. Adaptable to masks and other inhalation equipment. Some morphometric data on upper airways and tracheobronchial tree available. Husbandry and care well understood. Many pulmonary function characteristics are similar to human	Inconveniently large for manipulation. Requires large housing area. Expensive relative to many other species	Breeds vary considerably in cooperativeness
Bovine (mainly sheep)	Potentially similar to human in respiratory viral disease. Tracheal goblet cell density and mucus secreting glands of sheep similar to those of human. Sheep used as model in tracheal clearance studies. Pulmonary function measurement techniques available. Sheep used as model of allergic airway reactivity. Sheep have passive nature in laboratory permitting use of a variety of procedures on unanesthetized individuals	Requires relatively large housing space. Has tracheal bronchus	Sheep has very heavy wool coat
Rodents	Guinea pig useful as bronchoconstriction model. Morphometric airway models available. Widely used in modern inhalation studies. Gas (ozone) uptake in guinea pig nose may be similar to human. Data available on age-related changes in structure and function. Used as model for particle clearance. Used as infectivity models. Used as immunologic model. Literature available on lung biochemical events. Adapts to exercise and spontaneous activity studies	Pulmonary function measurements on awake animals are difficult but not impossible. Do not have well-developed respiratory bronchioles. Coprophagy can produce artifacts in metabolism and other types of studies. Nasopharynx anatomy not similar to human. Tendency toward confounding spontaneous respiratory infections (especially rats)	Guinea pig has unusually abundant bronchial smooth muscle. Rodents often show greater sensitivity to toxic gases than do larger animals. Short life spans, may be advantage or disadvantage depending on study

has turned attention to the ferret as a model in inhalation studies (Boyd and Mangos, 1981; Vinegar et al., 1985; Ellington et al., 1990; Maher and DeStefano, 2004; McLain et al., 2007). Also, the ferret's similarity to humans regarding viral infections has stimulated its use (Maher and DeStefano, 2004). The author's experience with female albino ferrets indicates that they are in fact relatively easy to handle in the laboratory and are willing runners on an exercise treadmill. Lung airway casts from ferrets exhibit very long tracheas in relation to the lung size, a monopodial airway branching system, and several orders of respiratory bronchioles. As shown in Figure 9.6, the tracheobronchial tree structure is slightly more similar to the human than is that of the dog. Further, because of the relatively large litter size and modest housing space requirements, this animal is valuable in experiments requiring large numbers of subjects. A review on the use of laboratory ferrets is available (Ball, 2006), and a dedicated issue of *Laboratory Animal Science* (35[3], 1985) has 18 papers on ferrets in research, including a compendium of about 600 references on ferret anatomy, biology, physiology, behavior, etc.

Nonhuman Primates

The mammalian order, Primate (top mammal), contains humans, apes (orangutans, chimpanzees, gorillas, and gibbons), monkeys (colobines, long-tailed, hand-tailed, and dog-faced), baboons, marmosets, tarsioids, lemurs, lorisoids, and tree shrews. Monkeys, and to a lesser extent baboons, are common subjects in toxicological studies. The great variety of monkeys found in nature theoretically provides the inhalation toxicologist with a large list of potential models. However, availability, cost, and other practical considerations have greatly limited the types actually used (Patterson and Carrion, 2005). Macaques (genus *Macaca*), medium sized monkeys with stout limbs, include several species that are used in biomedical research. The macaques include the bonnet monkey, stumptail monkey, rhesus monkey, Barbary apes, the cynomolgus monkey, and other species. The use of nonhuman primates in environmental pollution research was discussed by Castleman and colleagues (1979). These researchers, actively involved in inhalation studies, addressed the nature of the small airways and respiratory mechanical properties in the macaques. They concluded that the "structure and number of generations of respiratory bronchioles are similar between man and macaques," and that macaques are closer to humans in chest wall compliance than are dogs or small rodents. In addition, similarity to humans in many other respects has made nonhuman primates an essential resource for vaccine development (Waag et al., 1999; Phipps et al., 2004: Patterson and Carrion, 2005; Hobbs et al., 2006; Lawler et al., 2006). An issue of the *ILAR J.* (47[4], 2006) has 3 papers related to research with nonhuman primates.

Horses

Along with zebras and asses, the horses are in the genus *Equus* (equines). Their appeal in inhalation studies derives from two major considerations: their subgross pulmonary anatomy has been identified as similar to that of the human (McLaughlin et al., 1961a,b; Tyler and Julian, 1991) and they develop several respiratory disorders that are believed to be similar to those of humans (Garner et al., 1971; Mauderly, 1974a; Cook, 1976; McPherson et al., 1979). These two features are possibly related, in that mammals with similar lung anatomical characteristics are expected to exhibit similar pathophysiological phenomena. Equines vary widely in size and temperament, but they require freedom of head movement when awake. The pony and donkey have both been found manageable in the inhalation study setting (Garner et al., 1971; Albert et al., 1974; Mauderly, 1974a). And their size facilitates many procedures that are not feasible in small mammals.

Both pulmonary function and particle clearance data show wider variations from individual to individual than is seen in inbred laboratory animals. Thus, equines are best suited to studies in which individual pre-exposure parameters can be compared with corresponding postexposure measurements.

Bovids

The family Bovidae contains nearly 50 genera and over 100 species including antelopes, goats, sheep, oxen, buffalos, and domestic cattle. Although cattle and goats are used in biomedical research, the sheep is more common in the published inhalation toxicology literature. The passive, tolerant nature of the sheep, its manageable size, and availability, all contribute to its experimental value. Mawdesley-Thomas et al. (1970) selected the lamb from 12 examined mammals as a model for the effects of irritants on bronchial and tracheal glands. The ability of the sheep to tolerate bronchoscopic procedures led to its use as a model for study of the effects of air pollutants on tracheal clearance rates by Sackner et al. (1981). The goat has been used as a smoke injury model (Walker et al., 1981), and calves have been used in aerosol deposition and clearance studies (Jones and Bull, 1986; Diesel et al., 1991).

Rodents

The order Rodentia has nearly 30 nonextinct families including many that are commonly used in biomedical laboratories, such as true rats and mice, kangaroo rats and mice, hamsters, gerbils, and cavies (including guinea pigs). Their extensive use in research has led to an enormous literature covering their behavior, biochemistry, anatomy, physiology, gestation, growth, diseases, and husbandry. They are readily available from reputable suppliers, with specialized characteristics including specific genetic defects and resistance or susceptibility to specific diseases or classes of toxicants. Most laboratory rodents are easy to handle, house, and maintain. Rodents are commonly used in studies that require a large number of subjects.

Their primary disadvantage in inhalation studies derives from their short, relatively wide airways, tendency to lack respiratory bronchioles, and their tendency to harbor confounding pulmonary infections. Further, the guinea pig is known for its extreme bronchoconstrictive tendency, which some see as a disadvantage, but which was used to advantage by Amdur and associates (1958, 1978a,b) to rank and otherwise study inhaled bronchial irritants. Rats also suffer from a tendency to have particle clearance failure (particle overload) when high doses are delivered (Morrow, 1988; Mauderly and McCunney, 1996), which can produce unrealistic responses.

It is unnecessary to list all of the uses of rodents in inhalation studies because they have been used for essentially every type of endpoint including lethality, learning and behavioral alterations, infectivity after viral or bacterial challenge, teratogenicity, carcinogenicity, mutagenicity, fertility, immunology, bronchial reactivity, growth, metabolic alterations, respiratory tract clearance, etc. Of necessity, the listings in the Table 9.11 are not complete. The reader is referred to the *Handbook of Human Toxicology* (Massaro, 1997) and *Animal Models in Toxicology* (Gad, 2007a) for specific discussions regarding rodents in toxicological studies. The availablility of genetically modified (*transgenic* and *knockout*) rodents further increases the importance of rodent models in inhalation studies (Costa and Kodavanti, 2003; Bridges and Weaver, 2006). Because of their long history in toxicological research, the availability of useful genetic variants, and their many practical advantages, they will undoubtedly continue to be widely used.

Other Mammals

The order Lagomorpha contains only two families: One, the Leporidae, consists of nine genera which comprise the hares and rabbits. Although rabbits and hares are believed to be rodents by many people, zoologists place them in a separate order. Rabbits are used in research for many of the reasons that apply to rodents. Additionally, they are typically docile and conveniently larger than most rodents, and thus can be used to provide larger blood and tissue samples. As a result, rabbits are sometimes used in studies requiring serial hematologic evaluations, large numbers of pulmonary macrophages (or other cellular and noncellular material obtained using pulmonary lavage), and in studies of particle clearance from the respiratory tract. They are also useful as a model of ventilated newborns (O'Callaghan et al., 1992; Fok et al., 1997).

The domestic cat, a carnivore, has been widely available to investigators due to its familiarity and well-known husbandry. Uses in inhalation studies include studies of laryngeal reflexes (Szereda-Przestaszewska and Widdicombe, 1973), studies of mucociliary clearance (Adler et al. 1973), and studies of allergic asthma (Reinero et al., 2004).

Domesticated pigs *(Sus scrofa domesticus)* are widely used in training, testing, and medical research owing to their similarity to humans in many respects. Inhalation related uses include lung transplants (Warnecke et al., 2006), inhalation injury, including liquid ventilation therapy (Fitzpatrick et al., 1997), lung infection (Hensel et al., 1993), immunologic development (Butler et al., 2006), exercise physiology (McKirnan et al., 1986), etc. Because large pigs are difficult to handle, miniature variants have been developed (Koch et al., 2001; Gad et al., 2007a). Although they are invaluable models in biomedical research and testing, pigs are notorious for loud vocalizations and handling difficulties. Experts in handling these animals should be consulted prior to their use when fully awake in the laboratory.

The potential for introduction of new animal models is great in inhalation toxicology. This is largely due to specific shortcomings in each of the widely used models. However, the introduction of new animal models should not be attempted capriciously. One should adequately understand the normal anatomy, physiology, and their variabilities. The state of health should be rigidly controlled and known. Further, standardized methods for handling, exposing to materials of interest, and quantifying responses must be demonstrated.

10
Regulations and Guidelines

INTRODUCTION

Laws and regulations are usually effected in order to protect an interest. The interests of individuals, institutions, society, and laboratory animals are protected by legislation and regulation. Examples in the United States (U.S.) include: the Occupational Safety and Health Act; the Good Laboratory Practices Act; and the Animal Welfare Act. Both U.S. and international regulations of interest to toxicologists have been recently reviewed (Merrill, 2003; Gad, 2007b). Such regulations are often promulgated in response to outrage over real or perceived irresponsible behavior.

As an example, consider a lengthy quotation from the *Federal Register,* vol. 41, No. 225, Friday, November 19, 1976. This quote is from the introductory section of the Food and Drug Administration's (FDA's) *Proposed Regulations for Good Laboratory Practice (GLP).* Under the subheading "Statement of the Problem," eight types of fraud on the part of pharmaceutical firms and private testing laboratories are described.

> Recent FDA experiences have identified significant problems in the manner in which nonclinical laboratory studies are being performed. Deficiencies were found during inspections of the testing facilities of major pharmaceutical firms, inspections of several private contract testing facilities, and internal review of toxicity studies of color additives conducted by FDA. How widespread or serious the problems are is not known at this time. The agency is concerned, however, that significant deviations in the quality and integrity of reported data might be of a greater magnitude and scope than had previously been assumed. The following are the more important deficiencies in the animal testing procedures that have been observed:
>
> 1. Experiments were poorly conceived, carelessly executed, or inaccurately analyzed or reported. For example, FDA found: Original autopsy records for certain studies were either unavailable, or were apparently transcribed to new records several years after the autopsies; pathology reports submitted to the agency were inconsistent with the original autopsy records; microscopic examinations of tissue slides were conducted by more than one pathologist, each of whom came to different conclusions, yet only the conclusions favorable to the drug were submitted to the agency; in one long-term toxicity and carcinogenicity study, there was not a complete set of records for any single animal in the study, despite the fact that the

records were required for proper analysis of the study and were represented to the agency to exist; and, in the same long-term toxicity and carcinogenicity study, the protocol required daily physical examination of the animals, yet the agency was unable to find any records to indicate that these examinations were carried out.

2. Technical personnel were unaware of the importance of protocol adherence, accurate observations, accurate administration of the test substance, and accurate recordkeeping and record transcription. Examples of these deficiencies include: Certain employees of a firm were unable to explain the procedures used to record data; records of laboratory observations were neither dated nor signed; employees were unable to account for discrepancies between raw data and final reports submitted to the agency; animals were observed and recorded as normal for a variety of factors, including appearance, awareness, appetite, and thirst, when in fact the animals were dead; and drugs under study were administered to animals in a manner that made it impossible to determine how much, if any, of the required dosage was actually ingested by the animals.
3. Management did not assure critical review of the data or proper supervision of the personnel. For example, in one toxicity study involving rats, gross changes of tissue began to appear, yet management was not made aware of these alarming changes for approximately 4 to 8 months; and, in another study, a drug was determined to be a tumorigen, yet this information was not promptly given to the agency.
4. Studies were impaired by protocol designs that did not allow the evaluation of all available data. In one situation, protocols were discovered that called for looking at all the high-dose and control animals for tumorigenicity while the protocols did not require that all the low- and mid-dose animals be observed as well.
5. Assurance could not be given for the scientific qualifications and adequate training of personnel involved in the research study. At one firm, the reproduction and teratology studies were conducted and laboratory personnel were overseen by a senior scientist who did not have the proper qualifications or background to be conducting and supervising these critical studies. In another case, necropsies were being performed by people without the proper training, as was recognized by a senior scientist who reviewed the work.
6. There was a disregard for the need to observe proper laboratory, animal care, and data management procedures. Illustrations of these deficiencies are: Treatment and control animals were not properly identified; weighings of the animals were not accurately recorded; animals were fixed in toto and not necropsied for several months; one study was discontinued because a disease, unrelated to the drug under study, killed most of the animals in the study, yet none of the animal records contained any observations of symptoms of the disease; and a laboratory was sprayed and fogged with pesticides while the animals were present in the laboratory.
7. Sponsors failed to monitor adequately the studies performed in whole or in part by contract testing laboratories. For example: A contract laboratory failed to make a slide of lesions for histopathological examination, despite the fact that this was called for in the protocol, and management failed to adequately monitor the study; in one study done by a contract laboratory, although the management had serious questions about the conduct of

the study, they never questioned or exercised any control over the operating investigator; in another study done by a contract laboratory, FDA was told that animal tissues had been examined histopathologically when a review of the contract laboratory's original records indicates that these tissue samples were never even collected; and, in another study done by a contract laboratory the problem of autolysis of the animals was so extensive that the study should have been considered unacceptable, yet the final study report made no mention of this fact.

8. Firms failed to verify the accuracy and completeness of scientific data in reports of nonclinical laboratory studies in a systematic manner before submission to FDA. Examples of such failure include: Significant discrepancies found between gross observations on pathology sheets when compared with the individual pathology summaries to the agency; inconsistent progress reports of the same study submitted to FDA; and one firm submitted a study utilizing the wrong data and the wrong animal identification numbers which were easily discovered by the agency, yet management did not check the data used.

Thus, the GLPs were born. For a description of the U.S. and international GLPs, and their requirements, see Cwiertniewicz (2005).

Similarly, abuse of privilege was the impetus for strict federal regulations regarding the ethical review, monitoring, and conduct of research involving human subjects. Such abuse occurred primarily at the hands of physicians, both in Nazi Germany and in the United States. Apparently, the motivations of the abusers this time were not economic, but more political, social, and related to perceived needs for biomedical information. An excerpt from a committee report by Thomson et al. (1981), serves as an illustration of the studies that have shaped human protection legislation in the United States.

> In October of 1932, the United States Public Health Service began an experiment on 399 human subjects in Macon County, Alabama. The subjects were black, poor, and semiliterate; and they had syphilis. The aim of the experiment was to track the effects of syphilis on untreated black males. To encourage participation, the subjects were led to believe they were being treated. But they were not in fact treated; their symptoms were recorded at periodic physical examinations, and autopsies were performed after death. The public at large first learned of this experiment - now known as the Tuskegee Study - in 1972, when it was still in progress, since some of the subjects had not yet died.
>
> It is arguable that the treatment available for syphilis in 1932 was not particularly effective and that it was, itself, dangerous. Two facts remain: (1) treatment was still being withheld from the survivors in 1972, even though penicillin had become available for the treatment of syphilis in the 1940's; and (2) a treatment for syphilis was available in 1932, and the subjects were not informed that it was being withheld.
>
> There can be no doubt that public anger at this experiment played a major role in provoking the strict federal regulation of research on human subjects which was to be developed during the seventies.

The lesson is clear: those who do not behave in a responsible and ethical manner will stimulate restrictive legislation, and the legislation will be applied to the villain and hero alike.

GUIDELINES VERSUS REGULATIONS

Guidelines are voluntary standards usually originated by professional societies, published in a widely available form, and updated by task groups composed of active specialists in the field. Regulations, on the other hand, are mandatory, imposed by a governmental body, and often are not frequently revised, even in the face of relevant new knowledge. Guidelines and regulations are based on very different assumptions about the people to whom they apply. Guidelines are based upon a trust that the individual using them acts in good faith and welcomes the council and additional information that is provided by the guideline. Regulations are usually made on the assumptions that the affected persons would not spontaneously act in good faith, and in fact, require the threat of penalties to prevent them from causing harm. In some instances, guidelines are superior to regulations, and in other instances it is the other way around. Individuals whose main motivation is to provide some altruistic service should be given guidelines by knowledgeable specialists. Individuals whose main motivation is self-serving certainly require regulation.

The research scientist is entrusted by society with a unique leadership role in developing new knowledge. He or she is formally trained to fulfill that role by established scientists who have demonstrated their ability to develop scientific knowledge and to train others to do likewise. The responsibility for generating knowledge must be in the hands of dedicated individuals trained within a framework that preserves the lessons learned from past mistakes. This tradition of individual training, preserving and passing on critical knowledge, is essential to science in general and to inhalation toxicology in particular. A scientist should not rely on legislation for instructions on how to generate knowledge within their field of research. If an investigator is so poorly trained or inexperienced that he or she cannot decide how to do a study, they should not attempt that study. Nothing can substitute for a firm, in-depth understanding of the techniques to be used, the relevant phenomena, and the way in which their study adds to the accumulated scientific knowledge. In most cases guidelines are more valuable than regulations in the pursuit of knowledge. However, all applicable regulations must be scrupulously complied with.

PROTECTION OF LABORATORY PERSONNEL

Through the years, there have been intermittent attempts to improve the health of the working men and women of the United States. The result of this continuing effort was crystallized in the 91st Congress when Senator Harrison A. Williams (Democrat, New Jersey), Chairman of the Senate Labor Sub-Committee, collaborated with Congressman William Steiger (Republican, Wisconsin) on an occupational safety and health bill which went to President Richard Nixon's desk and was signed into law by him as the *Occupational Safety and Health Act* (Public Law 91-596) of 1970 on December 29, 1970. At the time of implementation, the National Safety Council had calculated that over 14,000 people were killed annually in on-the-job accidents and the President's Report on Occupational Safety and Health estimated that "there may be as many as 100,000 deaths per year from occupationally caused diseases."

The objective of the Occupational Safety and Health Act (the Act) is to assure safe and healthful working conditions for the employed men and women of the United States. The power to obtain this objective is a derivative of the Commerce Clause of the U.S. Constitution (Article I, Section 8) which empowers Congress to regulate commerce among the several states and to provide for the general welfare. The Occupational Safety and Health Act applies to all businesses affecting interstate commerce except those businesses regulated by another Federal agency, such as coal mines, metal mines, railroads, and atomic energy installations. In addition to the above indicated jurisdictional aspect of the Occupational Safety and Health Act, many individual states have established an Occupational Safety and Health Administration program under the Occupational Safety and Health Act based upon another element of the Act which provides for a state to assume the responsibility for the administration and enforcement of occupational safety and health (Section 2b). The Act created the *Occupational Safety and Health Administration* (OSHA) to set and enforce standards; the *National Institute for Occupational Safety and Health* (NIOSH) to provide information, recommendations, and training; and the *Occupational Health and Safety Review Commission*, to resolve disputes (Bingham, 1992).

The standards are not necessarily specific for application to laboratories but are generally applicable to workplaces. Therefore, where jurisdiction requirements are met and whether or not Federal or State occupational safety and health regulations apply, laboratories must meet the requirements of the applicable standards under the Occupational Safety and Health Act. For example, there are some standards that have been promulgated for the control of exposures to specifically identified carcinogens which require a medical surveillance program, protective clothing, respiration equipment, as well as other identified industrial hygiene control practices. Each laboratory, therefore, should review the applicable law in order to insure compliance.

Guidelines useful for the protection of laboratory personnel, on a substance by substance basis, are available in the form of publications such as the *NIOSH Pocket Guide to Chemical Hazards* (NIOSH, 2005), the *NIOSH / OSHA / DOE Occupational Health Guidelines* for chemical hazards (OSHA, 2007), Irving Sax's *Dangerous Properties of Industrial Materials, 10th Ed.* (Lewis, 2000), and the American Conference of Governmental Industrial Hygienists' *TLVs® and BEIs®* (ACGIH® 2007). For information on disease risks see *Zoonoses and Communicable Diseases Common to Man and Animals* (Acha and Szyfres, 2001, 2003). For studies conducted under Biosafety Levels 1 to 4 consult *Biosafety in Microbiological and Biomedical Laboratories (BMBL)* (CDC, 2007). In addition, the *Guide for Care and Use of Laboratory Animals* (ILAR, 1996) describes the elements of an occupational health and safety program for staff who work with laboratory animals.

PROTECTION OF RESEARCH SUBJECTS

Human Subjects

Research involving exposure of volunteer human subjects is necessary to an adequate understanding of inhalation risks. There are many examples of species differences with respect to response to toxicants, and it is therefore desirable to obtain data on human as well as nonhuman subjects. (Sometimes, even data obtained on one human population may not extrapolate well to other humans.) The problem of obtaining the needed information in an ethical fashion involves balancing the risks of scientific ignorance against those involved in experimentation.

Stimulus for the legal protection of human research subjects in modern times derives in part from the medical experimentation on prisoners during the Nazi regime in Germany. Shortly after World War II, the articles of the *Nuremberg Code* were drafted (Table 10.1). Although the code is not a legal standard, it provides general ethical guidelines to investigators involved in human experimentation. In the United States, these international articles have since been amended to cover subjects who are unable to give informed consent, and therefore include the provision for consent by guardians and conservators.

Table 10.1 The Nuremberg Code

1. The voluntary consent of the human subject is absolutely essential. The duty and responsibility for ascertaining the quality of the consent rests upon each individual who initiates, directs, or engages in the experiment. It is a personal duty and responsibility which may not be delegated to another with impunity[a].
2. The experiment should be such as to yield fruitful results for the good of society, unprocurable by other methods or means of study, and not random and unnecessary in nature.
3. The experiment should be so designed and based on the results of animal experimentation and a knowledge of the natural history of the disease or other problem under study that the anticipated results will justify the performance of the experiment.
4. The experiment should be so conducted as to avoid all unnecessary physical and mental suffering and injury.
5. No experiment should be conducted where there is an a priori reason to believe that death or disabling injury will occur; except, perhaps, in those experiments where the experimental physicians also serve as subjects.
6. The degree of risk to be taken should never exceed that determined by the humanitarian importance of the problem to be solved by the experiment.
7. Proper preparations should be made and adequate facilities provided to protect the experimental subject against even remote possibilities of injury, disability, or death.
8. The experiment should be conducted only by scientifically qualified persons. The highest degree of skill and care should be required through all stages of the experiment of those who conduct or engage in the experiment.
9. During the course of the experiment the human subject should be at liberty to bring the experiment to an end if he has reached the physical or mental state where continuation of the experiment seems to him to be impossible.
10. During the course of the experiment the scientist in charge must be prepared to terminate the experiment at any stage, if he has probable cause to believe, in the exercise of the good faith, superior skill and careful judgment required of him that a continuation of the experiment is likely to result in injury, disability or death in the experimental subject.

[a]Projects involving human subjects who are unable to give consent will require legally effective informed consent from guardians/conservators.

In the United States, in 1974, Federal regulations were promulgated by the then Department of Health, Education and Welfare for the purpose of "... safeguarding the rights and welfare of subjects at risk in activities supported under grants and contracts from the Department of Health, Education and Welfare (DHEW)." The U.S. Food and Drug Administration (FDA) has similar regulations regarding the testing of drugs and medical devices. The Office for Human Research Protections (OHRP) for the U.S. Department of Health and Human Services (HHS) implements regulations for conducting research with human subjects. Evolving policies of the OHRP can be found on the HHS website (www.hhs.gov/ohrp/assurances/).

State or other local legislation may also exist to cover experimentation using human subjects within a jurisdictional boundary. California enacted legislation in 1979 that pertains to all human medical experiments conducted within the state. These state laws include, for example, the requirement that each subject in a "medical investigation" be given a copy of an Experimental Subject's Bill of Rights (Table 10.2). Institutions such as universities may also impose additional unique local requirements. The rest of this section will cover HHS and FDA regulations for the protection of human research subjects in federally sponsored or regulated research. The regulations are not static, being periodically amended, supplemented, and modified.

Table 10.2 The Experimental Subject's Bill of Rights. Sample, Given to Subjects at The University of California, Irvine

Any person who is asked to consent as a subject in a medical experiment, or who is asked to consent on behalf of another, has the following rights:

1. To be told what the study is trying to find out.
2. To be told what will happen in the study and whether any of the procedures, drugs, or devices are different from what would be used in standard practice.
3. To be told about the risks, side effects, or discomforts of the things that may happen to me.
4. To be told if I can expect any benefit from participating and, if so, what the benefit might be.
5. To be told the other choices I have and how they may be better or worse than being in the study.
6. To be allowed to ask any questions concerning the study both before agreeing to be involved and during the course of the study.
7. To be told what sort of medical treatment is available if any complications arise.
8. To refuse to participate at all before or after the study is started. This decision will not affect my right to receive the standard medical care.
9. To receive a signed and dated copy of the consent form.
10. To be free of pressure when considering whether I wish to agree to be in the study.

The responsibilities for reviewing, approving, and monitoring human research rest in the hands of *Institutional Review Boards* (IRBs). These boards have specific requirements with respect to membership, performance of review, and record keeping. In addition, the Federal regulations clearly outline the process by which an investigator must obtain informed consent from each subject.

An IRB must have at least five members having sufficient diversity and expertise to review the research protocols of its institution. The board must not consist solely of members of one profession, must have at least one nonscientist and at least one member who is not otherwise affiliated with the institution. The main functions of an IRB are:

- To review, require modifications if necessary, and approve or disapprove the research
- To conduct continuing review, at least yearly, of the research and if necessary, suspend or terminate the project
- To report to institutional and Federal officials any serious or continuing noncompliance by the investigators or unexpected serious harm to subjects

When conducting a review of a research project the IRB must determine that all of the following are satisfied:

- Selection of subjects is equitable
- Risks are minimized
- Risks are reasonable in relation to the anticipated benefits to subjects and the importance of the knowledge to be gained
- Informed consent is sought, when appropriate, from each subject or his or her legally authorized representative
- Informed consent is appropriately documented
- When appropriate, the research plan includes monitoring of data collected to ensure the safety of subjects

- There are adequate provisions to protect the privacy of subjects and to maintain the confidentiality of data
- That additional regulations for the protection of fetuses, pregnant women, children, prisoners, those institutionalized as mentally disabled, and other vulnerable populations, are satisfied

Informed consent, unless specifically waived, must be obtained from each subject or his or her legal representative and must include the following elements:

- A statement that the activity involves research
- An explanation of the purposes, duration, and experimental procedures to be followed
- A description of any forseeable risks or discomforts to the subject
- A description of any benefits to the subject or others
- A disclosure of any alternate procedures, if any, that might be advantageous to the subject
- A statement on the extent to which confidentiality will be maintained
- An offer to answer any questions that the subject (or his or her representative) may have at any time about the research
- If risk is more than minimal, an explanation as to the availability of treatment or compensation should injury occur
- Who should be contacted should problems, injury, or questions arise
- A statement that participation is voluntary and that refusal to participate (or discontinue participation) will involve no penalty or loss of benefits to which the subject is otherwise entitled

This list contains only the minimum requirements, and several others may be required in specific instances. For example, any costs to the subject must be included in the informed consent.

Keeping up with the changing regulations can be a challenging task. One publication that is useful in this respect is *IRB: Ethics and Human Research* which is published six times per year.

Special legal aspects of human studies have been described by Kennedy (1963). Kennedy (legal staff member of the Los Angeles County Counsel's Office) analyzed the liability of an investigator performing human subject inhalation research. On the investigator's side is the clear need for such studies and the fact that the "... law permits a knowledgeable person to decide whether or not he would be willing to be a human guinea pig." He also stated that the pre-existence of a disease in a subject does not change the legal issues in any major way. The investigator is not absolved of liability if he or she is negligent in either the conduct of the experiments or in the provision of safeguards that are appropriate to the degree of risk. In this latter regard, Kennedy recommended a discussion between investigator and counsel regarding the worst possible situation that might arise in the research.

Dr. J. Hackney (formerly Chief of Environmental Health Service, Rancho Los Amigos Hospital Campus, University of Southern California, School of Medicine, Downey) was particularly active in performing human air pollution inhalation studies. In his laboratory a special set of ethical guidelines were adopted. These guidelines (Table 10.3), which may be more stringent than is always necessary, serve as an initial model for those entering the field of inhalation studies using human subjects.

Table 10.3 Ethical Guidelines Adopted by Rancho Los Amigos Hospital, for the Conduct of Human Inhalation Studies

A. Exposures

1. Pollutant concentrations are no greater than that might be experienced by similar subjects in their normal environment (community or industrial environment depending on purpose of study)
2. Exposures must be capable of producing only readily reversible health effects judging by all available information

B. Subjects

1. Prospective subjects must pass a physical examination and must give informed consent before participation
2. The benefit of participation to the individual subject or to society must be proportional to the risk incurred. Benefit to the individual may be financial (we try to minimize this aspect by supplying only modest compensation for the time spent), or it may be in the form of information about the subject's own health, i.e., his (or her) sensitivity or nonsensitivity to a given exposure. In this sense the experimental study is analogous to skin testing for allergic sensitivity or to cardiac testing under exercise stress
3. Each subject's condition is monitored continuously during his exposure. A trained observer monitors the electrocardiogram continuously and a physician and emergency resuscitation equipment are readily available
4. Investigators serve as the first subjects under any given experimental protocol.

Source: From Hackney, J.D., Linn, W.S., and Bell, K.A., personal communication, 1977.

Laboratory Animal Subjects

The use of laboratory animals in inhalation toxicology research is clearly essential for the development and continued refinement of knowledge on the effects and modes of action of airborne agents. As is true of research involving human subjects, ethical considerations arise in balancing the risks of scientific and medical ignorance with the risk of pain, distress, and discomfort for the animals involved in experimentation.

Perhaps the first law relating to animal protection was the *English Anti-Cruelty Law* enacted in 1822. Since that time, hundreds of societies have been organized throughout the world specifically for the protection of animals. In the United States, all states have laws that make cruelty toward animals a criminal offense, and police and sheriffs, or special officers are charged with enforcement.

In addition to the many laws created for the general protection of animals, the U.S. Congress, by passage of Public Law 89-544 (the *Laboratory Animal Welfare Act*) in 1966, established regulation of "... the transportation, sale and handling of dogs, cats and certain other animals intended to be used for purposes of research and experimentation ... ." The original Act was primarily concerned with dealers in nonhuman primates, dogs, cats, rabbits, guinea pigs, and hamsters, but it has been amended several times over the years and additionally applies to many other species used in research laboratories. The responsibility for enforcement and administration of the Animal Welfare Act (As found in the United States Code, Title 7, Chapter 54, Sections 2131-2159) rests with the U.S. Department of Agriculture. The passage of Public Law 99-158 by the U.S. Congress in November 1985, established additional guidelines which specifically apply to animal research funded by the Federal government. The Public Health Service Policy on Humane Care and Use of Laboratory Animals (Revised August, 2002) extends the scope of the Animal Welfare Act to all vertebrate animals, and is enforced and administered by the Office of Laboratory Animal Welfare (OLAW) at the National Institutes of Health under the Department of

Health and Human Services (DHHS). The Council for International Organizations of Medical Sciences (CIOMS) has also established principles for research involving animals. Institutional-level oversight of animal care and use in research is provided by the *Institutional Animal Care and Use Committee* (IACUC); the requirements for membership, record keeping, and review procedures of such committees are clearly defined by both the Animal Welfare Act/Regulations and Public Health Service Policy.

The scientific community, independently through the American Association for Laboratory Animal Science (AALAS), reviewed the ethics involved in the care and use of laboratory animals in 1959. Shortly thereafter, the AALAS appointed a Committee on Standards for the purpose of preparing guidelines for the care of laboratory animals. The U.S. Public Health Service (USPHS) provided the necessary financial support which led to publication in 1963 of the *Guide for Laboratory Animal Facilities and Care*. Since that time the *Guide* has been revised (1965, 1968, 1972, 1978, 1985, and 1996) and is now published by the National Academy Press, Washington, D.C. under the title *Guide for the Care and Use of Laboratory Animals* (ILAR, 1996). This guide is prepared by the Institute of Laboratory Animal Resources of the National Research Council. Prior to receiving research funds from the National Institutes of Health (NIH), an institution must maintain an assurance that it is committed to conform to the *Guide* and the *Principles for the Utilization and Care of Vertebrate Animals used in Testing, Research, and Training*. The *Principles* (Table 10.4)

Table 10.4 United States Government Principles for the Utilization and Care of Vertebrate Animals used in Testing, Research, and Training (Abbreviated)

1. The transportation, care, and use of animals should be in accordance with the Animal Welfare Act and other applicable Federal laws, guidelines, and policies[a].
2. Procedures involving animals should be designed and performed with due consideration of their relevance to human or animal health, the advancement of knowledge, or the good of society.
3. The animals selected for a procedure should be of an appropriate species and quality and the minimum number required to obtain valid results.
4. Proper use of animals, including the avoidance or minimization of discomfort, distress, and pain when consistent with sound scientific practices, is imperative.
5. Procedures with animals that may cause more than momentary or slight pain or distress should be performed with appropriate sedation, analgesia, or anesthesia. Surgical or other painful procedures should not be performed on unanesthetized animals paralyzed by chemical agents.
6. Animals that would otherwise suffer severe or chronic pain or distress that cannot be relieved should be painlessly killed at the end of the procedure or, if appropriate, during the procedure.
7. The living conditions of animals should be appropriate for their species and contribute to their health and comfort. Normally, the housing, feeding, and care of all animals used for biomedical purposes must be directed by a veterinarian or other scientist trained and experienced in the proper care, handling, and use of the species being maintained or studied.
8. Investigators and other personnel shall be appropriately qualified and experienced for conducting procedures on living animals.
9. Where exceptions are required in relation to the provision of these Principles, the decisions should not rest with the investigators directly concerned but should be made, with due regard to Principle II, by an appropriate review group such as an institutional animal care and use committee.

[a]For guidance throughout these Principles, the reader is referred to the *Guide for the Care and Use of Laboratory Animals* prepared by the Institute of Laboratory Animal Resources, National Academy of Sciences (ILAR, 1996).

relate to personnel, conduct of research, facilities, and transportation relating to animal research subjects. These *Principles* and the *Guide* in effect constitute standards to be followed by all who perform animal experimentation.

The *Guide* has steadily grown in scope, responding to both new information and the evolving views concerning the treatment of animals. The 1996 edition covers, among other items: veterinary care; occupational health; housing; exercise; restraint; pest and odor control; food; bedding; water; provisions for emergency, weekend, and holiday care; quarantine; disease control and treatment; separation of species; anesthesia and pain control; surgery; postsurgical care; euthanasia; personnel qualifications; ventilation; lighting; noise control; space recommendations; training programs; and relevant federal laws and policies. Table 8.1 is an excerpt from the *Guide*'s recommendations on housing space for various species.

Investigators involved in inhalation studies should be aware of, and responsible for, implementation of both the recommendations of the *Guide* and the *Principles*. To do otherwise may not only lead to conflict with peers and the public, but can also compromise the validity of one's study. As was pointed out in Chapter 5, the responses of animals to inhaled materials can be modified by the conditions of housing and care before, during, and after an exposure period. In chronic studies, it is advised that the exposure conditions conform to the extent possible with the conditions for housing set forth in the *Guide*. Furthermore, public sentiment, and hence the pressure for increasingly more rigid legislation, appears to be growing less sympathetic to the needs of the researcher. One's best protection in this regard is to assume a role of leadership in educating legislators and the public, and in promoting the welfare of research animals.

For further information on the care and management of laboratory animals, updates of the *Guide*, and the current version of *The Merck Veterinary Manual* (e.g., the 2006 50th Edition) should be consulted. In addition, current journals of interest to animal researchers include: *Lab Animal*; *Laboratory Animals*; *ILAR Journal*; and *Journal of the American Association for Laboratory Animal Science*.

INHALATION TOXICITY TESTING GUIDELINES

Although several agencies have been involved in publishing guidelines and mandatory inhalation testing procedures (Gross, 1981; Shoaf, 1994; Derelanko, 1995), the complexity and variety of inhalation exposures may make following such procedures impractical in pure research studies. Nevertheless, pesticide-testing guidelines of the U.S. Environmental Protection Agency (required by the U.S. Federal Insecticide, Fungicide, and Rodenticide Act, FIFRA), chemical testing guidelines (required by the Toxic Substances Control Act, TSCA), the Interagency Regulatory Liaison Group (IRLG), the Organization for Economic Co-operation and Development (OECD), and other entities have published guidelines that must be addressed in some studies (Pauluhn et al., 1988; Dorato and Wolff, 1991; Fogle, 1995; Whalan et al., 2006). Tables 10.5 and 10.6 describe some mandated tests for substances that are intended to expose human populations.

When fidelity to a promulgated testing regimen is not required, the inhalation toxicologist is free to design and conduct studies that will address specific problems/issues of interest. Fortunately, there is ample published information on guidelines for inhalation studies by qualified, experienced scientists.

For example, Donaldson et al. (2007) described recommended approaches to testing inhaled particles. They covered factors affecting particle toxicity, methods for characterizing exposures, and assessments of toxicity. Similarly, Gad (2006) provided recommendations for

Table 10.5 Proposed Types of Inhalation Studies Required Under Federal Law for Regulated Chemicals That Can Potentially Lead to Human Inhalation Risks

Acute tests

Purpose: LC_{50} (median lethal concentration), its statistical uncertainty (95% confidence interval), and the dose response curve and slope

Exposure: once for 4 h or longer, with 14 day postexposure observation

Animals: rat or approved alternate, minimum of 5 of each sex per dose level, controls not required

Limit test: if an airborne concentration of 5 mg/L of air for 4 h produces no mortality, additional acute inhalation testing is not required

Necropsy: all survivors with focus on respiratory tract

Subchronic tests

Purpose: find atmospheric concentration that produces no toxic effects

Exposure: 14, 28, or 90 days, for 6 or more hours per day, 5 or more days per week

Animals: rat or approved alternate, at least 10 animals of each sex per exposure level at time of termination. Controls required

Clinical tests: hematology, blood chemistry, urine analysis, and possibly other tests

Necropsy: complete gross examination of internal organs, organ weights, and histopathologic examination

Chronic tests

Purpose: define the toxic responses due to extended repeated exposures. Find the no-effect exposure concentration

Exposure: at least 1 year, lifetime for rodents: at least 6 h/day and at least 5 days/week

Animals: usually rats: at least 20 of each sex per level at start of exposure. Controls required

Clinical tests: required

Necropsy and histopathology: required

Table 10.6 Toxicity Test Protocol Battery

Acute test	**Chronic test**
Oral LD_{50}	12 months to lifetime oral, dermal, inhalation
Dermal LD_{50}	Lifetime oncogenicity
Inhalation LC	
Skin irritation	
Eye irritation	
Delayed neurotoxicity	
Dermal sensitization	
Subchronic tests	**Miscellaneous**
90-day oral	Toxicokinetics/metabolism
14-, 28- and 90-day dermal	Immunotoxicity
28- and 90-day inhalation	Behavioral
90-day neurotoxicity	Short-term
Teratology	Step sequence
Reproductive effects (2 and 3 generations)	
Perinatal	

Source: Adapted from Gross (1981).

safety assessment of inhaled medications, and Hahn (1999) gave guidance for conducting chronic carcinogen inhalation studies.

McClellan (1999) described the rationale and approaches to obtaining information in support of regulations and risk-assessment needs. In addition to describing the "risk paradigm" and some of the regulations that require obtaining risk-related data, he presented overviews of relevant epidemiologic, clinical, and toxicological studies (both in vivo and in vitro). McClellan described the relationships between experimental studies, and the setting of exposure criteria (such as the U.S. National Ambient Air Quality Standards; occupational Threshold Limit Values®; and standards for nearly 200 "hazardous air pollutants").

Folinsbee et al. (1997) presented a discussion of inhalation exposure techniques, and methods of measuring responses in human subjects. The authors emphasized reproducibility and standardization of assessment techniques, and cited a broad sample of published material on applicable methods, and successful studies.

Hobbs et al. (2006) addressed the special problems involved in designing, performing, and interpreting studies involving aerosolized bioterror agents, for the purpose of developing countermeasures (such as vaccines that require approval by the FDA). In this context, overcoming the normally lengthy FDA approval process involves performing animal inhalation studies that mimic not only the human exposure scenario, but also the expected human pathology, and response to new vaccines.

As described in previous chapters, the available literature on inhalation studies is rich and varied. The aerosol and gas generation and characterization methods, exposure methods, and techniques for designing and interpreting studies, have seen astonishing advances in the last 70 years. Still, new problems, with their new challenges to scientists, regulators, and medical and veterinary practitioners, will require significant advances in the future.

References

Abdel-Salam, M., Aerosol sampling methods in workplace and ambient environments, *J. Aerosol Med.*, 19: 434–455, 2006.

Abu-Musa, A., Nassar, A., and Usta, I., In vitro fertilization in two patients with Kartagner's syndrome and infertility, *Gynec. Obstr. Invest.*, 65: 29–31, 2008.

ACGIH® Technical Committee on Air Sampling Procedures, *Particle Size-Selective Sampling in the Workplace*, American Conference of Governmental Industrial Hygienists, Cincinnati, OH, 1985.

ACGIH®, *2007 TLVs® and BEIs®*, American Conference of Governmental Industrial Hygienists, Cincinnati, OH, 2007.

Acha, P.N. and Szyfres, B., *Zoonoses and Communicable Diseases Common to Man and Animals, 3rd Ed.*, Pan American Health Organization, Washington, DC, *Vol. I: Bacterioses and Mycoses* (2001), *Vol. II: Chlamydioses, Rickettsioses, and Viroses* (2003), and *Vol III: Parasitoses* (2003).

Adler, K.B., Wooten, O., and Dulfano, M.J., Mammalian respiratory mucociliary clearance, *Arch. Environ. Health*, 27: 364–369, 1973.

Aitken, R.J., Baldwin, P.E.J., Beaumont, G.C., Kenny, L.C., and Maynard, A.D., Aerosol inhalability in low air movement environments, *J. Aerosol Sci.*, 30: 613–626, 1999.

Alarie, Y., Sensory irritation by airborne chemicals, *Crit. Rev. Toxicol.*, 2: 299–363, 1973.

Albert, R.E., Berger, J., Sanborn, K., and Lippmann, M., Effects of cigarette smoke components on bronchial clearance in the donkey, *Arch. Environ. Health,* 29: 96–101, 1974.

Altman, P.L. and Dittmer, D.S., Development and growth, Part III, in *Biology Data Book, Vol. I., 2nd Ed.,* Altman, P.L. and Dittmer, D.S., Eds.. Federation of American Societies for Experimental Biology, Bethesda, MD, 1972, pp. 173–224.

Altman, P.L. and Dittmer, D.S., Eds., *Biology Data Book, Vol. III., 2nd Ed.,* Federation of American Societies for Experimental Biology, Bethesda, MD, 1974, pp. 1581–1587.

Amdur, M.O., The influence of aerosols upon the respiratory response of guinea pigs to sulfur dioxide, *Am. Ind. Hygiene Assoc. Quart.,* 18: 149–155, 1957.

Amdur, M.O. and Mead, J., Mechanics of respiration in unanesthetized guinea pigs, *Am. J. Physiol.,* 192: 364–368, 1958.

Amdur, M.O., Bayles, J., Ugro, V., and Underhill, D.W., Comparative irritant potency of sulfate salts, *Environ. Res.,* 16: 1–8, 1978a.

Amdur, M.O., Dubriel, M., and Creasia, D.A., Respiratory response of guinea pigs to low levels of sulfuric acid, *Environ. Res.,* 15: 418–423, 1978b.

Andersen, A.C., Ed., *The Beagle as an Experimental Dog,* Iowa State University Press, Ames, IA, 1970.

Andersen, A.C. and Goldman, M., Growth and development, in *The Beagle as an Experimental Dog,* Andersen, A.C., Ed., The Iowa State University Press, Ames, IA, 1970, Chapter 6.

Andersen, M.E., Toxicokinetic modeling and its applications in chemical risk assessment, *Toxicol. Lett.,*138: 9–27, 2003.

Ansborlo, E., Hengé-Napoli, M.H., Chazel, V., Gibert, R., and Guilmette, R.A., Review and critical analysis of available in vitro dissolution tests, *Health Phys.*, 77: 638–645, 1999.

Aris, R., Christian, D., Sheppard, D., and Balmes, J.R., Acid fog-induced bronchoconstriction, *Am. Rev. Respir. Dis.*, 141: 546–551, 1990.

Arnold, D.L., Charbonneau, S.M., Zawidzka, Z.Z., and Grice, H.C., Monitoring animal health during chronic toxicity studies, *J. Environ. Pathol. Toxicol.*, 1: 227–239, 1977.

Asgharian, B. and Moss, O.R., Particle suspension in a rotating drum chamber when the influence of gravity and rotation are both significant, *Aerosol. Sci. Technol.*, 17: 263–277, 1992.

Asgharian, B., Kelly, J.T., and Tewksbury, E.W., Respiratory deposition and inhalability of monodisperse aerosols in Long-Evans rats, *Toxicol. Sci.*, 71: 104–111, 2003.

Asgharian, B., A model of deposition of hygroscopic particles in the human lung, *Aerosol Sci. Technol.*, 38: 938–947, 2004.

Asgharian, B., Ménache, M.G., and Miller, F.J., Modeling age-related particle deposition in humans, *J. Aerosol Med.*, 17: 213–224, 2004.

Auerbach, O., Hammond, E.C., Kirman, D., and Garfinkel, L., Effects of cigarette smoking on dogs. 2. Pulmonary neoplasms, *Arch. Environ. Health,* 21: 754–768, 1970.

Austin, J.C., Cleaton-Jones, P.E., and Vieira, E.G., Design and use of an inhalation chamber for air pollution studies in small animals, *J. S. Afr. Vet. Assoc.*, 49: 235–238, 1978.

Avallone, E.A. and Baumeister, T., Eds., *Marks' Standard Handbook for Mechanical Engineers,10th Ed.*, McGraw-Hill, Columbus, OH, 1996.

Avol, E.L., Wightman, L.H., Linn, W.S., and Hackney, J.D., A movable laboratory for controlled clinical studies of air pollution exposure, *Air Pollut. Control Assoc. J.*, 29: 743–745, 1979.

Bair, W.J., Porter, N.S., Brown, D.P., and Wehner, A.P., Apparatus for direct inhalation of cigarette smoke by dogs, *J. Appl. Physiol.*, 26: 847–850, 1969.

Baker, G.L., Gupta, A., Clark, M.L., et al., Inhalation toxicity and lung toxicokinetics of C-60 fullerene nanoparticles and microparticles, *Toxicol. Sci.*, 101: 122–131, 2008.

Baker, M.A., Chapman, L.W., and Nathanson, M., Control of brain temperature in dogs: Effects of tracheostomy, *Respir. Physiol.*, 22: 325–333, 1974.

Baker, R.R., Temperature distribution inside a burning cigarette. *Nature (London)*, 247: 404–406, 1974.

Bakker-Woudenberg, I.A.J.M., Experimental models of pulmonary infection, *J. Microbiol. Mtds.*, 54: 295–313, 2003.

Balásházy, I., and Hofmann, W., Particle deposition in airway bifurcations: I. Inspiratory flow, *J. Aerosol Sci.*, 24: 745–772, 1993.

Balásházy, I., Hofmann, W., and Heistracher, T., Computation of local enhancement factors for the quantification of particle deposition patterns in airway bifurcations, *J. Aerosol Sci.*, 30: 185–203, 1999.

Ball, R.S., Issues to consider for preparing ferrets as research subjects in the laboratory, *ILAR J.*, 47: 348–357, 2006.

Banchero, N., Rostami, A., and Gimenez, M., Respiration in conscious dogs at rest and during exercise, *Respiration,* 37: 135–141, 1979.

Barlett, D. Jr., Faulkner, C.S., II, and Cook, K., Effect of chronic ozone exposure on lung elasticity in young rats, *J. Appl. Physiol.*, 37: 92–96, 1974.

Barrow, C.S. and Dodd, D.E., Ammonia production in inhalation chambers and its relevance to chlorine inhalation studies, *Toxicol. Appl. Pharmacol.*, 49: 89–95, 1979.

Barrow, C.S., Ed., *Toxicology of the Nasal Passages,* Hemisphere Publishing Corp., Washington, DC, 1986.

Barsocchi, A.T. and Knobel, R., Zero air: Selection of a high-volume preparation method, *Am. Lab.*, 12: 81–90, 1980.

Bates, D.V., The respiratory bronchiole as a target organ for the effects of dusts and gases, *J. Occup. Med.*, 15: 177–180, 1973.

Bates, D.V., *Respiratory Function in Disease, 3rd Ed.*, W.B. Saunders Co., Philadelphia, PA, 1989, Chapters 6, 7, 11.

Battista, S.P., Guerin, M.R., Gori, B.G., and Kensler, C.J., A new system for quantitatively exposing laboratory animals by direct inhalation, *Arch. Environ. Health,* 27: 376–382, 1973.

Beeckmans, J.M., The density of aggregated solid aerosol particles, *Ann. Occup. Hygiene*, 7: 299–305, 1964.

Bell, K.A., Linn, W.S., Hazucha, M., Hackney, J.D., and Bates, D.V., Respiratory effects of exposure to ozone plus sulfur dioxide in southern Californians and eastern Canadians, *Am. Ind. Hygiene Assoc. J.*, 38: 695–706, 1977.

Bell, K.A., Avol, E.L., Bailey, R.M., Kleinman, M.T., Landis, D.A., and Heisler, S.L., Design. operation and dynamics of aerosol exposure facilities for human subjects, in *Generation of Aerosols,* Willeke, K., Ed., Ann Arbor Science, Ann Arbor, MI, 1980, Chapter 23.

Bell, K.A. and Ho, A.T., Growth rate measurements of hygroscopic aerosols under conditions simulating the respiratory tract, *J. Aerosol Sci.,* 12: 247–254, 1981.

Bennett, W.D., Zeman, K.L., Kim, C.S., Variability of fine particle deposition in healthy adults: Effect of age and gender, *Am. J. Resp. Crit. Care Med.*, 153: 1641–1647, 1996.

Bennett, W.D., Zeman, K.L., Kim, C., and Mascarella, J., Enhanced deposition of fine particles in COPD patients spontaneously breathing at rest, *Inhal. Toxicol.*, 9: 1–14, 1997.

Bennett, W.D., Zeman, K.L., and Jarabek, A.M., Nasal contribution to breathing with exercise: Effect of race and gender, *J. Appl. Physiol.*, 95: 497–503, 2003.

Berglund, R.N. and Liu, B.Y.H., Generation of monodisperse aerosol standards, *Environ. Sci. Technol.,* 7: 147–153, 1973.

Bernstein, D.M. and Drew, R.T., The major parameters affecting temperature inside inhalation chambers. *Am. Ind. Hygiene Assoc. J.,* 41: 420–426, 1980.

Bianco, A., Gibb, F.R., Kilpper, R.W., Landman, S., and Morrow, P.E., Studies of tantalum dust in the lungs, *Radiology,* 112: 549–556, 1974.

Bice, D.E., Seagrave, J.C., and Green, F.H.Y., Animal models of asthma: Potential usefulness for studying health effects of inhaled particles, *Inhal. Toxicol.*, 12: 829–862, 2000.

Bide, R.W., Armour, S.J., and Yee, E., Allometric respiration/body mass data for animals to be used for estimates of inhalation toxicity to young adult humans, *J. Appl. Toxicol.*, 20: 273–290, 2000.

Bingham, E., The Occupational Health and Safety Act, in *Environmental and Occupational Medicine, 2nd Ed.,* Rom, W.N., Ed., Little, Brown & Co., Boston, MA, 1992, Chapter 112.

Birchall, A., Bailey, M.R., and James, A.C., LUDEP: A lung dose evaluation program, *Radiat. Protection Dosim.*, 38: 167–174, 1991.

Bittar, E.E., Ed., *Pulmonary Biology in Health and Disease*, Springer, New York, 2002.

Bland, R.D., Neonatal chronic lung disease in the post-surfactant era: Lessons learned from authentic animal models, *Biol. Neonate*, 88: 181–191, 2005.

Boecker, B.B., Aguilar, F.L., and Mercer, T.T., A canine inhalation exposure apparatus utilizing a whole-body plethysmograph, *Health Phys.,* 10: 1077–1089, 1964.

Bogdanffy, M.S. and Keller, D.A., Metabolism of xenobiotics by the respiratory tract, in *Toxicology of the Lung, 3rd Ed.*, Gardner, D.E., Crapo, J.D., and McClellan, R.O., Eds., Taylor & Francis, Philadelphia, PA, 1999, Chapter 3.

Boorman, G.A., Spontaneous lesions of importance for long-term toxicity testing in rats and mice, in *Scientific Considerations in Monitoring and Evaluating Toxicological Research,* Gralla, E.J., Ed., Hemisphere, Washington, DC, 1981, Chapter 6.

Bowen, S.E., Batis, J.C., Paez-Martinez, N., and Cruz, S.L., The last decade of solvent research in animal models of abuse: Mechanistic and behavioral studies, *Neurotox. Teratol.*, 28: 636–647, 2006.

Bowes, S.M. III, Frank, R., and Swift, D.L., The head dome: A simplified method for human exposures to inhaled air pollutants, *Am. Ind. Hygiene Assoc. J.*, 51: 257–260, 1990.

Boyd, M.R., Metabolic activation of pulmonary toxins, in *Mechanisms in Respiratory Toxicology, Vol. 2*, Witschi, H. and Nettesheim, P., Eds., CRC Press, Boca Raton, FL, 1982, Chapter 4.

Boyd, R.L. and Mangos, J.A., Pulmonary mechanics of the normal ferret, *J. Appl. Physiol., Respir. Environ. Exercise Physiol.,* 50: 799–804, 1981.

Brain, J.D., Knudson, D.E., Sorokin, S.P., and Davis, M.A., Pulmonary distribution of particles given by intratracheal instillation or by aerosol inhalation, *Environ. Res.,* 11: 13–33, 1976.

Brain, J.D and Corkery, G.C., The effect of increased particles on the endocytosis of radiocolloids by pulmonary macrophages in vivo: Competitive and toxic effects, in *Inhaled Particles IV,* Walton, W.H. and McGovern, B., Eds., Pergamon Press, Oxford, 1977, pp. 551–564.

Breysse, P.N. and Swift, D.L., Inhalability of large particles into the human nasal passage—in vivo studies in still air, *Aerosol Sci. Technol.*, 13: 459–464, 1990.

Bridges, J.P. and Weaver, T.E., Use of transgenic mice to study lung morphogenesis and function, *ILAR J.*, 47: 22–31, 2006.

Broday, D.M., Deposition of ultrafine particles at carinal ridges of the upper bronchial airways, *Aerosol Sci. Technol.*, 38: 991–1000, 2004.

Brooks, S.M., Gochfeld, M., Herzstein, J., Jackson, R.J., and Schenker, M.B., Eds., *Environmental Medicine*, Mosby, St. Louis, MO, 1995.

Brown, J.S., Particle inhalability at low wind speeds, *Inhal. Toxicol.*, 17: 831–837, 2005.

Brown, J.S., Wilson, W.E., and Grant, L.D., Dosimetric comparisons of particle deposition and retention in rats and humans, *Inhal. Toxicol.*, 17: 355–385, 2005.

Brown, M.G. and Moss, O.R., An inhalation exposure chamber designed for animal handling, *Lab. Anim. Sci.*, 31: 717–720, 1981.

Brunauer, S., Emmett, P.H., and Teller, E., Adsorption of gases in multimolecular layers., *J. Am. Chem. Soc.*, 60: 309–319, 1938.

Budyka, A.K. and Ogorodnikov, B.I., Aerosol filtration (aerosol sampling by fibrous filters), in *Aerosols Handbook: Measurement, Dosimetry, and Health Effects*, Ruzer, L.S., and Harley, N.H., Eds.,CRC Press, Boca Raton, FL, 2005, Chapter 17.

Burri, P.H. and Weibel, E.R., Ultrastructure and morphometry of the developing lung, in *Development of the Lung,* Hodson, W.A., Ed., Marcel Dekker, New York, 1977, Chapter 5.

Burri, P.H., Postnatal development and growth, in *The Lung: Scientific Foundations, 2nd Ed.*, Crystal, R.G., West, J.B., Weibel, E.R., and Barnes, P.J., Eds., Lippincott-Raven, Philadelphia, PA, 1997, Chapter 70.

Burtscher, H., Novel instrumentation for the characterization of ultrafine particles, *J. Aerosol Med.*, 15: 149–160, 2002.

Buskirk, E.R., Decrease in physical work capacity at high altitude, in *Biomedicine Problems of High Terrestrial Elevations*, Hegnauer, A.H., Ed., U.S. Army Res. Inst. Environ. Med., Natick, MA, 1969, pp. 204–222.

Bustad, L.K., Hegreberg, G.A., and Padgett, G.A., *Naturally Occurring Animal Models* of *Human Disease: A Bibliography,* Institute for Laboratory Animal Resources, National Academy of Sciences, Washington DC, 1975.

Butler, J.E., Sun, J.S., Wertz, N., and Sinkora, M., Antibody repertoire development in swine, *Devel. Compar. Immunol.*, 30: 199–221, 2006.

Butterworth, B.E., Predictive assays for mutagens and carcinogens, in *Scientific Considerations in Monitoring and Evaluating Toxicological Research,* Gralla, E.J., Ed., Hemisphere, Washington, DC, 1981, Chapter 10.

Calabrese, E.J., Hormesis: Implications for risk assessment, in *Inhalation Toxicology, 2nd Ed.*, Salem, H., and Katz, S.A., Eds., CRC Press, Boca Raton, 2006, Chapter 15.

Campbell, E.J.M., Motor pathways, in *Handbook of Respiration, Vol. I,* Fenn, W.O. and Rahn, H., Eds., American Physiological Society, Washington, DC, 1964, Chapter 21.

Cannon, G.J., and Swanson, J.A., The macrophage capacity for phagocytosis, *J. Cell Sci.*, 101: 907–913, 1992.

Cannon, W.C., Blanton, E.F., and McDonald, K.E., The flow-past chamber: An improved nose-only exposure system for rodents, *Am. Ind. Hygiene Assoc. J.*, 44: 923–928, 1983.

Cantor, J.O., Ed., *CRC Handbook of Animal Models of Pulmonary Disease*, CRC Press, Boca Raton, FL, 1989.

Caplan, K.J., Rotameter corrections for gas density, *Am. Ind. Hygiene Assoc. J.*, 46: B10–B16, 1985.

Cardoso, W.V., Lung morphogenesis, role of growth factors and transcription factors, in *The Lung: Development, Aging and the Environment*, Harding, R., Pinkerton, K.E., and Plopper, C. G., Eds., Elsevier, San Diego, CA, 2004, Chapter 1.

Carpenter, R.L., Barr, F.P., Leydig, R.L., and Rajala, R.E., Microprocessor-controlled inhalation system for repeated exposure of animals to aerosols, *Rev. Sci. Instrum.*, 50: 560–564, 1979.

Carpenter, R.L. and Beethe, R.L., Airflow and aerosol distribution in animal exposure facilities, in *Generation of Aerosols,* Willeke, K., Ed., Ann Arbor Science, Ann Arbor, MI, 1980, Chapter 22.

Castleman, W., Gillespie, J., Kosch, P., Schwartz, L., and Tyler, W., The role of nonhuman primates in environmental pollution research, in *Assessing Toxic Effects of Environmental Pollutants,* Lee, S.D. and Mudd, J.B., Eds., Ann Arbor Science, Ann Arbor, MI, 1979, Chapter 2.

CDC (Centers for Disease Control and Prevention), *Biosafety in Microbiological and Biomedical Laboratories (BMBL), 5th Ed.*, Centers for Disease Control and Prevention, Washington, DC, 2007.

CDC/NIOSH (Centers for Disease Control and Prevention/National Institute for Occupational Safety and Health), *Guidance for Protecting Building Environments from Airborne Chemical, Biological, or Radiological Attacks*, DHHS (NIOSH) Publication No. 2002-139, NIOSH Publications, Cincinnati, 2002.

CDC/NIOSH (Centers for Disease Control and Prevention/National Institute for Occupational Safety and Health), *Biosafety in Microbiological and Biomedical Laboratories (BMBL), 5th Ed.*, U.S. Government Printing Office, Washington, DC, 2007.

Chalupa, D.C., Morrow, P.E., Oberdörster, G., Utell, M.J., and Frampton, M.W., Ultrafine particle deposition in subjects with asthma, *Environ. Health Persp.*, 112: 879–882, 2004.

Chang, M.C., Kim, S., and Sioutas, C., Experimental studies on particle impaction and bounce: Effects of substrate design and material, *Atmos. Environ.*, 33: 2313–2322, 1999.

Cheng, Y.S., Yeh, H.C., and Kanapilly, G.M., Collection efficiencies of a point-to-plane electrostatic precipitator, *Am. Ind. Hygiene Assoc. J.*, 42: 605–610, 1981.

Cheng, Y.S., Hansen, G.K., Su, Y.F., Yeh, H.C., and Morgan, K.T., Deposition of ultrafine aerosols in rat nasal molds, *Toxicol. Appl. Pharmacol.*, 106: 222–233, 1990.

Cheng, Y.S. and Moss, O.R., Inhalation exposure systems, *Toxicol. Methods*, 5: 161–197, 1995.

Cheng, Y.S., Denuder systems and diffusion batteries, in *Air Sampling Instruments, 9th Ed.*, Cohen, B.S. and McCammon, C.S. Jr., Eds., ACGIH®, Cincinnati, OH, 2001, Chapter 19.

Churg, A. and Vedal, S., Carinal and tubular airway particle concentrations in the large airways of non-smokers in the general population: Evidence for high particle concentration at airway carinas, *Occup. Environ. Med.,* 53: 553–558, 1996.

Clarke, R.W., Coull, B., Reinisch, U., et al., Inhaled concentrated ambient particles are associated with hematologic and bronchoalveolar lavage changes in canines, *Environ. Health Perspect.,* 108: 1179–1187, 2000.

Clyde, W.A. Jr., Experimental models for study of common respiratory viruses, *Environ. Health Perspect.*, 35: 107–112, 1980.

Cobourn, W.G., Husar, R.B., and Husar, J.D., Continuous in-situ monitoring of ambient particulate sulfur using flame photometry and thermal analysis, *Atmos. Environ.*, 12: 89–98, 1978.

Coffin, D.L., Gardner, D.E., Holzman, R.S., and Wolock, F.J., Influence of ozone on pulmonary cells, *Arch. Environ. Health,* 16: 633–636, 1968.

Coffin, D.L. and Stokinger, H.E., Biological effects of air pollutants, in *Air Pollution, Vol. 2, 3rd Ed.*, Stern. A.C., Ed., Academic Press. New York, 1977, Chapter 5.

Cohen, B.S., Xiong, J.Q., and Li, W., The influence of charge on the deposition behavior of aerosol particles with emphasis on singly charged nanometer sized particles, in *Aerosol Inhalation: Recent Research Fontiers*, Marijnissen, J.C.M. and Gradon, L., Eds., Kluwer Academic Publishers, Dordrecht, 1996, pp. 153–164.

Cohen, B.S. and McCammon, C.S. Jr., Eds., *Air Sampling Instruments, 9th Ed.*, ACGIH®, Cincinnati, OH, 2001.

Cohen, M.D., Pulmonary immunotoxicology, in *Toxicology of the Lung, 4th Ed.*, Gardner, D.E., Ed., Taylor & Fancis (CRC Press), Boca Raton, FL, 2006, Chapter 9.

Committee on Medical and Biologic Effects of Environmental Pollutants, *Vapor-Phase Organic Pollutants,* National Academy of Sciences, Washington, DC, 1976, p. 305.

Committee on Sulfur Oxides, *Sulfur Oxides,* National Academy of Sciences, Washington, DC, 1978, pp. 136 and 157.

Conn, C.A., Green, F.H.Y., and Nikula, K.J., Animal models of pulmonary infection in the compromised host: Potential usefulness for studying health effects of inhaled particles, *Inhal. Toxicol.*, 12: 783–827, 2000.

Cook, W.R., Chronic bronchitis and alveolar emphysema in the horse, *Vet. Rec.,* 99: 448–451, 1976.

Cooper, G., Langer, G., and Rosinski, J., Submicron aerosol losses in aluminized mylar bags, *J. Appl. Meteorol.*, 18: 57–68, 1979.

Corn, M. and Esmen, N.A., Aerosol generation, in *Handbook on Aerosols,* Dennis, R., Ed., U.S. Energy Research and Development Administration, TID-26608, Springfield, IL, 1976, Chapter 2.

Cornelius, C.E., Animal models—a neglected medical resource, *N. Engl. J. Med.,* 281: 934–944, 1969.

Costa, D.L., Tepper, J.S., and Raub, J.A., Intrepretations and limitations of pulmonary function testing in small laboratory animals, in *Comparative Biology of the Normal Lung*, Parent, R.A., Ed., CRC Press, Boca Raton, FL, 1991, Chapter 22.

Costa, D.L., Air pollution, in *Casarett and Doull's Essentials of Toxicology*, Klaassen, C.D. and Watkins, J.B. III., Eds., McGraw-Hill, New York, 2003, Chapter 28.

Costa, D.L. and Kodavanti, U.P., Toxic responses of the lung to inhaled pollutants: Benefits and limitations of lung-disease models, *Toxicol. Lett.*, 140 (*Special Issue*): 195–203, 2003.

Covert, D., Wiedensohler, A., and Russell, L., Particle charging and transmission efficiencies of aerosol charge neutralizers, *Aerosol Sci. Technol.*, 27: 206–214, 1997.

Cox, C.S. and Wathes, C.M., Eds., *Bioaerosols Handbook*, Lewis Publishers (CRC Press), Boca Raton, FL, 1995.

Crapo, J.D., Miller, F.J., Smolko, E.D., Graham, J.A., and Hayes, A.W., Eds., *Extrapolation of Dosimetric Relationships for Inhaled Particles and Gasses*, Academic Press, San Diego, CA, 1989.

Creasia, D.A., Nettesheim, P., and Hammons, A.S., Impairment of deep lung clearance by influenza virus infection, *Arch. Environ. Health,* 26: 197–201, 1973.

Crofton, J.W. and Douglas, A., *Respiratory Diseases, 2nd Ed.*, Blackwell Scientific, Oxford, 1975, Chapters. 31 and 52.

Crook, B., Inertial samplers: Biological perspectives, in *Bioaerosols Handbook,* Cox, C.S. and Wathes, C.M., Eds., Lewis Publishers, Boca Raton, FL, 1995, Chapter 9.

Crosfill, M.L. and Widdicombe, J.G., Physical characteristics of the chest and lungs and the work of breathing in different mammalian species, *J. Physiol.*, 158: 1–14, 1961.

Crystal, R.G., Ed., *The Biochemical Basis of Pulmonary Function,* Marcel Dekker, New York, 1976, p. 28.

Crystal, R.G., West, J.B., Weibel, E.R., and Barnes, P.J., Eds., *The Lung: Scientific Foundations, 2nd Ed.* (2 vols.), Lippincott-Raven, Philadelphia, PA, 1997.

Cuddihy, R.G. and Boecker, B.B., Controlled administration of respiratory tract burdens of inhaled radioactive aerosols in beagle dogs, *Toxicol. Appl. Pharmacol.*, 25: 597–605, 1973.

Cuddihy, R.G., McClellan, R.O., and Griffith, W.C., Variability in target organ deposition among individuals exposed to toxic substances, *Toxicol. Appl. Pharmacol.,* 49: 179–187, 1979.

Cwiertniewicz, J., Introduction to the Good Laboratory Practice Regulations, *Lab. Anim.*, 34(3): 29–32, 2005.

Dahl, A.R., Metabolic characteristics of the respiratory tract, in *Concepts in Inhalation Toxicology 2nd Ed.*, McClellan, R.O., and Henderson, R.F., Eds., Taylor & Francis, Washington, DC, 1995, Chapter 7.

Dai, Y.T., Juang, Y.J., Wu, Y.Y., Breysse, P.N., and Hsu, D.J., In vivo measurements of inhalability of ultralarge aerosol particles in calm air by humans, *J. Aerosol Sci.*, 37: 967–973, 2006.

Dasgupta, P.K., Raabe, O.G., Duvall, T.R., and Tarkington, B.K., Generation and characterization of sodium sulfite aerosols for applications in inhalation toxicologic research, *Am. Ind. Hygiene Assoc. J.,* 41: 660–665, 1980.

Dauterman, W.C., Metabolism of toxicants: Phase II reactions, in *Introduction to Biochemical Toxicology*, Hodgson, E. and Guthrie, F.E., Eds., Elsevier, New York, 1980, Chapter 5

Davies, C.N., Definitive equations for the fluid resistance of spheres, *Proc. Phys. Soc. Lond.*, 57: 259–270, 1945.

Davies, C.N., The entry of aerosols into sampling tubes and heads, *Brit. J. Appl. Phys. Ser. 2,* 1: 921–932, 1968.

Dennison, J.E., Andersen, M.E., and Yang, R.S.H., Pitfalls and related improvements of in vivo gas uptake pharmacokinetic experimental systems, *Inhal. Toxicol.*, 17: 539–548, 2005.

Derelanko, M.J., Regulatory toxicology in the United States: An overview, in *CRC Handbook of Toxicology,* Derelanko, M.J. and Hollinger, M.A., Eds., CRC Press, Boca Raton, FL, 1995, Chapter 18.

Derelanko, M.J. and Hollinger, M.A., Eds., *CRC Handbook of Toxicology*, CRC Press, Boca Raton, FL, 1995.

DiBerardinis, L.J., Baum, J.S., First, M., Gatwood, G.T., and Seth, A.K., *Guidelines for Laboratory Design: Health and Safety Considerations, 3rd Ed.*, Wiley Interscience, New York, 2001.

Diesel, D.A., Lebel, J.L., and Tucker, A., Pulmonary particle deposition and airway mucociliary clearance in cold-exposed calves, *Am. J. Vet. Res.,* 52: 1665–1671, 1991.

Dixon, R.L., Problems in extrapolating toxicity data for laboratory animals to man, *Environ. Health Perspect.,* 13: 43–50, 1976.

DOE (U.S. Department of Energy), *Nuclear Air Cleaning Handbook,* DOE-HDBK-1169-2003, U.S. Department of Energy, Washington, DC, 2003.

Donaldson, K., and Borm, P., Eds., *Particle Toxicology*, CRC Press, Boca Raton, 2007.

Donaldson, K., Faux, S., Borm, P.J.A., and Stone, V., Approaches to the toxicological testing of particles, in *Particle Toxicology*, Donaldson, K., and Borm, P., Eds., CRC Press, Boca Raton, 2007, Chapter 17.

Dorato, M.A. and Wolff, R.K., Inhalation exposure technology, dosimetry, and regulatory issues, *Toxicol. Pathol.*, 19: 373–383, 1991.

Dorman, D.C., Brenneman, K.A., McElveen, A.M., Lynch, S.E., Roberts, K.C., and Wong, B.A., Olfactory transport: A direct route of delivery of inhaled manganese phosphate to the rat brain, *J. Toxicol. Environ. Health Part A*, 65: 1493–1511, 2002.

Dorman, R.G., Filtration, in *Aerosol Science,* Davies, C.N., Ed., Academic Press, New York, 1966, Chapter 8.

Doty, R.L., Cometto-Muñiz, J.E., Jalowayski, A.A., Dalton, P., Kendal-Reed, M., and Hodgson, M., Assessment of upper respiratory tract and ocular irritative effects of volatile chemicals in humans, *Crit. Revs. Toxicol.*, 34: 85–142, 2004.

Doyle, G.J., Bekowies, P.J., Winer, A.M., and Pitts, J.N. Jr., Charcoal-adsorption air purification system for chamber studies investigating atmospheric photochemistry, *Environ. Sci. Technol.,* 11: 45–51, 1977.

Drew, R.T. and Laskin, S., Environmental inhalation chambers, in *Methods of Animal Experimentation*, Gay, W.I., Ed., Academic Press. New York. 1973, Chapter 1.

Drinker, P. and Hatch, T., *Industrial Dust*, McGraw-Hill, New York, 1936.

Dubin, S. and Westcott, R.J., Functional residual capacity of normal unanesthetized Beagle dogs, *Am. J. Vet. Res.,* 30: 2027–2030, 1969.

Dubin, S.E. and Morrison, G., A face mask and mouthpiece for respiratory studies in unanesthetized Beagle dogs, *J. Appl. Physiol.,* 27: 104–105, 1969.

Dungworth, D.L., Schwartz, L,.W., Tyler, W.S., and Phalen, R.F., Morphological methods for evaluation of pulmonary toxicity in animals, *Annu. Rev. Pharmacol. Toxicol.,* 16: 381–399, 1976.

Dungworth, D., Kimmerle, G., Lewkowski, J., McClellan, R., and Stöber, W., Eds., *Inhalation Toxicology: The Design and Interpretation of Inhalation Studies and Their Use in Risk Assessment*, Springer-Verlag, New York, 1988.

Dungworth, D.L., Histopathological end-points in the respiratory tract, in *Respiratory Toxicology and Risk Assessment*, Jenkins, P.G., Kayser, D., Muhle, H., Rosner, G., and Smith, E.M., Eds., Wissenschaftlicke Verlagsgesellschaft mbH, Stuttgart, Germany, 1994, pp. 233–249.

Dungworth, D.L., Hahn, F.F., and Nikula, K.J., Noncarcinogenic responses of the respiratory tract to inhaled toxicants, in *Concepts in Inhalation Toxicology, 2nd Ed.*, McClellan, R.O., and Henderson, R.F., Eds., Taylor & Francis, Weshington, DC, 1995, Chapter 18.

Ellington, B., McBride, J.T., and Stokes, D.C., Effects of corticosteroids on postnatal lung and airway growth in the ferret, *J. Appl. Physiol.,* 68: 2029–2033, 1990.

Emery, J.L., The postnatal development of the human lung and its implications for lung pathology, *Respiration*, (*Suppl.*), 27: 41–50, 1970.

EPA (U.S. Environmental Protection Agency), *Development of Statistical Distributions or Ranges of Standard Factors Used in Exposure Assessments*, EPA/600/8-85/010, Office of Health and Environmental Assessment, Washington, DC, 1985.

Esmen, N.A., Adhesion and aerodynamic resuspension of fibrous particles, *J. Environ. Eng.—ASCE,* 122: 379–383, 1996.

Esparza, D.C., Schum, G.M., and Phalen, R.F., A latex-sponge collar for partial-body plethysmography using anesthetized rodents, *Lab. Anim. Sci.,* 29: 652–655, 1979.

Evans, G.W. and Jacobs, S.V., Air pollution and human behavior, *J. Social Issues,* 37: 95–125, 1981.

Everett, K. and Hughes, D., *A Guide to Laboratory Design,* Butterworths, Boston, MA, 1981.

Farrell, B.P., Kerr, H.D., Kulle, T.J., Sauder, L.R., and Young, J.L., Adaptation in human subjects to the effects of inhaled ozone after repeated exposure, *Am. Rev. Respir. Dis.,* 119: 725–730, 1979.

Fehrenbach, H., Animal models of chronic obstructive pulmonary disease: Some critical remarks, *Pathobiol.,* 70: 277–283, 2002.

Fenn, W.O., The phagocytosis of solid particles I. Quartz, *J. Gen. Physiol.,* 3: 439–464, 1921.

Fenn, W.O., The phagocytosis of solid particles IV. Carbon and quartz in solutions of varying acidity, *J. Gen. Physol.,* 5: 311–325, 1923.

Ferin, J. and Leach, L.J., The effect of SO_2 on lung clearance of TiO_2 particles in rats, *Am. Ind. Hygiene Assoc. J.,* 34: 260–263, 1973.

Ferin, J. and Leach, L.J., Horizontal airflow inhalation exposure chamber, in *Generation of Aerosols,* Willeke, K., Ed., Ann Arbor Science, Ann Arbor, 1980, Chapter 25.

Ferron, G.A., The size of soluble aerosol particles as a function of the humidity of the air: Application to the human respiratory tract. *J. Aerosol Sci.,* 8: 251–267, 1977.

Ferron, G.A., Kreyling, W.G., and Haider, B., Inhalation of salt aerosol particles: II Growth and deposition in the human respiratory tract, *J. Aerosol Sci.,* 19: 611–631, 1988.

Ferron, G.A. and Busch, B., Deposition of hygroscopic aerosol particles in the lungs, in *Aerosol Inhalation: Recent Research Frontiers,* Marijnissen J.C.M. and Gradón, L. Eds., Kluwer Academic Publishers, Dordrecht, 1996, pp. 143–152.

Filov, V.A., Golubev, A.A., Liublina, E.I., and Tolokontsev, N.A., *Quantitaitive Toxicology,* John Wiley & Sons, New York, 1979, Chapter 2.

Findeisen, W., Uber das Absetzen kleiner, in der Luft suspendierter Teilchen in der menschliehen Lunge bei der Atmung, *Pflug. Arch. Ges. Physiol. Mensch. Tiere,* 236: 367–379, 1935.

Finlay, W.H., *The Mechanics of Inhaled Pharmaceutical Aerosols: An Introduction,* Academic Press, New York, 2001, Chapter 4.

Fiserova-Bergerova, V., Extrapolation of physiological parameters for physiologically based simulation models, *Toxicol. Lett.* 79: 77–86, 1995.

Fisher, A.B., Normal and pathologic biochemistry of the lung, *Environ. Health Perspect.,* 16: 3–9, 1976.

Fisher, A.B., Lung biochemistry and intermediary metabolism for concepts in inhalation toxicology, in *Concepts in Inhalation Toxicology, 2nd Ed.,* McClellan, R.O., and Henderson, R.F., Eds., Taylor & Francis, Washington, DC, 1995, Chapter 6.

Fitzpatrick, J.C., Jordan, B.S., Salman, N., Williams, J., Cioffi, W.G., and Pruitt, B.A., The use of perfluorocarbon-associated gas exchange to improve ventilation and decrease mortality after inhalation injury in a neonatal swine model, *J. Pediat. Surg.,* 32: 192–196, 1997.

Flavin, M., MacDonald, M., Dolovich, M., Eng, P., Coates, G., and O'Brodovich, H., Aerosol delivery to the rabbit lung with an infant ventilator, *Pediatr. Pulmonol.,* 2: 35–39, 1986.

Fogle, H.C., Environmental Protection Agency: TSCA, in *CRC Handbook of Toxicology,* Derelanko, M.J. and Hollinger, M.A., Eds., CRC Press, Boca Raton, FL, 1995, Chapter 19.

Fok, T.F., Al-Essa, M., Monkman, S., et al., Pulmonary deposition of salbutamol aerosol delivered by metered dose inhaler, jet nebulizer and ultrasonic nebulizer in mechanically ventilated rabbits, *Pediat. Res.,* 42: 721–727, 1997.

Folinsbee, L.J., Wagner, J.A., Borgia, J.F., Drinkwater, B.L., Gliner, J. A., and Bedi, J.F., Eds., *Environmental Stress, Individual Human Adaptations,* Academic Press, New York, 1978.

Folinsbee, L.J., Kim, C.S., Kehrl, H.R., Prah, J.D., and Devlin, R.B., Methods in human inhalation toxicology, in *Handbook of Human Toxicology*, Massaro, E.J., Ed., CRC Press, Boca Raton, FL, 1997, Chapter 14.

Fox, J.G., Clinical assessment of laboratory rodents on long-term bioassay studies, *J. Environ. Pathol. Toxicol.*, 1: 199–226, 1977.

Frampton, M.W. and Utell, M.J., Clinical studies of airborne pollutants, in *Toxicology of the Lung, 3rd Ed.*, Gardner, D.E., Crapo, J.D., and McClellan, R.O., Eds.,Taylor & Francis, Philadelphia, PA, 1999, Chapter 16.

Frank, N.R., Yoder, R.E., Yokoyama, E., and Speizer, F.E., The diffusion of $^{35}SO_2$, from tissue fluids into lungs following exposure to dogs to $^{35}SO_2$, *Health Phys.,* 13: 31–39, 1967.

Fraser, D.A., Bales, R.E., Lippmann, M., and Stokinger, H.E., *Exposure Chambers for Research in Animal Inhalation*, Public Health Monogr., No. 57, U.S. Department of Health. Education and Welfare, U.S. Government Printing Office, Washington, DC, 1959.

Freeman, G., Crane, S.C., Furiosi, N.J., Stephens, R.J, Evans, M.J., and Moore, W.D., Covert reduction in ventilatory surface in rats during prolonged exposure to subacute nitrogen dioxide, *Am. Rev. Respir. Dis.*, 106: 563–579, 1972.

Frostling, H., A rotating drum for the study of toxic substances in aerosol form, *Aerosol Sci.,* 4: 411–419, 1973.

Fuchs, N.A., *The Mechanics of Aerosols*, Dover Publications Inc., New York, 1964.

Fuchs, N.A. and Sutugin, A.G., Generation and use of monodisperse aerosols, in *Aerosol Science,* Davies, C.N., Ed., Academic Press, New York, 1966, Chapter 1.

Fuchs, N.A. and Sutugin, A.G., *Highly Dispersed Aerosols,* Ann Arbor Science, Ann Arbor, MI, 1970.

Fuchs, N.A., Latex aerosols—caution!, *Aerosol Sci.*, 4: 405–410, 1973.

Fuchs, N.A., Sampling of aerosols, *Atmos. Environ.*, 9: 697–707, 1975.

Fulwyler, M.J., Electronic separation of biological cells by volume, *Science,* 150: 910–911, 1965.

Gad, S.C., Safety assessment of therapeutic agents administered by the respiratory route, in *Toxicology of the Lung, 4th Ed.*, Gardner, D.E., Ed., CRC Press, Boca Raton, FL, 2006, Chapter 7.

Gad, S.C., Ed., *Animal Models in Toxicology, 2nd Ed.*, CRC Press (Taylor & Francis), Boca Raton, FL, 2007a.

Gad, S.C., Laws and regulations governing animal care and use in research, in *Animal Models in Toxicology, 2nd Ed.,* Gad, S.C., Ed., CRC Press (Taylor & Francis), Boca Raton, FL, 2007b, Chapter 15.

Gad, S.C., Dincer, Z., Svendsen, O., and Skaanild, M.T., The minipig, in *Animal Models in Toxicology 2nd Ed.*, Gad, S.C., Ed., CRC Press (Taylor & Francis), Boca Raton, FL, 2007, Chapter 10.

Ganong, W.F., *Review of Medical Physiology, 19th Ed.*, Appleton & Lange, Stamford, CT, 1999, Chapters 14, 34.

Garay, S.M., Pulmonary function testing, in *Environmental and Occupational Medicine, 2nd Ed.*, Rom, W.N., Ed., Little, Brown & Co., Boston, MA, 1992, Chapter 15.

Gardner, D.E., Ed., Miller, F.J., Illing, J.W., and Kirtz, J.M., Increased infectivity with exposure to ozone and sulfuric acid, *Toxicol. Lett.,* 1: 59–64, 1977.

Gardner, D.E., Crapo, J.D., and Massaro, E.J., Eds., *Toxicology of the Lung*, Raven Press, New York, 1988.

Gardner, D.E. and Kennedy, G.L. Jr., Methodologies and technology for animal inhalation toxicology studies, in *Toxicology of the Lung, 2nd Ed.,* Gardner, D.E., Crapo, J.D., and McClellan, R.O., Eds., Raven Press, New York, 1993, Chapter 1.

Gardner, D.E., Crapo, J.D. and McClellan, R.O., Eds., *Toxicology of the Lung, 3rd Ed.*, Taylor and Francis, Philadelphia, PA, 1999.

Gardner, D.E., Ed., *Toxicology of the Lung, 4th Ed.*, CRC Press, Boca Raton, FL, 2006.

Garner, H.E., Rosborough, J.P., Amend, J.F., and Hoff, H.E., The grade pony as a new laboratory model in cardiopulmonary physiology, *Cardiovas. Res. Cent. Bull.,* 9: 91–104, 1971.

Gehr, P. and Heyder, J., Eds., *Particle–Lung Interactions*, Marcel Dekker, New York, 2000.

Gerde, P., Animal models and their limitations: On the problem of high-to-low dose extrapolations following inhalation exposures, *Exper. Toxicol. Pathol.*, 57 (Suppl. 1): 143–146, 2005.

Ghio, A.J. and Huang, Y.C.T., Exposure to concentrated ambient particles (CAPs): A review, *Inhal. Toxicol.,* 16: 53–59, 2004.

Gilboa, A. and Silberberg, A., Characterization of epithelial mucus and its function in clearance by ciliary propulsion, in *Air Pollution and the Lung,* Aharonson, E.F., Ben-David, A., Klingberg, M.A., and Kaye, M., Eds., John Wiley & Sons, New York, 1976, pp. 49–62.

Gilmour, M.I., Park, P., Doerfler, D., and Selgrade, M.K., Factors that influence the suppression of pulmonary antibacterial defenses in mice exposed to ozone, *Exp. Lung Res.*, 19: 299–314, 1993.

Giordano, A.M., Holsclaw, D., and Litt, M., Mucus rheology and mucociliary clearance: Normal physiologic state. *Am. Rev. Respir. Dis.,* 118: 245–250, 1978.

Girod, S., Zahm, J.M., Plotkowski, C., Beck, G., and Puchelle, E., Role of the physicochemical properties of mucus in the protection of the respiratory epithelium, *Eur. Respir. J.*, 5: 477–487, 1992.

Goldstein, E., Tyler, W.S., Hoeprich, P.D., and Eagle, C., Ozone and the antibacterial defense mechanisms of the murine lung. *Arch. Intern. Med.,* 127: 1099–1102, 1971.

Goldstein, E., Bartlema, H.C., van der Ploeg, M., van Duijn, P., van der Stap, J.G.M.M., and Lippert, W., Effect of ozone on lysosomal enzymes of alveolar macrophages engaged in phagocytosis and killing of inhaled *Staphylococcus aureus. J. Infect. Dis.,* 138: 299–311, 1978.

Gong, H. Jr., Linn, W.S., Terrell, S.L., et al., Altered heart-rate variability in asthmatic and healthy volunteers exposed to concentrated ambient coarse particles, *Inhal. Toxicol.*, 16: 335–343, 2004.

Gordon, C.J., Role of environmental stress in the physiological response to chemical toxicants, *Environ Res.* 92: 1–7, 2003.

Gradón, L. and Orlicki, D., Deposition of inhaled aerosol particles in a generation of the tracheobronchial tree, *J. Aerosol Sci.*, 21: 3–19, 1990.

Granito, S., *Calculated Retention of Aerosol Particles in the Rat Lung*, M.S. dissertation, University of Chicago, Chicago, IL, 1971.

Green, G.M., Patterns of bacterial clearance in murine influenza, in *Antimicrobial Agents and Chemotherapy,* American Society for Microbiology, Washington, DC, 1965, pp. 26–29.

Green, G.M. and Goldstein, E.A., A method of quantitating intrapulmonary bacterial inactivation in individual animals, *J. Lab. Clin. Med.,* 68: 669–677, 1966.

Griffin, B. and Arch, B., *Laboratory Design Guide: For Clients Architects, and Their Design Team: The Laboratory Design Process from Start to Finish,* Architectural Press, Boston, 1998.

Gross, E.A., and Morgan, K.T., Architecture of nasal passages and larynx, in *Comparative Biology of the Normal Lung*, Parent, R.A., Ed., CRC Press, Boca Raton, FL, 1991, Chapter 2.

Gross, P. and Westrick, M., The permeability of lung parenchyma to particulate matter, *Am. J. Pathol.*, 30: 195–213, 1954.

Gross, S.B., Regulatory guidelines for inhalation toxicity testing, in *Inhalation Toxicology and Technology,* Leong, B.K.J., Ed., Ann Arbor Science, Ann Arbor, MI, 1981, pp. 279–298.

Grover, R.F., Adaptation to high altitude, in *Environmental Stress*, Folinsbee, L.J., Wagner, J.A., Borgia, J.F., Drinkwater, B.L., Gliner, J.A., and Bedi, J.F., Eds., Academic Press, New York, 1978, pp. 325–334.

Gunderson, E.C. and Anderson, C.C., *Development and Validation of Methods for Sampling and Analysis of Workplace Toxic Substances*, Department of Health and Human Services, NIOSH Publ. No. 80-133, Cincinnati, OH, 1980.

Gunn, R., The statistical electrification of aerosols by ionic diffusion, *J. Colloid Sci.*, 10: 107–119, 1955.

Guyton, A.C., Analysis of respiratory patterns in laboratory animals, *Am. J. Physiol.,* 150: 78–83, 1947.

Guyton, A.C., Aviation, high altitude, and space physiology, in *Texbook of Medical Physiology, 8th Ed.*, Guyton, A.C., Ed., W.B. Saunders, Philadelphia, PA, 1991, Chapter 43.

Hackney, J.D., Linn, W.S., Buckley, R.D., Pedersen, E.E., Karuza, S.K., Law, D.C., and Fischer, D.A., Experimental studies on human health effects of air pollutants: I. Design considerations, *Arch. Environ. Health,* 30: 373–378, 1975.

Haefeli-Bleuer, B., and Weibel, E.R., Morphometry of the human pulmonary acinus, *Anat. Rec.*, 220: 401–414, 1988.

Hahn, E.W. and Wester, R.C., *The Biomedical Use of Ferrets in Research*, Marshall Research Animals, North Rose, NY, 1969.

Hahn, F.F., Carcinogenic responses of the lung to inhaled toxicants, in *Concepts in Inhalation Toxicology, 2nd Ed.*, McClellan, R.O. and Henderson, R.F., Eds., Taylor & Francis, Washington, DC, 1995, Chapter 12.

Hahn, F.F., Chronic inhalation bioassays for respiratory-tract carcinogenesis, in *Toxicology of the Lung, 3rd Ed.*, Gardner, D.E., Crapo, J.D., and McClellan, R.O., Eds., Taylor & Francis, Philadelphia, 1999, PA, Chapter 9.

Hanna, M.G. Jr., Nettesheim, P., and Gilbert, J.R., Eds., *Inhalation Carcinogenesis*, A.E.C. Symp. Set. No. 18, U.S. Atomic Energy Commission, Division of Technical Information, CONF 691001, National Technical Information Service, Springfield, VA, 1970.

Harbison, R.D., Teratogens, in *Casarett and Doull's Toxicology, 2nd Ed.*, Doull, J., Klaassen, C.D., and Amdur, M.O., Eds., Macmillan, New York, 1980, Chapter 8.

Harding, R., Pinkerton, K.E., and Plopper, C.G., Eds., *The Lung: Development, Aging and the Environment*, Elsevier, San Diego, CA, 2004.

Hartings, J.M., and Roy, C.J., The automated bioaerosol exposure system: Preclinical platform development and a respiratory dosimetry application with nonhuman primates, *J. Pharmacol. Toxicol. Methods*, 49: 39–55, 2004.

Hasani, A., Vora, H., Pavia, D., Agnew, J.E., and Clarke, S.W., No effect of gender on lung mucociliary clearance in young healthy adults, *Respir. Med.*, 88: 697–700, 1994.

Haskins, S., Pascoe, P.J., Ilkiw, J.E., Fudge, J., Hopper, K., and Aldrich, J., Reference cardiopulmonary values in normal dogs, *Compar. Med.*, 55: 156–161, 2005.

Hatch, T. and Choate, S.P., Statistical description of the size properties of non-uniform particulate substances. *J. Franklin Inst.*, 207: 369–387, 1929.

Hayatdavoudi, G., Crapo, J.D., Miller, F.J., and O'Neil, J.J., Factors determining degree of inflation in intratracheally fixed rat lungs, *J. Appl. Physiol. Respir. Environ. Exercise Physiol.*, 48: 389–393, 1980.

Heinemann, H.O. and Fishman, A.P., Nonrespiratory functions of mammalian lung, *Physiol. Rev.*, 49: 1–46, 1969.

Henderson, R.F., Use of bronchoalveolar lavage to detect respiratory tract toxicity of inhaled material, *Exptl. Toxicol. Pathol.*, 57 (Suppl. 1): 155–159, 2005.

Henry, M.C., Spangler, J., Findlay, J., and Ehrlich, R., Effects of nitrogen dioxide and tobacco smoke on retention of inhaled bacteria, in *Inhaled Particles III. Vol. 1,* Walton, W.H., Ed., Unwin, Surrey, England, 1970, p. 527.

Hensel, A., Windt, H., Stockhofezurwieden, N., Lodding, H., Koch, W., and Petzoldt, K., A porcine aerosol infection model for studying dose-dependent effects caused by actinobacilluspleuropneumoniae bacteria, *J. Aerosol Med.,* 6: 73–88, 1993.

Hering, S.V., Impactors, cyclones, and other particle collectors, in *Air Sampling Instruments, 9th Ed.*, Cohen, B.S., and McCammon, C.S. Jr., Eds., ACGIH®, Cincinnati, OH, 2001, Chapter 14.

Hesketh, H.E., *Fine Particles in Gaseous Media.* Ann Arbor Science, Ann Arbor, MI, 1977.

Hickey, A.J., Ed., *Inhalation Aerosols: Physical and Biological Basis for Therapy, 2nd Ed.*, Informa Healthcare, U.S.A., New York, 2007.

Hinds W.C., *Aerosol Technology: Properties, Behavior and Measurement of Airborne Particles, 2nd Ed.*, Wiley-Interscience, New York, 1999.

Hinds W.C. and Kennedy N.J., An ion generator for neutralizing concentrated aerosols. *Aerosol Sci. Technol.*, 32: 214–220, 2000.

Hinds, W.C., Ashley, A. Kennedy, N.J., and Bucknam, P., Conditions for cloud settling and Rayleigh-Taylor instability, *Aerosol. Sci. Technol.*, 36: 1128–1138, 2002.

Hinners, R.G., Burkart, J.K., and Contner, G.L., Animal exposure chambers in air pollution studies, *Arch. Environ. Health,* 13: 609–615, 1966.

Hobbs, C.H., Dorman, D.C., Griffin, D.E., et al., *Overcoming Challenges to Develop Countermeasures Against Aerosolized Bioterrorism Agents.* National Academies Press, Washington, DC, 2006.

Hocking, W.G. and Golde, D.W., The pulmonary-alveolar macrophage, *N. Engl. J. Med.,* 301: 580–639, 1979.

Hodkinson, J.R., The optical measurement of aerosols, in *Aerosol Science*, Davies, C.N., Ed., Academic Press. New York, 1966 Chapter 10.

Hodson, W.A., Ed., *Development of the Lung,* Marcel Dekker, New York, 1977.

Hoffmann, D. and Wynder, E.L., Chamber development and aerosol dispersion, in *Inhalation Carcinogenesis*, Hanna, M.G. Jr., Nettesheim, P., and Gilbert, J.R., Eds., U.S. Atomic Energy Commission, Oak Ridge, TN, 1970, pp. 173–191.

Hofmann, W., Morawska, L., and Bergmann, R., Environmental tobacco smoke deposition in the human respiratory tract: Differences between experimental and theoretical approaches, *J. Aerosol. Med.*, 14: 317–326, 2001.

Holma, B., Lung clearance of mono- and di-disperse aerosols determined by profile scanning and whole-body counting—a study on normal and SO_2 exposed rabbits, *Acta Med. Scand.*, 5 (Suppl. 473): 1–102, 1967.

Holma, B., The acute effects of cigarette smoke on the initial course of lung clearance in rabbits, *Arch. Environ. Health,* 18: 171–173, 1969.

Hopke, P.K., Advances in monitoring methods for airborne particles, *Aerosols Handbook: Measurement Dosimetry, and Health Effects*, in Ruzer, L.S. and Harely, N.H., Eds., CRC Press, Boca Raton, FL, 2005, Chapter 3.

Horsfield, K., Dart, G., Olson, D. E., Filley, G., and Cumming, G., Models of the human bronchial tree, *J. Appl. Physiol.,* 31: 207–217, 1971.

Horsfield, K., Kemp, W., and Phillips, S., An asymmetrical model of the airways of the dog lung, *J. Appl. Physiol.,* 52: 21–26, 1982.

Hoskins, J.A., Brown, R.C., Cain, K., et al., The construction and validation of a high containment nose-only rodent inhalation facility, *Ann. Occup. Hygiene*, 41: 51–61, 1997.

Hubal, E.A.C., Fedkiw, P.S., and Kimbell, J.S., Mass-transport models to predict toxicity of inhaled gasses in the upper respiratory tract, *J. Appl. Physiol.* 80: 1415–1427, 1996.

Hyde, D.M., Bolender, R.P., Harkema, J.R., and Plopper, C.G., Morphometric approaches for evaluating pulmonary toxicity in mammals: Implication for risk assessment, *Risk Analysis*, 14: 293–302, 1994.

Hyde, D.M., Tyler, N.K., and Plopper, C.G., Morphometry of the respiratory tract: Avoiding the sampling, size, orientation, and reference traps, *Toxicol. Pathol.*, 35: 41–48, 2007.

Hyman, A.L., Spannhake, E.W., and Kadowitz, P.J., Prostaglandins and the lung, in *Lung Disease, State* of *the Art,* Murray, J.F., Ed., American Lung Association, New York, 1979, pp. 229–254.

Ichioka, M., Model experiments on absorbability of the airway mucous membrane of SO_2 and NO_2 gases, *Bull. Tokyo Med. Dent. Univ.,* 19: 361–375, 1972.

ICRP (International Commission on Radiatiological Protection, Task Group of Committee 2), *Human Respiratory Tract Model for Radiological Protection*, Publication 66, Pergamon Press, New York, 1994.

ILAR (Institute of Laboratory Animal Resources), *Guide for the Care and Use of Laboratory Animals*, National Research Council, National Academy Press, Washington, DC, 1996.

Ishizu, Y., Ohta, K., and Okada, T., Changes in the particle size and the concentration of cigarette smoke through the column of a cigarette, *J. Aerosol Sci*., 9: 25–29, 1978.

Jacobson, A.R. and Morris, S.C., The primary air pollutants—viable particles their occurrence, sources and effects, in *Air Pollution, Vol. 1, 3rd Ed.,* Stern, A.C., Ed., Academic Press, New York, 1976, Chapter 4.

Jarabek, A.M., The application of dosimetry models to identify key processes and parameters for default dose–response assessment approaches, *Toxicol. Lett.*, 79: 171–184, 1995.

Jarabek, A.M., Asgharian, B., and Miller, F.J., Dosimetric adjustments for interspecies extrapolation of inhaled poorly soluble particles (PSP), *Inhal. Toxicol.*, 17: 317–334, 2005.

Jeffery, P.K., Comparative morphology of the airways in asthma and chronic obstructive pulmonary-disease, *Am. J. Respir. Crit. Care Med*., 150 (Suppl.): S6–S13, 1994.

Jenkins, P.G., Kayser, D., Muhle, H., Rosner, G., and Smith, E.M., Eds., *Respiratory Toxicology and Risk Assessment,* Wissenschaftliche Verlagsgesellschaft mbH, Stuttgart, Germany, 1994.

Jenkins, R.A., Guerin, M.R., and Tomkins, B.A., *The Chemistry of Environmental Tobacco Smoke: Composition and Measurement, 2nd Ed*., Lewis Publishers (CRC Press), Boca Raton, FL, 2000.

Ji J.H., Bae G.N., and Hwang J., Characteristics of aerosol charge neutralizers for highly charged particles, *J. Aerosol Sci.*, 35: 1347–1358, 2004.

Jones, C.D.R. and Bull, J.R., Deposition and clearance of radiolabelled monodisperse polystyrene spheres in the calf lung, *Res. Vet. Sci.,* 42: 82–91, 1986.

Jones, J.H. and Longworth, K.E., Gas exchange at rest and during exercise in mammals, in *Comparative Biology of the Normal Lung*, Parent, R.A., Ed., CRC Press, Boca Raton, FL, 1991, Chapter 19.

Kacergis, J.B., Jones, R.B., Reeb, C.K., Turner, W.A., Ohman, J.L., Ardman, M.R., and Paigen, B., Air quality in an animal facility: Particulates, ammonia, and volatile organic compounds, *Am. Ind. Hygiene Assoc. J.*, 57: 634–640, 1996.

Kane, L.E., Barrow, C.S., and Alarie, Y., A short-term test to predict acceptable levels of exposure to airborne sensory irritants, *Am. Ind. Hygiene Assoc. J.*, 40: 207–229, 1979.

Karg, E., Tuch, Th., Ferron, G.A., et al., Design operation and performance of whole body chambers for long-term aerosol exposure of large experimental animals, *J. Aerosol. Sci.*, 23:279–290, 1992.

Keith, C.H. and Derrick, J.C., Measurement of the particle size distribution and concentration of cigarette smoke by the "conifuge", *J. Colloid Sci.* 15: 340–356, 1960.

Kennedy, H.W., Legal aspects of human exposure to atmospheric pollutants, *Arch. Environ. Health,* 6: 99–112, 1963.

Kenett, R. and Thyregod, P., Aspects of statistical consulting not taught by academia, *Stat. Neerl.*, 60: 396–411, 2006.

Kenoyer, J.L., Phalen, R.F., and Davis, J.R., Particle clearance from the respiratory tract as a test of toxicity: Effect of ozone on short and long term clearance, *Exp. Lung Res.,* 2: 111–120, 1981.

Kilburn, K.H., Model systems for the biological effects of air pollutants, in *Air Pollution and the Lung,* Aharonson, E.F., Ben-David, A., and Klingberg, M.A., Eds., John Wiley & Sons, New York, 1976, pp. 221–247.

Kim, C.S. and Kang, T.C., Comparative measurement of lung deposition of inhaled fine particles in normal subjects and patients with obstructive airway disease, *Am. J. Respir. Crit. Care Med.*, 155: 899–905, 1997.

Kim, C.S. and Hu, S.C., Regional deposition of inhaled particles in human lungs: Comparison between men and women, *J. Appl. Physiol.*, 84: 1834–1844, 1998.

Kim, W.D., Lung mucus: A clinician's view, *Eur. Respir. J.*, 10: 1914–1917, 1997.

Kim, Y., Furr, R., Gibbs, L.M., and Teng, L.S., Health and safety challenges in start-up of a multi-discipline/open lab research facility, *Chem. Health Safety*, 12: 24–29, 2005.

Kimbell, J.S. and Miller, F.J., Regional respiratory-tract absorption of inhaled reactive gases, in *Toxicology of the Lung, 3rd Ed.*, Gardner, D.E., Crapo, J.D., and McClellan, R.O., Eds., Taylor & Francis, Philadelphia, PA, 1999, Chapter 19.

Kimbell, J.S. and Subramaniam, R.P., Use of computational fluid dynamics models for dosimetry of inhaled gases in the nasal passages, *Inhal. Toxicol.,* 13: 325–334, 2001.

King, M., Zahm, J.M., Pierrot, D., Vaquez-Girod, S., and Puchelle, E., The role of mucus gel viscosity, spinnability, and adhesive properties in clearance by simulated cough, *Biorheology*, 26: 737–745, 1989.

Kinsara, A.A., Tompson, R.V., and Loyalka, S.K., Computational flow and aerosol concentration profiles in lung bifurcations, *Health Phys.*, 64: 13–22, 1993.

Kirk, W.P., Rehnberg, B.F., and Morken, D.A., Acute lethality in guinea pigs following respiratory exposure to Kr-85, *Health Phys.,* 28: 275–284, 1975.

Klaassen, C.D. and Doull, J., Evaluation of safety: Toxicologic evaluation, in *Casarett and Doull's Toxicology,* Doull, J., Klaassen, C.D., and Amdur, M.O., Eds., Macmillan, New York, 1980, Chapter 2.

Kleinman, M.T., Bailey, R.M., Chang, Y.T.C., et al., Exposures of human volunteers to a controlled atmospheric mixture of ozone, sulfur dioxide and sulfuric acid, *Am. Ind. Hygiene Assoc. J.*, 42: 61–69, 1981.

Kleinman, M.T. and Phalen, R.F., Exposure atmosphere generation and characterization, in *Methods in Inhalation Toxicology,* Phalen, R.F., Ed., CRC Press, Boca Raton, FL, 1997, Chapter 4.

Kliment, V., Libich, J., and Kaudersova, V., Geometry of guinea pig respiratory tract and application of Landahl's model of deposition of aerosol particles, *J. Hygiene Epidemiol. Microbiol. Immunol.,* 16: 107–114, 1972.

Koch, W., Generation of exposure atmospheres, in *Respiratory Toxicology and Risk Assessment,* Jenkins, P.G., Kayser, D., Muhle, H., Rosner G., and Smith, E.M., Eds. Wissenschaftliche Verlagsgesellschaft mbH, Stuttgart, Germany, 1994, pp. 49–58.

Koch, W., Windt, H., Walles, M., Borlak, J., and Clausing, P., Inhalation studies with the Gottingen minipig, *Inhal. Toxicol.,* 13: 249–259, 2001.

Kodavanti, U.P., and Costa, D.L., Rodent models of susceptibility: What is their place in inhalation toxicology? *Respir. Physiol.,* 128: 57–70, 2001.

Koenig, J.Q. and Luchtel, D.L., Respiratory responses to inhaled toxicants, in *Handbook of Human Toxicology,* Massaro, E.J., Ed., CRC Press, Boca Raton, FL, 1997, Chapter 13.

Kotov, Y.A., Electric explosion of wires as a method for preparation of nanopowders, *J. Nanoparticle Res.,* 5: 539–550, 2003.

Kotrappa, P. and Moss, O.R., Production of relatively monodisperse aerosols for inhalation experiments by aerosol centrifugation, *Health Phys.,* 21: 531–535, 1971.

Krasovskii, G.N., Extrapolation of experimental data from animals to man, *Environ. Health Persp.,* 13: 51–58, 1976.

Kreyling, W.G., and Scheuch, G., Clearance of particles deposited in the lungs, in *Particle–Lung Interactions,* Gehr, P. and Heyder, J., Eds., Marcel Dekker, New York, 2000, Chapter 7.

Kreyling, W.G., Semmler-Behnke, M., and Möller, W., Ultrafine particle–lung interactions: Does size matter? *J. Aerosol Med.,* 19: 74–83, 2006.

Kreyling, W.G., Möller, W., Semmler-Behnke, M., and Oberdörster, G., Particle dosimetry: Deposition and clearance from the respiratory tract and translocation towards extra-pulmonary sites, in *Particle Toxicology,* Donaldson, K. and Borm, P., Eds., CRC Press, Boca Raton, FL, 2007, Chapter 3.

Kromback, F., Münzing, S., Allmeling, A.M., Gerlack, J.T., Behr, J., and Dörger, M., Cell size of alveolar macrophages: An interspecies comparison, *Environ, Health Persp.,* 105 (Suppl. 5): 1261–1263, 1997.

Kurucz, I. and Szelenyi, I., Current animal models of bronchial asthma, *Current Pharm. Design,* 12: 3175–3194, 2006.

Lab Animal, *Lab Animal Buyers Guide,* http://guide.labanimal.com/guide/index.html (accessed 12/6/2007); Also, *Lab Animal,* December issue.

Lai, A.C.K., Investigation of electrostatic forces on particle deposition in a test chamber, *Indoor Built Environ.,* 15: 179–186, 2006.

Lai, Y.L., Comparative ventilation of the normal lung, in *Comparative Biology of the Normal Lung,* Parent, R.A., Ed., CRC Press, Boca Raton, FL, 1991, Chapter 17.

Lambert, L.B., Singer, T.M., Boucher, S.E., and Douglas, G.R., Detailed review of transgenic rodent mutation assays, *Mutation Res.-Rev. Mutation Res.,* 590: 1–280, 2005.

Lambertsen, C.J., Anoxia, altitude and acclimatization, in *Medical Physiology 11th Ed.,* Bard, P., Ed., Mosby, St. Louis, MO, 1961, Chapter 42.

Landahl, H.D., On the removal of air-borne droplets by the human respiratory tract. I. The lung, *Bull. Math. Biophys.,* 12: 43–56, 1950.

Landahl, H.D. and Tracewell, T.N., An investigation of cigarette smoke as an aerosol with special reference to retention in lungs in human subjects, *Trans. Ill. State Acad. Sci.,* 50: 213–220, 1957.

Landahl, H.D., The effect of gravity, hygroscopicity, and particle size on the amount and site of deposition of inhaled particles, with particular reference to hazard due to airborne viruses, in *Assessment of Airborne Particles,* Mercer, T.T., Morrow, P.E., and Stöber, W., Eds., Charles C. Thomas, Springfield, IL, 1972, Chapter 21.

Lange, C.F. and Finlay, W.H., Liquid atomizing: Nebulizing and other methods of producing aerosols, *J. Aerosol Med.,* 19: 28–35, 2006.

Laskin, S., Kuschner, M., and Drew, R.T., Studies in pulmonary carcinogenesis, in *Inhalation Carcinogenesis,* Hanna, M.G. Jr., Nettesheim, P., and Gilbert, J.R., Eds., U.S. Atomic Energy Commission, Oak Ridge, TN, 1970, pp. 321–351.

Laskin, S., Kuschner, M., Sellakumar, A., and Katz, G.V., Combined carcinogen-irritant animal inhalation studies, in *Air Pollution and the Lung,* Aharonson, E.F., Ben-David, A., and Klingberg, M.A., Eds., John Wiley & Sons, New York, 1976, pp. 190–213.

Lawler, J.V., Endy, T.P., Hensley, L.E., et al., Cynomolgus macaque as an animal model for severe acute respiratory syndrome, *Plos Med.*, 3: 677–686, 2006.

Leach, L.J., Spiegl, C.J., Wilson, R H., Sylvester, G.E., and Lauterbach, K.E., A multiple chamber exposure unit designed for chronic inhalation studies, *Am. Ind. Hygiene Assoc. J.*, 20: 13–22, 1959.

Lee, J.W. and Goo, J.H., Numerical simulation of air flow and inertial deposition of particles in a bifurcating channel of square cross section, *J. Aerosol Med.*, 5: 131–154, 1992.

Lehrer, R.I., Ferrari, L.G., Patterson-Delafield, J., and Sorrell, T., Fungicidal activity of rabbit alveolar and peritoneal-macrophages against *Candida albicans*, *Infect. Immun.*, 28: 1001–1008, 1980.

Leong, B.K.J., The current state of chamber design and inhalation toxicology instrumentation, *Proc. 7th Annu. Conf. Environ. Toxicol.*, available from National Technical Information Service as AMRL-TR-76-125, Springfield, VA., 1976, pp. 141–150.

Leong, B.K.J., Coombs, J.K., Sabaitis, C.P., Rop, D.A., and Aaron, C.S., Quantitative morphometric analysis of pulmonary depositon of aerosol particles inhaled via intratracheal nebulization, intratracheal instillation or nose-only inhalation in rats, *J. Appl. Toxicol.*, 18: 149–160, 1998.

Leung, K., Louca, E., Gray, M., Tipples, G., and Coates, A.L., Use of the next generation pharmaceutical impactor for particles size distribution measurements of live viral aerosol vaccines, *J. Aerosol Med.*, 18: 414–426, 2005.

Levine, M.D., Reisch, M.L., and Thurlbeck, W.M., Automated measurement of internal surface area of human lung, *IEEE Trans. Biomed. Eng.,* 17: 254–262, 1970.

Lewis, R.J. Sr., *Sax's Dangerous Properties of Industrial Materials, 6th Ed.* (3 vols.), John Wiley & Sons, New York, 2000.

Lewis, S.M., and Carraway, J.H., Animal models of human disease, *Lab. Animals*, 21(1): 22–29, 1992.

Lidén, G. and Harper, M., The need for an international sampling convention for inhalable dust in calm air, *J. Occup. Environ. Hygiene*, 3: D94–D101, 2006.

Likens, S.A. and Mauderly, J.L., *Respiratory Measurements in Small Laboratory Mammals: A Literature Review*, Inhalation Toxicology Research Institute Report, LF-68, 1979, (available from National Technical Information Service, Springfield, VA).

Lin, A.C., Size matters: Regulating nanotechnology, *Havard Environ. Law Rev.*, 31: 349–408, 2007.

Lindstedt, S.L. and Schaeffer, P.J., Use of allometry in predicting anatomical and physiological parameters of mammals, *Lab. Animals*, 36:1–19, 2002.

Lippmann, M. and Albert, R.E., A compact electric-motor driven spinning disc aerosol generator, *Am. Ind. Hygiene Assoc. J.,* 28: 501–506, 1967.

Lippmann, M., Experimental inhalation studies—equipment and procedures, in *Inhalation Carcinogenesis*, Hanna. M.G. Jr., Nettesheim. P., and Gilbert. J.R., Eds., U.S. Atomic Energy Commission, Oak Ridge, TN., 1970, pp. 55–76.

Lippmann, M., Aerosol exposure methods, in *Generation of Aerosols.* Willeke, K., Ed., Ann Arbor Science, Ann Arbor, MI, 1980, Chapter 21.

Lippmann, M., Ed., *Environmental Toxicants: Human Exposures and Their Health Effects*, Van Nostrand Reinhold, New York, 1992.

Lippmann, M., Gordon, T., and Chen, L.C., Effects of subchronic exposure to concentrated ambient particles (CAPs) in mice: I. Introduction, objectives and experimental plan, *Inhal. Toxicol.*, 17: 177–187, 2005.

Liu, B.Y.H. and Pui, D.Y.H., Electrical neutralization of aerosols. *J. Aerosol Sci.*, 5: 465–472, 1974.

Lodge, J.P. Jr. and Chan, T.L., Eds., *Cascade Impactor: Sampling and Data Analysis*, American Industrial Hygiene Association, Akron, OH, 1986.

Loffert, D.T., Ikle, D., and Nelson, H.S., A comparison of commercial jet nebulizers, *Chest*, 106: 1788–1793, 1994.

Longest, P.W. and Oldham, M.J., Mutual enhancements of CFD modeling and experimental data: A case study of 1-μm particle deposition in a branching airway model, *Inhal. Toxicol.*, 18: 761–771, 2006.

Longest, P.W., Vinchurkar, S., and Martonen, T., Transport and deposition of respiratory aerosols in models of childhood asthma, *J. Aerosol Sci.*, 37: 1234–1257, 2006.

Loosli, C.G., Interalveolar communications in normal and in pathologic mammalian lungs, *Arch. Pathol.*, 24: 743–776, 1937.

MacKenzie, K.M. and Hoar, R.M., Developmental toxicology, in *CRC Handbook of Toxicology*, Derelanko, M.J. and Hollinger, M.A., Eds., CRC Press, Boca Raton, FL, 1995, Chapter 12.

Macklem, P.T., Airway obstruction and collateral ventilation, *Physiol. Rev.*, 51: 368–436, 1971.

Maher, J.A. and DeStefano, J., The ferret: An animal model to study influenza virus, *Lab. Animals*, 33(9): 50–53, 2004.

Ma-Hock, L., Gamer, A.O., Landsiedel, R., et al., Generation and characterization of test atmospheres with nanomaterials, *Inhal. Toxicol.*, 19: 833–848, 2007.

Mannix, R.C., Phalen, R.F., Kenoyer, J.L., and Crocker, T.T., Effect of sulfur dioxide–sulfate exposure on rat respiratory tract clearance, *Am. Ind. Hygiene Assoc. J.*, 43: 679–685, 1982.

Mariassy, A.T., Epithelial cells of trachea and bronchi, in *Comparative Biology of the Normal Lung*, Parent, R.A., Ed., CRC Press, Boca Raton, FL, 1991. Chapter 6.

Maricq, M.M., Xu, N., and Chase, R.E., Measuring particulate mass emissions with the electrical low pressure impactor, *Aerosol Sci. Technol.*, 40: 68–79, 2006.

Marijnissen, J.C.M. and Gradón, L., Eds., *Aerosol Inhalation: Recent Research Frontiers*, Kluwer Academic Publishers, Dordrecht, Netherlands, 1996.

Marple, V.A., Lui, B.Y.H., and Rubow, K.L., A dust generator for laboratory use, *Am. Ind. Hygiene Assoc. J.*, 39: 26–32, 1978.

Martin, H.B., The effects of aging on the alveolar pores of Kohn in the dog, *Am. Rev. Respir. Dis.*, 88: 773–778, 1963.

Martonen, T.B., Analytical model of hygroscopic particle behavior in human airways, *Bull. Math. Biol.*, 44: 425–442, 1982.

Martonen, T.B., Deposition patterns of cigarette smoke in human airways, *Am. Ind. Hygiene Assoc. J.*, 53: 6–18, 1992.

Martonen, T.B., Zhang, Z., and Yang, Y., Interspecies modeling of inhaled particle deposition patterns, *J. Aerosol Sci.*, 23: 389–406, 1992.

Martonen, T.B. and Zhang, Z., Deposition of sulfate acid aerosols in the developing human lung, *Inhal. Toxicol.*, 5: 165–187, 1993.

Martonen, T.B., Zhang, Z., and Yang, Y., Interspecies modeling of inhaled gases, *Inhal. Toxicol.*, 7: 1125–1139, 1995.

Martonen, T.B., Ed., *Medicinal Applications of Computer Modeling: Respiratory System,* WIT Press, Southampton, United Kingdom, 2001.

Massaro, E.J., Ed., *Handbook of Human Toxicology*, CRC Press, Boca Raton, FL, 1997.

Mauderly, J.L., Nenno, W.C., and Morrison, G.A., Stocks for holding unanesthetized dogs in the standing position. *Lab. Anim. Sci.*, 21: 263–266, 1971.

Mauderly, J.L., Steady-state carbon monoxide-diffusing capacity of unanesthetized Beagle dogs, *Am. J. Vet. Res.*, 33: 1485–1491, 1972.

Mauderly, J.L. and Tesarek, J.E., Effects of current aerosol exposure methods on the pulmonary function of beagle dogs and Syrian hamsters, in *Inhalation Toxicology Research Institute Annual Report 1972 to 1973*, McClellan. R.O., Ed., LF-46, Lovelace Foundation, Albuquerque, NM, 1973, pp. 163–168.

Mauderly, J.L., Evaluation of the grade pony as a pulmonary function model, *Am. J. Vet. Res.*, 35: 1025–1029, 1974a.

Mauderly, J.L., Influence of sex and age on the pulmonary function of the unanesthetized beagle dog, *J. Gerontol.*, 29: 282–298, 1974b.

Mauderly, J.L., Respiration of F344 rats in nose-only inhalation exposure tubes, *J. Appl. Toxicol.*, 6: 25–30, 1986.

Mauderly, J.L., Comparisons of respiratory function responses of laboratory animals and humans, in *Inhalation Toxicology: The Design and Interpretation of Inhalation Studies and Their Use in Risk Assessment*, Dungworth, D., Kimmerle, G., Lewkowski, J., McClellan, R., and Stöber, W., Eds., Springer-Verlag, New York, 1988, Chapter 15.

Mauderly, J.L., Assessment of pulmonary function and the effects of inhaled toxicants, in *Concepts in Inhalation Toxicology, 2nd Ed.*, McClellan, R.O. and Henderson, R.F., Eds., Taylor & Francis, Washington, DC, 1995, Chapter 13.

Mauderly, J.L. and McCunney, R.J., Eds., *Particle Overload in the Rat Lung and Lung Cancer: Implications for Human Risk Assessment*, Taylor & Francis, Washington, DC, 1996.

Mauderly, J.L., Lung overload: The dilemma and opportunities for resolution, in *Particle Overload in the Rat Lung and Lung Cancer: Implications for Human Risk Assessment,* Mauderly, J.L. and McCunney, R.J., Eds., Taylor & Francis, Washington, DC, 1996a, pp. 1–28.

Mauderly, J.L., Usefulness of animal models for predicting human responses to long-term inhalation of particles, *Chest,* 109 (Suppl.): S65–S68, 1996b.

Mauderly, J.L., Animal models for the effect of age on susceptibility to inhaled particulate matter, *Inhal. Toxicol.*, 12: 863–900, 2000.

Mautz, W.J., Phalen, R.F., McClure, T.R., and Bufalino, C., A rodent treadmill for inhalation toxicological studies and respirometry, *J. Appl. Physiol.*, 58: 673–679, 1985.

Mautz, W.J., Kleinman, M.T., Phalen, R.F., and Crocker, T.T., Effects of exercise exposure on toxic interactions between inhaled oxidant and aldehyde air pollutants, *J. Toxicol. Environ. Health,* 25: 165–177, 1988.

Mautz, W.J., Animal monitoring, in *Methods in Inhalation Toxicology*, Phalen, R.F., Ed., CRC Press, Boca Raton, FL, 1997, Chapter 6.

Mautz, W.J., Exercising animal models in inhalation toxicology: Interactions with ozone and formaldehyde, *Environ. Res.*, 92: 14–26, 2003.

Mawdesley-Thomas, L.E., Healey, P., and Barry, D.H., Experimental bronchitis in animals due to sulphur dioxide and cigarette smoke. An automated quantitative study, in *Inhaled Particles III, Vol. I,* Walton, W.H., Ed., Unwin Brothers, Surrey, England, 1970, pp. 509–515.

May, K.R., The Collison nebulizer: Description, performance and application, *Aerosol Sci.*, 4: 235–243, 1973.

Maynard, A.D., Overview of methods for analyzing single ultrafine particles, in *Ultrafine Particles in the Atmosphere*, Brown, L.M., Collings, N., Harrison, R.M., Maynard, A.D., and Maynard, R.L., Eds., Imperial College Press, London, 2000, Chapter 3.

McAllen, S.J., Chiu, S.P., Phalen, R.F., and Rasmussen, R.E., Effect of in vivo ozone exposure on in vitro pulmonary alveolar macrophage mobility, *J. Toxicol. Environ. Health*, 7: 373–381, 1981.

McBride, J.T., Architecture of the tracheobronchial tree, in *Comparative Biology of the Lung*, Parent, R.A., Ed., CRC Press, Boca Raton, FL, 1991. Chapter 5.

McClellan, R.O. and Henderson, R.F., Eds., *Concepts in Inhalation Toxicology, 2nd Ed.*, Taylor & Francis, Washington, DC, 1995.

McClellan, R.O., Developing risk assessments for airborne materials, in *Toxicology of the Lung, 3rd Ed.,* Gardner, D.E., Crapo, J.D., and McClellan, R.O., Eds., Taylor & Francis, Philadelphia, PA, 1999, Chapter 20.

McClure, T.R., Diller, P.J., and Phalen, R.F., An airway perfusion apparatus for whole lung fixation, *Lab. Anim. Sci.,* 32: 195–196, 1982.

McJilton, C., Frank, R., and Charlson, R., Role of relative humidity in the synergistic effect of a sulfur dioxide–aerosol mixture on the lung, *Science,* 182: 503–504, 1973.

McJilton, C.E., Frank, R., and Charlson, R.J., Influence of relative humidity on functional effects of an inhaled SO_2 aerosol mixture, *Am. Rev. Respir. Dis.,* 113: 163–169, 1976.

McKinney, W. and Frazer, D., Computer-controlled ozone inhalation exposure system, *Inhal. Toxicol.*, 20: 43–48, 2008.

McKirnan, M.D., White, F.C., Guth, B.D., Longhurst, J.C., and Bloor, C.M., Validation of a respiratory mask for measuring gas-exchange in exercising swine, *J. Appl. Physiol.*, 61: 1226–1229, 1986.

McLain, D.E., Chandra, S., and Gad, S.C., The ferret, in *Animal Models in Toxicology, 2nd Ed.*, Gad, S.C., Ed., CRC Press, Boca Raton, FL, 2007, Chapter 7.

McLaughlin, R.F. Jr., Tyler, W.S., and Canada, R.O., Subgross pulmonary anatomy in various mammals and man. *JAMA*, 175: 148–151, 1961a.

McLaughlin, R.F. Jr., Tyler, W.S., and Canada, R.O., A study of the subgross pulmonary anatomy in various mammals, *Am. J. Anat.*, 108: 149–165, 1961b.

McPherson, E.A., Lawson, G.H.K., Murphy, J.R., Nicholson, J.M., Breeze, R.G., and Pirie, H.M., Chronic obstructive pulmonary disease (COPD) in horses: Aetiological studies: Responses to intradermal and inhalation antigenic challenge. *Equine Vet. J.*, 11: 159–166, 1979.

Medinsky, M.A., Bond, J.A., Schlosser, P.M., and Morris, J.B., Mechanisms and models for respiratory-tract uptake of volatile organic chemicals, in *Toxicology of the Lung, 3rd Ed.*, Gardner, D.E., Crapo, J.D., and McClellan, R.O., Eds., Taylor & Francis, Philadelphia, PA, 1999, Chapter 17.

Meessen, H., Die pathomorphologie der diffusion und perfusion, *Verh. Dtsch. Ges. Pathol.*, 44: 98–127, 1960.

Ménache, M.G., Miller, F.J., and Raabe, O.G., Particle inhalability curves for humans and small laboratory animals, *Ann. Occup. Hygiene*, 39: 317–328, 1995.

Ménache, M.G., Hofmann, W., Ashgarian, B., and Miller, F.W., Airway geometry models of children's lungs for use in dosimetry modeling, *Inhal. Toxicol.*, 20: 101–126, 2008.

Mercer, R.R. and Crapo, J.D., Architecture of the acinus, in *Comparative Biology of the Normal Lung*, Parent, R.A., Ed., CRC Press, Boca Raton, FL, 1991, Chapter 10.

Mercer, T.T., On the role of particle size in the dissolution of lung burdens, *Health Phys.*, 13: 1211–1221, 1967.

Mercer, T.T., Goddard, R.F., and Flores, R.L., Output characteristics of three ultrasonic nebulizers, *Ann. Allergy,* 26: 18–27, 1968a.

Mercer, T.T., Tillery, M.I., and Chow, H.Y., Operating characteristics of some compressed-air nebulizers, *Am. Ind. Hygiene Assoc. J.,* 29: 66–78, 1968b.

Mercer. T.T., Morrow, P.E., and Stober, W., Eds., *Assessment of Airborne Particles*, Charles C. Thomas, Springfield, IL, 1972.

Mercer, T.T., *Aerosol Technology in Hazard Evaluation*, Academic Press, New York, 1973.

Merrill, R.A., Regulatory toxicology, in *Casarett and Doull's Essentials of Toxicology,* Klaassen, C.D. and Watkins, J.B. III, Eds., McGraw-Hill, New York, 2003, Chapter 34.

Miller, F.J., Illing J.W., and Gardner, D.E., Effect of urban ozone levels on laboratory-induced respiratory infections. *Toxicol. Lett.,* 2: 163–169, 1978.

Miller, F.J. and Menzel, D.B., Eds., *Fundamentals of Extrapolation Modeling of Inhaled Toxicants: Ozone and Nitrogen Dioxide*, Hemisphere Publishing Corp., Washington, DC, 1984.

Miller, F.J., Dosimetry of inhaled gasses, in *Respiratory Toxicology and Risk Assessment*, Jenkins, P.G., Kayser, D., Muhle, H., Rosner, G., and Smith, E.M., Wissenschaftliche Verlagsgesellschaft mbH, Stuttgart, Germany, 1994, pp. 111–144.

Miller, F.J., Ed., *Nasal Toxicity and Dosimetry of Inhaled Xenobiotics: Implications for Human Health*, Taylor & Francis, Washington, DC, 1995.

Miller, W.S., *The Lung*, Charles C. Thomas, Springfield, IL, 1947.

Mitchell, G.D., A review of permeation tubes and permeators, *Sepn. Purif. Methods.*, 29: 119–128, 2000.

Mitchell, J.P., Aerosol generation for instrument calibration, in *Bioaerosols Handbook*, Cox, C.S. and Wathes, C.M., Eds., Lewis Publishers (CRC Press), Boca Raton, FL, 1995a, Chapter 7.

Mitchell, J.P., Particle size analyzers: Practical procedures and laboratory techniques, in *Bioaerosols Handbook*, Cox, C.S. and Wathes, C.M., Eds., Lewis Publishers (CRC Press), Boca Raton, FL, 1995b, Chapter 8.

Mitzner, W., Collateral ventilation, in *The Lung: Scientific Foundations, 2nd Ed.*, Crystal, R.G., West, J.B., Weibel, E.R., and Barnes, P.J., Eds., Lippincott-Raven, Philadelphia, PA, 1997, Chapter 104.

Montgomery, M.R., Anderson, R.E., and Mortenson, G.A., A compact, versatile inhalation exposure chamber for small animal studies, *Lab. Anim. Sci.,* 26: 461–464, 1976.

Morgan, M.S. and Frank, R., Uptake of pollutant gases by the respiratory system, in *Respiratory Defense Mechanisms, Part I,* Brain, J.D., Proctor, D.F., and Reid, L.M., Eds., Marcel Dekker, New York, 1977, Chapter 6.

Morris, K.J., Modern microscopic methods for bioaerosol analysis, in *Bioaerosols Handbook*, Cox, C.S. and Wathes, C.M., Eds., Lewis Publishers (CRC Press), Boca Raton, FL, 1995, Chapter 11.

Morrow, P.E. and Mercer, T.T., A point-to-plane electrostatic precipitator for particle size sampling, *Am. Ind. Hygiene Assoc. J.,* 25: 8–14, 1964.

Morrow, P.E., Clearance kinetics of inhaled particles, in *Respiratory Defense Mechanisms, Part II,* Brain, J.D., Proctor, D.F., and Reid, L.M., Eds., Marcel Dekker, New York, 1977, Chapter 14.

Morrow, P.E., Possible mechanisms to explain dust overloading of the lungs, *Fund. Appl. Toxicol.,* 10: 369–384, 1988.

Moskal, A., Makowski, L., Sosnowski, T.R., and Gradón, L., Deposition of fractal-like aerosol aggregates in a model of human nasal cavity, *Inhal. Toxicol.,* 18: 725–731, 2006.

Moss, O.R. and Cheng, Y.S., Generation and characterization of test atmospheres: Particles and droplets, in *Concepts in Inhalation Toxicology, 2nd Ed.,* McClellan, R.O. and Henderson, R.F., Eds., Taylor & Francis, Washington DC, 1995, Chapter 4.

Mossberg, B., Human tracheobronchial clearance by mucociliary transport and cough, *Eur. J. Respir. Dis.* (*Suppl. 107*) 61: 51–58, 1980.

Muggenberg, B.A. and Mauderly, J.L., Cardiopulmonary function of awake, sedated, and anesthetized beagle dogs, *J. Appl. Physiol.,* 37: 152–157, 1974.

Muggenburg, B.A., Tilley, L., and Green, F.H.Y., Animal models of cardiac disease: Potential usefulness for studying health effects of inhaled particles, *Inhal. Toxicol.,* 12: 901–925, 2000.

Murphy, D.J., Safety pharmacology of the respiratory system—techniques and the study design, *Drug Devel. Res.,* 32: 237–246, 1994

Murray, M.J. and Driscoll, K.E., Immunology of the respiratory system, in *Comparative Biology of the Normal Lung,* Parent, R.A., Ed., CRC Press, Boca Raton, FL, 1991, Chapter 37.

Nadithe, V., Rahamatalla, M., Finlay, W.H., Mercer, J.R., and Samuel, J., Evaluation of nose-only aerosol inhalation chamber and comparison of experimental results with mathematical simulation of aerosol deposition in mouse lungs, *J. Pharm. Sciences,* 92: 1066–1076, 2003.

Narcisco, S.P., Nadziejko, E., Chen, L.C., Gordon, T., and Nadziejko, C., Adaptation to stress induced by restraining rats and mice in nose-only inhalation holders, *Inhal. Toxicol.,* 15: 1133–1143, 2003.

Nathan, C.F., Secretory products of macrophages, *J. Clin. Invest.,* 79: 319–326, 1987.

National Oceanic and Atmospheric Administration, National Aeronautics and Space Administration and U.S. Air Force, *U.S. Standard Atmosphere, 1976,* NOAA-S/T 76-1562, Washington DC, 1976.

NCRP (National Council on Radiation Protection and Measurements), *Deposition Retention and Dosimetry of Inhaled Radioactive Substances,* NCRP SC 57-2 Report, National Council on Radiation Protection and Measurements, Bethesda, MD, 1997.

Nelson, G.O., *Controlled Test Atmospheres: Principles and Techniques,* Ann Arbor Science, Ann Arbor, MI, 1971.

Netter, F.H., *Respiratory System, Vol. 7,* CIBA Collection of Medical Illustrations, CIBA Pharmaceutical Corporation, Summit, NJ, 1979.

Nettesheim, P., Hanna, M.G. Jr., and Deatherage, J.W. Jr., Eds., *Morphology of Experimental Respiratory Carcinogenesis,* A.E.C. Symp. Ser. No. 21, U.S. Atomic Energy Commission, Division of Technical Information, CONF-700501, National Technical Information Service, Springfield, VA, 1970.

Newton, P.E., Inhalation toxicology, in *CRC Handbook of Toxicology,* Derelanko, M.J. and Hollinger, M.A., Eds., CRC Press, Boca Raton, FL, 1995, Chapter 5.

Nikula, K.J. and Green, F.H.Y., Animal models of chronic bronchitis and their relevance to studies of particle-induced disease, *Inhal. Toxicol,* (*Suppl. 4*), 12: 123–153, 2000.

NIOSH (National Institute for Occupational Safety and Health), *NIOSH Pocket Guide to Chemical Hazards,* National Institute for Occupational Safety and Health, Cincinnati, OH, 2005.

NIOSH (National Institute for Occupational Safety & health), *NIOSH Manual of Analytical Methods, 2nd Ed.,* National Institute for Occupational Safety and Health, Cincinnati, OH, http://www.cdc.gov/niosh/nmam/ (accessed 1/2/2006).

Oberdörster, G., Pulmonary effects of inhaled ultrafine particles, *Internat. Arch. Occup. Environ. Health,* 74: 1–8, 2001.

Oberdörster, G., Sharp, Z., Atudorei, V., et al., Translocation of inhaled ultrafine particles to the brain, *Inhal. Toxicol.*, 16: 437–445, 2004.

O'Callaghan, C., Hardy, J., Stammers, J., Stephenson, T.J., and Hull, D., Evaluation of techniques for delivery of steroids to lungs of neonates using a rabbit model, *Arch. Dis. Childhood*, 67: 20–24, 1992.

Ochs, M., Nyengaard, J.R., Jung, A., et al., The number of alveoli in the human lung, *Am. J. Respir. Crit. Care Med.*, 169: 120–124, 2004.

Oldham, M.J., Phalen, R.F., Schum, G.M., and Daniels, D.S., Predicted nasal and tracheobronchial particle deposition efficiencies for the mouse, *Ann. Occup. Hygiene*, 38 (Suppl 1): 135–141, 1994.

Oldham, M.J., Phalen, R.F., and Heistracher, T., Computational fluid dynamic predictions and experimental results for particle deposition in an airway model, *Aerosol Sci. Technol.*, 32: 61–71, 2000.

Oldham, M.J. and Phalen, R.F., Dosimetry implications of upper tracheobronchial airway anatomy in two mouse varities, *Anat. Rec.*, 268: 59–65, 2002.

Oldham, M.J., Phalen, R.F., Robinson, R.J., and Kleinman, M.T., Performance of a portable whole-body mouse exposure system, *Inhal. Toxicol.*, 16: 657–662, 2004

Oldham, M.J., Challenges in validating CFD-derived inhaled aerosol deposition predictions, *Inhal. Toxicol.*, 18: 781–786, 2006.

Olson, H., Betton, G., Robinson, D., et al., Concordance of the toxicity of pharmaceuticals in humans and animals, *Regul. Toxicol. Parmacol.*, 32: 56–67, 2000.

OSHA (Occupational Health and Safety Administration), *NIOSH/OSHA/DOE Health Guidelines,* Occupational Health and Safety Administration, Washington, DC, http://www.osha.gov/SLTC/healthguidelines/index.html (accessed 1/8/2008).

Page, N.P., Chronic toxicity and carcinogenicity guidelines, *J. Environ. Pathol. Toxicol.*, 1: 161–182, 1977.

Parent, R.A., Ed., *Comparative Biology of the Normal Lung*, CRC Press, Boca Raton, FL, 1991.

Patrick, G. and Stirling, C., The retention of particles in large airways of the respiratory tract, *Proc. R. Soc. London, Ser. B*, 198: 455–462, 1977.

Patterson, J.L. and Carrion, R. Jr., Demand for nonhuman primate resources in the age of biodefense, *ILAR J.*, 46: 15–22, 2005.

Patterson, R. and Kelly, J. F., Animal models of the asthmatic state. *Annu. Rev. Med.*, 25: 53–68, 1974.

Pattle, R.E., Surface lining of lung alveoli, *Physiol. Rev.*, 45: 48–79, 1965.

Pauluhn, J., Machemer, L., Kimmerle, G., and Eben, A., Methodological aspects of the determination of the acute inhalation toxicity of spray-can ingredients, *J. Appl. Toxicol.*, 8: 431–437, 1988.

Pauluhn, J., Validation of an improved nose-only exposure system for rodents, *J. Appl. Toxicol.*, 14: 55–62, 1994a.

Pauluhn, J., Species differences: Impact on testing, in *Respiratory Toxicology and Risk Assessment*, Jenkins, P.G., Kayser, D., Muhle, H., Rosner, G., and Smith, E.M., Eds., Wissenschaftliche Verlagsgesellschaft mbH, Stuttgart, Germany, 1994b, pp. 145–173.

Pauluhn, J., Overview of testing methods used in inhalation toxicity: From facts to artifacts, *Toxicol. Lett.*, 140 (Special Issue S1): 183–193, 2003.

Pauluhn, J., Animal models of respiratory allergy, in *Toxicology of the Lung, 4th Ed.*, Gardner, D.E., Ed., CRC Press, Boca Raton, FL, 2006, Chapter 10.

Pavia, D., Bateman, J.R.M., and Clarke, S.W., Deposition and clearance of inhaled particles, *Bull. Eur. Physiopathol. Respir.*, 16: 335–336, 1980.

Pavia, D., Acute respiratory infections and mucociliary clearance, *Eur. J. Respir. Dis.*, 71: 219–226, 1987.

Perry, R.H. and Chilton, C.H., *Chemical Engineer's Handbook, 5th Ed.*, McGraw-Hill, New York, 1973.

Phalen, R.F., Evaluation of an exploded-wire aerosol generator for use in inhalation studies. *Aerosol Sci.*, 3: 395–406, 1972.

Phalen, R.F. and Morrow, P.E., Experimental inhalation of metallic silver, *Health Phys.*, 24: 509–518, 1973.

Phalen, R.F., Yeh, H.C., Raabe, O.G., and Velasquez, D.J., Casting the lungs in situ, *Anat. Rec.*, 177: 255–263, 1973.

Phalen, R.F., Yeh, H.C., Schum, G.M., and Raabe, O.G., Application of an idealized model to morphometry of the mammalian tracheobronchial tree, *Anat. Rec.*, 190: 167–176, 1978a.

Phalen, R.F., Crocker, T.T., Wilson, A.F., and Kenoyer, J.L., Effect of dose rate on ozone toxicity, paper read at Annual American Industrial Hygiene Association Conference, Los Angeles, May 1978b.

Phalen, R.F., Kenoyer, J.L., Crocker, T.T., and McClure, T.R., Effects of sulfate aerosols in combination with ozone on elimination of tracer particles inhaled by rats, *J. Toxicol. Environ. Health,* 6: 797–810, 1980.

Phalen, R.F. and Oldham, M.J., Tracheobronchial structure as revealed by casting techniques, *Am. Rev. Respir. Dis.*, 128 (Suppl.): S1–S4, 1983.

Phalen, R.F., *Inhalation Studies: Foundations and Techniques*, CRC Press, Boca Raton, FL, 1984.

Phalen, R.F., Crocker, T.T., McClure, T.R., and Tyler, N.K., Effect of ozone on mean linear intercept in the lung of young beagles, *J. Toxicol. Environ. Health*, 17: 285–296, 1986.

Phalen, R.F., Schum, G.M., and Oldham, M.J., The sensitivity of an inhaled aerosol tracheobronchial models to input parameters, *J. Aerosol Med.*, 3: 271–282, 1990.

Phalen, R.F., Oldham, M.J., and Dunn-Rankin, D., Inhaled particle mass per unit body mass per unit time, *Appl. Occup. Environ. Hygiene*, 7: 246–252, 1992.

Phalen, R.F., Oldham, M.J., Mannix, R.C., and Schum, G.M. Cigarette smoke deposition in the tracheobronchial tree: Evidence for colligative effects. *Aerosol. Sci. Technol.*, 20: 215–226, 1994a.

Phalen, R.F., Kleinman, M.T., Mautz, W.J., and Drew, R.T., Inhalation exposure methodology, in *Respiratory Toxicology and Risk Assessment,* Jenkins, P.G., Kayser, D., Muhle, H., Rosner, G., and Smith, E.M., Eds., Wissenschaftliche Verlagsgesellschaft mbH, Stuttgart, Germany, 1994b, pp. 59–82.

Phalen, R.F., Inhalation exposure methods, in *Methods in Inhalation Toxicology*, Phalen, R.F., Ed., CRC Press, Boca Raton, FL, 1997a, Chapter 5.

Phalen, R.F., Cleaning the air, in *Methods in Inhalation Toxicology,* Phalen, R.F., Ed., CRC Press, Boca Raton, FL, 1997b, Chapter 3.

Phalen, R.F., and Oldham, M.J., Modeling particle deposition as a function of age, *Respir. Physiol.*, 128: 119–130, 2001.

Phalen, R.F., *The Particulate Air Pollution Controversy: A Case Study and Lessons Learned,* Kluwer Academic Publishers, Boston, MA, 2002, Chapter 5.

Phalen, R.F., Oldham, M.J., and Nel, A.E., Tracheobronchial particle dose considerations for in-vitro toxicology studies, *Toxicol. Sci.*, 92: 126–132, 2006.

Phillips, C.G. and Kaye, S.R., On the asymmetry of bifurcations in the bronchial tree, *Respir. Physiol.*, 107: 85–98, 1997.

Phipps, A.J., Premanandan, C., Barnewall, R.E., and Lairmore, M.D., Rabbit and nonhuman primate models of toxin-targeting human anthrax vaccines, *Microbiol. Molec. Biol. Rev.*, 68: 617–629, 2004.

Phipps, R.J. and Richardson, P.S., The effects of irritation at various levels of the airway upon tracheal mucus secretion in the cat, *J. Physiol.,* 261: 563–581, 1976.

Pich, J., Theory of aerosol filtration by fibrous and membrane filters, in *Aerosol Science,* Davies, C.N., Ed., Academic Press, New York, 1966, Chapter 9.

Pinkerton, K.E., Gehr, P., and Crapo, J.D., Architecture and cellular composition of the air–blood barrier, in *Comparative Biology of the Normal Lung*, Parent, R.A., Ed., CRC Press, Boca Raton, FL, 1991, Chapter 11.

Pinkerton, K.E. and Green, F.Y.H., Normal aging of the lung, in *Comparative Biology of the Normal Lung*, Parent, R.A., Ed., CRC Press, Boca Raton, FL, 1991, Chapter 15.

Pitot, H.C. III and Dragan, Y.P., Chemical carcinogenesis, in *Casarett and Doull's Essentials of Toxicology*, Klaassen, C.D. and Watkins, J.B. III, Eds., McGraw Hill, New York, 2003, Chapter 8.

Plopper, C.G., and Pinkerton, K.E., Overview of the diversity in the respiratory system of mammals, in *Comparative Biology of the Lung*, Parent, R.A., Ed., CRC Press, Boca Raton, FL, 1991, Chapter 1.

Plopper, C.G., and Fanucchi, M.V., Development of airway epithelia, in *The Lung: Development, Aging and the Environment*, Harding, R., Pinkerton, K.E., and Plopper, C.G., Eds., Elsevier, San Diego, CA, 2004, Chapter 2.

Port, C.D., Ketels, K.V., Coffin, D.L., and Kane, P., A comparative study of experimental and spontaneous emphysema, *J. Toxicol. Environ. Health,* 2: 589–604, 1977.

Poynter, D. and Spurling, N.W., Some cardiac effects of beta-adrenergic stimulants in animals, *Postgrad. Med. J. Suppl.,* 47: 21–24, 1971.

Prasad, S.B., Rao, S.V., Mannix, R.C., and Phalen, R.F., Effects of pollutant atmospheres on surface receptors of pulmonary macrophages, *J.Toxicol. Environ. Health,* 24: 385–402, 1988.

Preston, R.J. and Hoffmann, G.R., Genetic toxicology, in *Casarett and Doull's Essentials of Toxicology,* Klaassen, C.D., and Watkins, J.B. III, Eds., McGraw-Hill, New York, 2003, Chapter 9.

Pritchard, J.N., Jefferies, J., and Black, A. Sex differences in the regional deposition of inhaled particles in the 2.5–7.5 µm size range, *J. Aerosol Sci.,* 17: 385–389, 1986.

Prodi, V., A condensation aerosol generator for solid monodisperse particles, in *Assessment of Airborne Particles,* Mercer, T.T., Morrow, P.E., and Stöber, W., Eds., Charles C. Thomas, Springfield, IL, 1972, Chapter 9.

Raabe, O.G., The dilution of monodisperse suspensions for aerosolization, *Am. Ind. Hygiene Assoc. J.,* 29: 439–443, 1968.

Raabe, O.G. and Newton, G.J., Development of techniques for generating monodisperse aerosols with the Fulwyler droplet generator, *Fission Product Inhalation Program Annual Report 1969-1970,* LF 43, Lovelace Foundation, Albuquerque, NM, 1970, pp. 13–17.

Raabe, O.G., Bennick, J.E., Light, M.E., Hobbs, C.H., Thomas, R.L., and Tillery, M.I., An improved apparatus for acute inhalation exposure of rodents to radioactive aerosols, *Toxicol. Appl. Pharmacol.,* 26: 264–273, 1973.

Raabe, O.G. and Yeh, H.C., Principles for inhalation exposure systems using concurrent flow spirometry, *J. Aerosol Sci.,* 7: 233–243, 1976.

Raabe, O.G., Yeh, H.C., Newton, G.J., Phalen, R.F., and Velasquez, D.J., Deposition of inhaled monodisperse aerosols in small rodents, in *Inhaled Particles IV,* Walton, W.H., Ed., Pergamon Press, Oxford, 1977, pp. 3–21.

Raabe, O.G., Generation and characterization of aerosols, in *Inhalation Carcinogenesis,* Hanna, M.G. Jr., Nettesheim, P., and Gilbert, J.R., Eds., U.S. Atomic Energy Commission, AEC Symp. Ser. 18, Springfield, VA, 1979, pp. 123–172.

Raabe, O.G., Generation and characterization of power plant fly ash, in *Generation of Aerosols,* Willeke, K., Ed., Ann Arbor Science, Ann Arbor, MI, 1980, Chapter 11.

Raabe, O.G., Al-Bagati, M.A., Teague, S.V., and Rasolt, A., Regional deposition of inhaled monodisperse coarse and fine aerosol particles in small laboratory animals, *Ann Occup. Hygiene,* 32 (Suppl. 1): 53–63, 1988.

Rall, D.P., Relevance of animal-experiments to humans, *Environ. Health Perspect.,* 32: 297–300, 1979.

Rasmussen, R.E., Tissue acquisition and processing, in *Methods in Inhalation Toxicology,* Phalen, R.F., Ed., CRC Press, Boca Raton, FL, 1997, Chapter 7.

Redman, T.K., Rudolph, K., Barr, E.B., Bowen, L.E., Muggenburg, B.A., and Bice, D.E., Pulmonary immunity to ragweed in a Beagle dog model of allergic asthma, *Exper. Lung Res.,* 27: 433–451, 2001.

Reid, L., The lung: Its growth and remodeling in health and disease, *Am. J. Roentgenol.* 129: 777–788, 1977.

Reinero, C.R.N., Decile, K.C., Berghaus, R.D., et al., An experimental model of allergic asthma in cats sensitized to house dust mite or Bermuda grass allergen, *Intl. Arch. Allergy Immunol.,* 135: 117–131, 2004.

Reischl, P., Mautz, W.J., and Phalen, R.F., The effective dose concept extended to breathing pattern kinetics in exercising dogs exposed to ozone-containing atmospheres, *Fed. Proc.,* Abstr. 8656: 1750, 1982.

Rennard, S.I. and Spurzen, J.R., Methods for evaluating the lung in human subjects, in *Toxicology of the Lung, 4th Ed.,* Gardner, D.E., Ed., Taylor & Francis, Boca Raton, FL, 2006, Chapter 1.

Roberts, C.R. and Wong, S.J., Metabolic heat loads: Go figure, *Lab Animals,* 27(4): 36–37, 1998.

Roehlich, F. Jr. and Rodgers, S.J., Development of an air purifier for personnel cubicles, *Am. Ind. Hygiene Assoc. J.,* 37: 586–589, 1976.

Rollins, J.P., Ed., *Compressed Air and Gas Handbook, 5th Ed.*, Prentice-Hall, Englewood Cliffs, NJ, 1988.

Rom, W.N., Ed., *Environmental and Occupational Medicine, 2nd Ed.*, Little, Brown & Co., Boston, MA, 1992.

Roosevelt, J.W., The anatomy of the thorax and lungs in relation to certain points in physical diagnosis, *Med. Rec.*, 37: 201–204, 1890.

Roth, C., Ferron, G.A., Karg, E., et al., Generation of ultrafine particles by spark discharging, *Aerosol Sci. Technol.*, 38: 228–235, 2004.

Rotton, J., Barry, T., Frey, J., and Soler, E., Air pollution and interpersonal attraction, *J. Appl. Social Psychol.*, 8: 57–71, 1978.

Rotton, J., Frey, J., Barry, T., Milligan, M., and Fitzpatrick, M., The air pollution experience and physical aggression, *J. Appl. Social Psychol.*, 9: 397–412, 1979.

Roy, C.J., Hale, M., Hartings, J.M., Pitt, L., and Duniho, S., Impact of inhalation exposure modality and particle size on the respiratory deposition of ricin in balb/c mice, *Inhal. Toxicol.*, 15: 619–638, 2003.

Ruzer, L.S. and Harley, N.H., Eds., *Aerosols Handbook: Measurement, Dosimetry, and Health Effects*, CRC Press, Boca Raton, FL, 2005.

Sackner, M.A., Dougherty, R.L., Chapman, G.A., et al., Effects of brief and intermediate exposures to sulfate submicron aerosols and sulfate injections on cardiopulmonary function of dogs and tracheal mucous velocity of sheep, *J. Toxicol. Environ. Health,* 7: 951–972, 1981.

Sahebjami, H., Aging of the normal lung, in *Comparative Biology of the Normal Lung,* Parent, R.A., Ed., CRC Press, Boca Raton, FL, 1991, Chapter 21.

Saibene, F., Mognoni, P., Lafortuna, C.L., and Mostardi, R., Oronasal breathing during exercise, *Pflug. Arch. Euro. J. Physiol.*, 378: 65–69, 1978.

Salathé, M., O'Riordan, T.G., and Wanner, A., Mucociliary clearance, in *The Lung: Scientific Foundations, 2nd Ed.*, Crystal, R.G., West, J.B., Weibel, E.R., and Barnes, P.J., Eds., Lippincott-Raven, Philadelphia, PA, 1997, Chapter 177.

Salem, H. and Katz, S.A., Eds., *Inhalation Toxicology, 2nd Edn.*, Taylor & Francis, Boca Raton, FL, 2006.

Salsburg, D.S., Statistics and toxicology: An overview, in *Scientific Considerations in Monitoring and Evaluating Toxicologic Research,* Gralla, E.J., Ed., Hemisphere, Washington, DC, 1981, Chapter 7.

Saltzman, B.E. and Wartburg, A.F. Jr., Absorping tube for removal of interfering SO_2 in analysis of atmospheric oxidants, *Anal. Chem.,* 37: 779–782, 1965.

Savin, W.M. and Adams, W.C., Effects of ozone inhalation on work performance and V_{O2max}, *J. Appl. Physiol.*, 46: 309–314, 1979.

Schaper, M., Development of a database for sensory irritants and its use in establishing occupational exposure limits, *Am. Ind. Hygiene Assoc. J.*, 54: 488–544, 1993.

Scheimberg, J., McShane, O.P., Carson, S., Swann, H.E., and Blair, A.M., Inhalation of a powdered aerosol medication by non-human primates in individual space-type exposure helmets (Abstr.), *Toxicol. Appl. Pharmacol.,* 25: 478–479, 1973.

Scherle, W., A simple method for volumetry of organs in quantitative stereology, *Mikroskopie*, 26: 57–60, 1970.

Schiller-Scotland, C.F., Hlawa, R., Gebhart, J., Wönne, R., and Heyder, J., Total depositon of aerosol particles in the respiratory tract of children during spontaneous and controlled mouth breathing. *J. Aerosol Sci.*, 23 (Suppl. 1), S457–S460, 1992.

Schleien, B., Slaback, L.A. Jr., and Birky, B.K., Eds., *Handbook of Health Physics and Radiological Health, 3rd Ed.*, Williams & Wilkins, Baltimore, MD, 1998, Chapter 12.

Schlesinger, R.B. and Lippmann, M., Particle deposition in the trachea: In vivo and in hollow casts, *Thorax,* 31: 678–684, 1976.

Schlesinger, R.B., Halpern, M., Albert, R.E., and Lippmann, M., Effect of chronic inhalation of sulfuric acid mist upon mucociliary clearance from the lungs of donkeys, *J. Environ. Pathol. Toxicol.*, 2: 1351–1367, 1979.

Schlesinger, R.B., Particle deposition in model systems of human and experimental animals, in *Generation of Aerosols,* K. Willeke, Ed., Ann Arbor Science, Ann Arbor, 1980, pp. 553–575.

Schlesinger, R.B., and McFadden, L.A., Comparative morphometry of the upper bronchial tree in 6 mammalian species, *Anat. Rec.*, 199: 99–108, 1981.

Schlesinger, R.B., Comparative deposition of inhaled aerosols in experimental animals and humans: A review, *Toxicol. Environ. Health,* 15: 197–214, 1985.

Schlesinger, R.B., Ben-Jebria, A., Dahl, A.R., Snipes, M.B., and Ultman, J., Disposition of inhaled toxicants, in *Handbook of Human Toxicology,* Massaro, E.J., Ed., CRC Press, Boca Raton, FL, 1997, Chapter 12.

Schreider, J.P. and Hutchens, J.O., Particle deposition in the guinea pig respiratory tract, *J. Aerosol Sci.*, 10: 599–607, 1979.

Schreider, J.P. and Raabe, O.G., Anatomy of the nasal-pharyngeal airway of experimental animals, *Anat. Rec.*, 200: 195–205, 1981.

Schum, G.M. and Phalen, R.F., Modeling hygroscopic particle growth in human lung airways, *Ann. Occup. Hygiene*, 41 (Suppl. 1): 60–64, 1997.

Schwetz, B.A., Monitoring problems in teratology, in *Scientific Considerations in Monitoring and Evaluating Toxicological Research,* Gralla, E.J., Ed., Hemisphere, Washington, DC, 1981, Chapter 11.

Segal, R.A., Martonen, T.B., Kim, C.S., and Schearer, M., Computer simulations of particle deposition in the lungs of chronic obstructive pulmonary disease patients, *Inhal. Toxicol.*, 14: 705–720, 2002.

Shapiro, S.D., Transgenic and gene-targeted mice as models for chronic obstructive pulmonary disease, *Eur. Respir. J.*, 29: 375–378, 2007.

Shoaf, C.R., U.S. EPA inhalation testing guidelines, in *Respiratory Toxicology and Risk Assessment*, Jenkins, P.G., Kayser, D., Muhle, H., Rosner, G., and Smith, E.M., Eds., Wissenschaftliche Verlagsgesellschaft mbH, Stuttgart, Germany, 1994, pp. 11–25.

Silver, S.D., Constant flow gassing chambers: Principles influencing design and operation, *J. Lab. Clin. Med.*, 31: 1153–1161, 1946.

Sinclair, D. and LaMer, V.K., Light scattering as a measure of particle size in aerosols, *Chem. Rev.*, 44: 245–267, 1949.

Smaldone, G.C., Foster, W.M., Oriordan, T.G., Messina, M.S., Perry, R.J., and Langenback, E.G., Regional impairment of mucociliary clearance in chronic obstructive pulmonary disease, *Chest*, 103: 1390–1396, 1993.

Smith, S.W. and Spurling, N.W., Device for exposing the respiratory tract of the rat to medicinal aerosols, *Lab. Pract.*, 23: 717–721, 1974.

Snipes, M.B., McClellan, R.O., Mauderly, J.L., and Wolff, R.K., Retention patterns for inhaled particles in the lung: Comparisons between laboratory animals and humans in chronic exposures, *Health Phys.*, 57 (Suppl. 1): 69–78, 1989.

Soderholm, S.C., Proposed international conventions for particle size-selective sampling, *Ann. Occup. Hygiene*, 33: 301–320, 1989.

Spiegelman, J.R., Hanson, G.D., Lazarus, A., Bennett, R.J., Lippmann, M., and Albert, R.E., Effect of acute sulfur dioxide exposure on bronchial clearance in the donkey, *Arch. Environ. Health,* 17: 321–326, 1968.

Sprent, P., Some problems of statistical consultancy (with discussion), *J. Roy. Stat. Soc. Series A—Gen.*, 133: 139–165, 1970.

Stahlhofen W., Scheuch G., and Bailey M.R., Investigations of retention of inhaled particles in the human bronchial tree, *Radiat. Prot. Dosim.*, 60: 311–319, 1995.

Stanescu, D.C., Pattijn, J., Clement, J., and van de Woestijne, K.P., Glottis opening and airway resistance, *J. Appl. Physiol.,* 32: 460–466, 1972.

Stara, J.F. and Kello, D., Relationship of long-term animal studies to human disease, in *Assessing Toxic Effects of Environmental Pollutants,* Lee, S.D. and Mudd, J.B., Eds., Ann Arbor Science, Ann Arbor, MI, 1979, Chapter 4.

Stavert, D.M., Reischl, P., and Murdock, L.C., A refrigerated treadmill apparatus for exercising dogs, *J. Appl. Physiol.,* 52: 495–499, 1982a.

Stavert, D.M., Reischl, P., and O'Loughlin, B.J., A respiratory mask for resting and exercising dogs, *J. Appl. Physiol.,* 52: 500–504, 1982b.

Stavert, D.M., Archuleta, D.C., Behr, M.J., and Lehnert, B.E., Relative acute toxicities of hydrogen fluoride, hydrogen chloride, and hydrogen bromide in nose- and pseudo-mouth-breathing rats, *Fund. Appl. Toxicol.*, 16: 636–655, 1991.

Stefaniak, A.B., Guilmette, R.A., Day, G.A., Hoover, M.D., Breysse, P.N., and Scripsick, R.C., Characterization of phagolysosomal simulant fluid for study of beryllium aerosol particle dissolution, *Toxicol. In Vitro*, 19: 123–134, 2005.

Stern, S.T. and McNeil, S.E., Nanotechnology safety concerns revisited, *Toxicol. Sci.*, 101: 4–21, 2008.

Stevens, R.K., Dzubay, T.G., Russwurm, G., and Rickel, D., Sampling and analysis of atmospheric sulfates and related species, *Atmos. Environ.*, 12: 55–68, 1978.

Stewart, B.W., LeMesurier, S.M., and Lykke, A.W.J., Correlation of biochemical and morphological changes induced by chemical injury to the lung. *Chem. Biol. Interact.*, 26: 321–338, 1979.

Stinson, S.F. and Loosli, C.G., The effect of synthetic smog on voluntary activitiy of CD-1 mice, in *Animals as Monitors of Environmental Pollutants,* National Academy of Sciences, Washington, DC, 1979, pp. 233–239.

Stöber, W., On the theory of silicosis, *Arch. Environ. Health*, 16: 706–708, 1968.

Stöber, W. and Flachsbart, H., Size-separating precipitation of aerosols in a spinning spiral duct, *Environ. Sci. Technol.*, 3: 1280–1296, 1969.

Stöber, W., Morrow, P.E., Koch, W., and Morawietz, G., Alveolar clearance and retention of inhaled insoluble particles in rats simulated by a model inferring macrophage particle load distributions, *J. Aerosol Sci.*, 25: 975–1002, 1994.

Stokinger, H.E., Ozone toxicology: A review of research and industrial experience: 1954–1964, *Arch. Environ. Health,* 10: 719–731, 1965.

Stossel, T.P., The mechanism of phagocytosis. *J. Reticuloendothel. Soc.*, 19: 237–245, 1976.

Strom, L., Generation of monodisperse aerosols by means of a disintegrated jet of liquid, *Rev. Sci. Instrum.*, 40: 778–782, 1969.

Stuart, B.O., Willard, D.H., and Howard, E.B., Studies of inhaled radon daughters, uranium ore dust, diesel exhaust and cigarette smoke in dogs and hamsters, in *Inhaled Particles III,* Walton, W.H., Ed., Unwin Brothers Ltd, Surrey, England, 1971, pp. 543–560.

Stuart, B.O., Selection of animal models for evaluation of inhalation hazards in man, in *Air Pollution and the Lung,* Aharonson, E.F., Ben-David, A., and Klingberg, M.A., Eds., John Wiley & Sons, New York, 1976, pp. 268–288.

Sweeney, T.D., Skornik, W.A., Brain, J.D., Hatch, V., and Godleski, J.J., Chronic-bronchitis alters the pattern of aerosol deposition in the lung, *Am. J. Respir. Crit. Care Med.*, 151: 482–488, 1995.

Szereda-Przestaszewska, M. and Widdicombe, J.G., Reflex effects of chemical irritation of the upper airways on the laryngeal lumen in cats, *Respir. Physiol.*, 18: 107–115, 1973.

Talavera, J., Kirschvink, N., Schuller, S., et al., Evaluation of respiratory function by barometric whole-body plethysmography in healthy dogs, *Vet. J.*, 172; 67–77, 2006.

Tarroni, G., Melandri, C., Prodi, V., DeZaiacomo, T., Formignani, M., and Bassi, P., An indication on the biological variability of aerosol total deposition in humans. *Am. Ind. Hygiene Assoc. J.*, 41: 826–831, 1980.

TGLD (Task Group on Lung Dynamics), ICRP Committee II, Deposition and retention models for internal dosimetry of the human respiratory tract, *Health Phys.*, 12: 173–207, 1966.

Thilenius, O.G. and Vial, C.B., Chronic tracheostomy in dogs, *J. Appl. Physiol.*, 18: 439–440, 1963.

Thomas, A.A., Low ambient pressure environments and toxicity, *Arch. Environ. Health.*, 11: 316–322, 1965.

Thomas, R.G. and Lie, R., *Procedures and Equipment Used in Inhalation Studies on Small Animals*, U.S. Atomic Energy Commission Research and Development Report, Lovelace Foundation Report No. LF-11, Albuquerque, NM, 1963.

Thomas, R.G. and Morgan, B.N., *The Effects of Age, Weight, Strain and Anesthesia Upon the Breathing Patterns in Two Strains of Rats*, U.S. Atomic Energy Commission Research and Development Report, Lovelace Foundation Report LF-42, Albuquerque, NM, 1970.

Thomas, R.L. and Roe, L.J., Corprophagy in laboratory animals: A preliminary study to assess potential influences of excreta ingestion on radionuclide retention patterns, *Am. Ind. Hygiene Assoc. J.,* 35: 741–747, 1974.

Thompson, T. and Schuster, C.R., *Behavioral Pharmacology,* Prentice-Hall, Englewood Cliffs, NJ, 1968.

Thomson, J.J., Chodosh, S., Fried, C., Goodman, D.S., Wax, M.L., and Wilson, J.Q., Regulations governing research on human subjects: Academic freedom and the Institutional Review Board, *Academe,* December: 358–370, 1981.

Thurlbeck, W.M., Postnatal growth and development of the lung, *Am. Rev. Respir. Dis.,* 111: 803–844, 1975.

Thurlbeck, W.M., Structure of the lungs, in *International Review of Physiology: Respiratory Physiology II, Vol. 14,* Widdicombe, J.C., Ed., University Park Press, Baltimore, MD, 1977, Chapter 1.

Tillery, M.I., Wood, G.O., and Ettinger, H.J., Generation and characterization of aerosols and vapors for inhalation experiments, *Environ. Health Perspect.,* 16: 25–40, 1976.

Timbrell, V., An aerosol spectrometer and its applications, in *Assessment of Airborne Particles,* Mercer, T.T., Morrow, P.E., and Stöber, W., Eds., Charles C. Thomas, Springfield, IL, 1972, Chapter 15.

Tran, L. and Kuempel, E., Biologically based lung dosimetry and exposure–dose–response models for poorly soluble inhaled particles, in, *Particle Toxicology,* Donaldson, K. and Borm, P., Eds., CRC Press, Boca Raton, FL, 2007, Chapter 20.

Tsai, C.J. and Lu, H.C., Design and evaluation of a plate-to-plate thermophoretic precipitator, *Aerosol Sci. Technol.,* 22: 172–180, 1995.

Tsuji, J.S., Maynard, A.D., Howard, P.C., et al., Research strategies for safety evaluation of nanomaterials, Part IV: Risk assessment of nanoparticles, *Toxicol. Sci.,* 89: 42–50, 2006.

Tsujino, I., Kawakami, Y., and Kaneko, A., Comparative simulation of gas transport in airway models of rat, dog, and human, *Inhal. Toxicol.,* 17: 475–485, 2005.

Tucker, A.D., Wyatt, J.H., and Undery, D., Clearance of inhaled particles from alveoli by normal interstitial drainage pathways, *J. Appl. Physiol.,* 35: 719–732, 1973.

Tyler, W.S. and Julian, M.D., Gross and subgross anatomy of lungs, pleura, connective tissue septa, distal airways and structural units, in *Comparative Biology of the Normal Lung,* Parent, R.A., Ed., CRC Press, Boca Raton, FL, 1991, Chapter 4.

Urch, B., Brook, J.R., Wasserstein, D., et al., Relative contributions of $PM_{2.5}$ chemical constituents to acute arterial vasoconstriction in humans, *Inhal. Toxicol.,* 16: 345–352, 2004.

Utell, M.J., Morrow, P.E., Hyde, R.W., and Schreck, R.M., Exposure chamber for studies of pollutant gases and aerosols in human subjects: Design considerations, *J. Aerosol Sci.,* 15: 219–221, 1984.

Utell, M.J. and Frampton, M.W., Toxicologic methods: Controlled human exposures, *Environ. Health. Persp.,* 108(Suppl.4): 605–613, 2000.

Valberg, P.A. and Blanchard, J.D., Pulmonary macrophage physiology: Origin, motility, endocytosis, in *Comparative Biology of the Normal Lung,* Parent, R.A., Ed., CRC Press, Boca Raton, FL, 1991, Chapter 36.

Van Dingenen, R., Raes, F., and Vanmarcke, H., Molecule and aerosol particle wall losses in smog chambers made of glass, *J. Aerosol Sci.,* 20: 113–122, 1989.

Vedal, S., Rich, K., Brauer, M., White, R., and Petkau, J., Air pollution and cardiac arrhythmias in patients with implantable cardioverter defibrillators, *Inhal. Toxicol.,* 16: 353–362, 2004.

Viau, A. and Robinson, K., Reproductive toxicology testing of inhaled pharmaceuticals and biotechnology products, in *Toxicology of the Lung, 4th Ed.,* Gardner, D.E., Ed., Taylor & Francis, Boca Raton, FL, 2006, Chapter 12.

Vick, A., Wolff, R., Koester, A., Reams, R., Deaver, D., and Heidel, S., A 6-month inhalation study to characterize the toxicity, pharmacokinetics and pharmacodynamics of human insulin inhalation powder (HIIP) in Beagle dogs, *J. Aerosol Med.,* 20: 112–126, 2007.

Vincent, J.H., Mark, D., Miller, B.G., Armbruster, L., and Ogden, T.L., Aerosol inhalability at higher windspeeds, *J. Aerosol Sci.,* 21: 577–586, 1990.

Vincent, J.H., *Aerosol Science for Industrial Hygienists*, Pergamon (Elsevier Science) Tarrytown, NY, 1995.

Vinegar, A., Sinnett, E.E., Kosch, P.C., and Miller, M.L., Pulmonary physiology of the ferret and its potential as a model for inhalation toxicology, *Lab. Anim. Sci.*, 35: 246–250, 1985.

Vogelweid, C.M., Hill, J.B., Shea, R.A., and Johnson, D.B., Earthquakes and building design: A primer for the laboratory animal professional, *Lab. Animals*, 34(7): 35–42, 2005.

von Smoluchowski, M., A mathematical theory of the kinetics of coagulation of colloidal solutions, *Z. Phys. Chem.*, 92: 129–168, 1917.

Waag, D.M., Byrne, W.R., Estep, J., Gibbs, P., Pitt, M.L.M., and Banfield, C.M., Evaluation of cynomolgus (*Macaca fascicularis*) and rhesus (*Macaca mulatta*) monkeys as experimental models of acute Q fever after aerosol exposure to phase I *Coxiella burnetii*, *Lab. Anim. Sci.*, 49: 634–638, 1999.

Wada, H., Toluene and temporal discrimination behavior in the rat, *Neurotox. Teratol.,* 19: 399–403, 1997.

Walker, H.L., McLeod, C.G., and McManus, W.F., Experimental inhalation injury in the goat, *J.Trauma,* 21: 962–964, 1981.

Walters, R.B., Phalen, R.F., Mannix, R.C., and Smart, G.L., An aerosol and gas aging line suitable for use in inhalation toxicology research, *Am. Ind. Hygiene Assoc. J.*, 43: 218–222, 1982.

Walton, W.H. and Prewett, W.C., The production of sprays and mists of uniform drop size by means of spinning disc type sprayers, *Proc. Phys. Soc., Ser B,* 62: 341–350, 1949.

Warheit, D.B., Carakostas, M.C., Hartsky, M.A., and Hansen, J.F., Development of a short-term inhalation bioassay to assess pulmonary toxicity of inhaled particles: Comparisons of pulmonary responses to carbonyl iron and silica, *Toxicol. Appl. Pharmacol.*, 107: 350–368, 1991.

Warheit, D.B., Borm, P.J.A., Hennes, C., and Lademann, J., Testing strategies to establish the safety of nanomaterials: Conclusions of an ECETOC workshop, *Inhal. Toxicol.*, 19: 631–643, 2007.

Warnecke, G., Avsar, M., Morancho, M., et al., Preoperative low-dose irradiation promotes long-term allograft acceptance and induces regulatory T cells in a porcine model of pulmonary transplantation, *Transplantation*, 82: 93–101, 2006.

Wehner, A.P., Craig, D.K., and Stuart, B.O., An aerosol exposure system for chronic inhalation studies with rodents. *Am. Ind. Hygiene Assoc. J.*, 33: 483–487, 1972.

Wehner, A.P., Dagle, G.E., and Cannon, W.C., Development of an animal model, techniques, and an exposure system to study the effects of asbestos cement dust inhalation, *Environ. Res.,* 16: 393–407, 1978.

Weibel, E.R., *Morphometry of the Human Lung,* Springer-Verlag, Berlin, 1963a.

Weibel, E.R., Principles and methods for the morphometric study of the lung and other organs., *Lab. Invest.,* 12: 131–155, 1963b.

Weibel, E.R., Morphometrics of the Lung, *in Respiration, Vol. 1,* Fenn, W.O. and Rahn, H., Eds., American Physiological Society, Washington, DC, 1964, Chapter 7.

Weibel, E.R. and Gil, J., Structure–function relationships at the alveolar level, in *Bioengineering Aspects of the Lung,* West. J.B., Ed., Marcel Dekker, New York, 1977, Chapter 1.

Weibel, E.R., Hsia, C.C.W., and Ochs, M., How much is there really? Why stereology is essential in lung morphometry, *J.Appl. Physiol.*, 102: 459–467, 2007.

Weil, C.S., Guidelines for experiments to predict degree of safety of a material for man, *Toxicol. Appl. Pharmacol.,* 21: 194–199, 1972.

Weinberg, A.D., Dimen, E., Borzelleca, J.F., and Harris, L.S., Weight and activity in male mice after daily inhalation of cannabis smoke in an automated smoke exposure chamber, *J. Pharm. Pharmacol.*, 29: 477–481,1977

Weiss, B. and Rahill, A., Applications of behavioral measures to inhalation toxicology, in *Concepts in Inhalation Toxicology, 2nd Ed.*, McClellan, R.O. and Henderson, R.F., Eds., Taylor & Francis, Washington, DC, 1995, Chapter 17.

Wen, J. and Wexler, A.S., Thermophoretic sampler and its application in ultrafine particle collection, *Aerosol Sci. Technol.*, 41: 624–629, 2007.

Whalan, J.E., Foureman, G.L., and Vandenberg, J.J., Inhalation risk assement at the Environmental Protection Agency, in *Inhalation Toxicology, 2nd Ed.*, Salem, H. and Katz, S.A., Eds., CRC Press (Taylor & Francis), Boca Raton, FL, 2006, Chapter 1.

Willeke, K., Ed., *Generation of Aerosols and Facilities for Exposure Experiments,* Ann Arbor Science, Ann Arbor, MI, 1980.

Willeke, K. and Baron, P.A., Eds., *Aerosol Measurement: Principles, Techniques, and Applications,* Van Nostrand Reinhold, New York, 1993.

Wilson, A.F., Fairshter, R.D., Gillespie, J.R., and Hackney, J., Evaluation of abnormal lung-function. *Annu. Rev. Pharmacol. Toxicol.,* 16: 465–486, 1976.

Wine, R.I., *Statistics for Scientists and Engineers,* Prentice-Hall, Englewood Cliffs, NJ, 1964.

Witschi, H., Extrapolation: What do the animals tell us?, in *Respiratory Toxicology and Risk Assessment,* Jenkins, P.G., Kayser, D., Muhle, H., Rosner, G., and Smith, E.M., Eds., Wissenschaftliche Verlagsgesellschaft mbH, Stuttgart, Germany, 1994, pp. 341–358.

Witschi, H.R. and Last, J.A., Toxic responses of the respiratory system, in *Casarett and Doull's Essentials of Toxicology,* Klaassen, C.D. and Watkins, J.B. III., Eds., Mc Graw-Hill, New York, 2003, Chapter 15.

Wolff, R.K., Mucociliary function, in *Comparative Biology of the Normal Lung*, Parent, R.A., Ed., CRC Press, Boca Raton, FL, 1991, Chapter 35.

Wong, B.A., Generation and characterization of gases and vapors, in *Concepts in Inhalation Toxicology, 2nd Ed.*, McClellan, R.O. and Henderson, R.F., Eds., Taylor & Francis, Washington, DC, 1995, Chapter 3.

Wong, B.A. and Moss, O.R., Application of a building management system to automate a 90-day chloroform inhalation study, *Toxicol. Methods*, 6: 53–64, 1996.

Wong, B.A., Inhalation exposure systems design, methods, and operation, in *Toxicology of the Lung, 3rd Ed.*, Gardner, D.E., Crapo, J.D., and Mc Clellan, R.O., Eds., Taylor & Francis, Philadelphia, PA, 1999, Chapter 1.

Wood, R.W., Stimulus properties of inhaled substances, *Environ. Health Perspect.,* 26: 69–76, 1978.

Wright, B.M., A new dust-feed mechanism, *J. Sci. Instrum.*, 27: 12–15, 1950.

Yeh, H.C., A theoretical study of electrical discharging of self-charging aerosols, *J. Aerosol Sci.*, 7: 343–349, 1976.

Yeh, H.C., Newton, G.J., Raabe, O.G., and Boor, D.R., Self-charging of ^{198}Au-labeled monodisperse gold aerosols studied with a miniature electrical mobility spectrometer, *J. Aerosol Sci.*, 7: 245–248, 1976.

Yeh, H.C., Schum, G.M., and Duggan, M.T., Anatomic models of the tracheobronchial and pulmonary regions of the rat, *Anat. Rec.,* 195: 483–492, 1979.

Yeh, H.C. and Schum, G.M., Models of human lung airways and their application to inhaled particle deposition, *Bull. Math. Biol.,* 42: 461–480, 1980.

Yeh, H.C., Cheng, Y.S., and Kanapilly, G.M., Electrical aerosol analyzer: Data reduction for high altitude or reduced pressure. *Atmos. Environ.*, 15: 713–718, 1981.

Yeh, H.C., Newton, G.J., Barr, E.B., Carpenter, R.L., and Hobbs, C.H., Studies of the temporal and spatial distribution of aerosols in mult-tiered inhalation exposure chambers, *Am. Ind. Hygiene Assoc. J.*, 47: 540–545, 1986.

Yeh, H.C., Electrical techniques, in *Aerosol Measurement: Principles, Techniques, and Applications*, Willeke, K. and Baron, P.A., Eds., Van Nostrand Reinhold, New York, 1993, Chapter 18.

Young, J.Y., Ed., *Particle Sampling Using Cascade Impactors: Some Practical Application Issues*, American Industrial Hygiene Association, Fairfax, VA, 1995.

Yu, C.P. and Xu, G.B., Predicted deposition of diesel particles in young humans. *J. Aerosol Sci.* 18: 419–429, 1987.

Yung, C.N., De Witt, K.J., and Keith T.G. Jr., Three-dimensional steady flow through a bifurcation, *ASME J. Biomech. Engr.*, 112: 189–197, 1990.

Zebel, G., Coagulation of aerosols, in *Aerosol Science*, Davies, C.N., Ed., Academic Press, New York, 1966, Chapter II.

Zhang, L. and Yu, C.P., Empirical equations for nasal deposition of inhaled particles in small laboratory animals, *Aerosol Sci. Technol.*, 19: 51–56, 1993.

Zitnik, L.A., Schwartz, L.W., Mc Quillen, N.K., Zee, Y.C., and Osebold, J.W., Pulmonary changes induced by low-level ozone: Morphological observations, *J. Environ. Pathol. Toxicol.,* 1: 365–376, 1978.

Zoetis, T. and Hurtt, M.E., Species comparison of lung development, *Birth Defects Res.*, 68: 121–124, 2003.

Zorychta, E. and Richardson, J.B., Innervation of the lung, in *Comparative Biology of the Normal Lung*, Parent, R.A., Ed., CRC Press, Boca Raton, FL, 1991, Chapter 14.

Zosky, G.R. and Sly, P.D., Animal models of asthma, *Clin. Exptl. Allergy*, 37: 973–988, 2007.

Index